Climate Change Management

Series editor

Walter Leal Filho, Hamburg, Germany

For further volumes:
http://www.springer.com/series/8740

CELA

ALFA

FINANCED BY
THE EUROPEAN UNION

Walter Leal Filho · Fátima Alves · Sandra Caeiro
Ulisses M. Azeiteiro
Editors

International Perspectives on Climate Change

Latin America and Beyond

Springer

Editors
Walter Leal Filho
International Climate Change Information Programme
Hamburg University of Applied Sciences
Hamburg
Germany

Sandra Caeiro
DCeT
Universidade Alberta Campus do Taguspark
Porto Salvo
Portugal

Fátima Alves
Ulisses M. Azeiteiro
Universidade Alberta
Porto
Portugal

ISSN 1610-2010 ISSN 1610-2002 (electronic)
ISBN 978-3-319-35572-6 ISBN 978-3-319-04489-7 (eBook)
DOI 10.1007/978-3-319-04489-7
Springer Cham Heidelberg New York Dordrecht London

Printed on acid-free paper

Springer is part of Springer Science+Business Media (www.springer.com)

Preface

The need for climate change adaptation is becoming more and more pressing, as a noticeable increase in both the frequency and intensity of extreme events is now being registered.

Adaptation, a process which entails anticipation of the adverse effects of climate change and the appropriate action needed to prevent or minimize the damage it causes, is now a priority for many countries across the world. Indeed, by pursuing appropriate adaptation strategies, it may be possible to cope with extreme events on the one hand, and protect human lives and property on the other. International cooperation in the field of climate change adaptation is very important, since climate change impacts transcend borders of individual states—such as with river basins—and their impacts vary considerably across regions.

Although there are many interesting and useful experiences on climate change adaptation across the world, only a few initiatives are ever documented or disseminated. This book addresses this need, since it provides concrete examples of concrete adaptation measures taking place in Latin America and in other parts of the world. Prepared in the context of the project "Climate Change Transfer Centers in Europe and in Latin America" (CELA), partly funded by the EU ALFA III Program, this books contains a wealth of information and experiences, illustrating how climate change adaptation is being implemented in practice.

Part I of this book deals with the impacts of change in regions and specific geographical areas, with examples predominantly from Latin America, meeting the need for analyses and projects which investigate the specific impacts of climate change in that part of the world. Part II tackles the management of climate change impacts and introduces projects, initiatives, and adaptation tools. Part III of the book emphasizes the role of policy-making and the contribution of information, communication, and stakeholders' involvement.

Examples of works documented in this book include how to use scarce water resources more efficiently; how agriculture can be adapted to future climate conditions and extreme weather events; how to develop drought-resistant crops, and the important role to be played by information, education, and communication on climate change.

We hope this book will provide a valuable support to international climate change adaptation efforts, and the ability of countries and regions to cope with the multiple impacts of climate change across populations, economic sectors, and regions around the world.

Summer 2014

Walter Leal Filho
Fátima Alves
Sandra Caeiro
Ulisses M. Azeiteiro

Contents

Part I
Impacts of Change Impacts in Regions and Geographical Areas

Chapter 1
Regional Vulnerability of Agro-Environmental Processes Facing Climate Change. Latin American Adaptation Agendas

Pablo Torres-Lima and Rey Acosta-Barradas

Abstract Agricultural systems have begun to proactively address vulnerability and climate risks as challenges and opportunities to develop climate change adaptation measures that protect the natural resources and the ecosystems services on which they depend. However, in most regional conditions, the level of understanding of local changes in hazard exposure caused by climate change, the geographic and spatial distribution of vulnerability, as well as underlying socioeconomic factors remains limited. Here, it is analyzed how in regional agro-environmental processes underline the need for a convergence among the assessments of local climate risk, agricultural vulnerability and adaptive capacities with national and international policy guidance and climate change science, particularly in Latin America.

Keywords Vulnerability • Agro-environmental processes • Climate change • Adaptation • Institutional systems

Introduction

Currently, several reports exist that describe and present different scenarios relating to climate change impacts, as a basis for analyzing the vulnerability of ecosystems, social groups and institutions, whilst also offering an evaluation of current and future options in terms of mitigation and adaptation processes. Comprehensive assessment concerning the vulnerability of agro-environmental processes including agricultural production systems must refer to different temporal and spatial scales, as well

P. Torres-Lima (✉) · R. Acosta-Barradas
Universidad Autónoma Metropolitana, Xochimilco, Calz. del Hueso 1100,
Col. Villa Quietud, Coyoacán, 04960 México, D. F., México
e-mail: ptorres@correo.xoc.uam.mx

R. Acosta-Barradas
e-mail: acostarey@gmail.com

W. Leal Filho et al. (eds.), *International Perspectives on Climate Change*,
Climate Change Management, DOI: 10.1007/978-3-319-04489-7_1,

as diverse levels of theoretical and methodological understanding (Ostrom 2009). This is particularly so as the concept of vulnerability refers to human-environmental interactions. Agro-environmental vulnerability to climate change, understood as a function of biophysical and socio-economic variables and varying factors that comprise agricultural activities must offer the possibility of conceptual definition within a spatially determined typology, particularly in relation to the fundamental attribute of vulnerability itself, which consists of adaptive capacity.

This paper aims to review the vulnerability of regional agro-environmental processes facing climate change and present the current situation as a crucial opportunity for transformation; concerning not only green house gas reduction, but also aimed at reducing general vulnerability. It emphasizes the importance of diagnosing and analyzing the vulnerability of agriculture to climate change, whilst considering an approach to adaptation based on ecosystems. Factors are presented that refer to the institutional response to agro-environmental processes and the design of mechanisms in agricultural policy that imply adaptation.

Regional Vulnerability of Agro-Environmental Processes in Latin America

As regional vulnerability reveals the differential effects of climate on society, it is imperative to study the causes and distribution of climate change impacts on agro-productive systems (IUCN 2008) regarding the complex interaction of environmental, social, economic and political factors involved in each region or geographic area. In Latin America food availability is threatened by climate variability in the short term and climate change in the medium and long terms. Changes in temperature and rainfall patterns will affect different regions and countries to differing degrees, and will lead to smaller yields, affect adversely the quality of food, influence the distribution of pests and the virulence of diseases that impact crops and livestock, affect the storage and distribution of food, the control of pests and diseases that cause post-harvest losses, and affect the incomes and living conditions of vulnerable segments of the population, especially in rural areas (IICA 2012). For instance, the vulnerability of farmers in Mexico and Argentina is highly related to their access to natural and material resources, defining their potential to react with flexibility to changing environmental, economic and institutional situations (Eakin et al. 2009a).

Agricultural productivity is affected by climate change that causes variations in the availability and quality of natural resources, as well as changes in temperature, precipitation, winds and water availability for crop growth and reproduction. Thus agro-ecosystems are directly related to the positive or negative climate effects (IUCN 2008). It is essential to develop methodologies, indicators and studies that take into account perspectives and strategies aimed towards the adaptation of agriculture based on the biodiversity, ecosystem services and rural communities on which it depends. Global agriculture contributes between 17 and 32 % of total emissions of greenhouse gases (GHG), including those resulting from changes in land use. Polluting emissions from agriculture in developing countries increased by 32 % between 1990 and 2005, and this trend is expected to continue in order to meet the food demands of a growing population. Moreover, considering the total global potential for mitigating GHG agricultural emissions; 74 % relates to these same countries (Kok et al. 2010). In terms of the promotion of agricultural practices that are friendly to the environment, the project "Mitigation of Climate Change in Agriculture" (Varming et al. 2010) considers that agro-ecosystems have significant potential for mitigating climate change by means of carbon sequestration at ground level, corresponding to 89 % of total mitigation capacity (Bellarby et al. 2008).

In Mexico, most agricultural and forestry systems are critically dependent on climate. It is estimated that the annual mean temperature for 2060 will increase between 1.1 and 3 °C whereas the average precipitation decrease from −3 to −15 %, according to the GCM model (INE 2009). Because the varied topography in Mexico imposes climatic differences in different regions of the country, it is estimated that there will be a decrease in arable land and crop yields due to climate change, implying challenges to alleviating poverty and to ensuring the food supply and well-being of local and regional populations. It is considered that a 10 % reduction in crop productivity will cause an additional 2 % of emigration population rate to the United States. This means that as a result of the decline in agricultural production for the year 2080, climate change alone will result in the emigration of between 1.4 and 6.7 millions of Mexicans, representing from 2 to 10 % of the current population within the age range of 15–65 years (Feng et al. 2010).

Due to the margin of error concerning these data and questions related to their value for application in decision-making processes as well as their implementation in public policy, it is important to expand studies and research that evaluate the impact of potential climate change scenarios on agro-environmental processes and crop yield in specific localities and regional frameworks (Lobelland and Ortiz-Monasterio 2007). The existence of studies using models that simulate cultivation in future climate scenarios for Latin America, where it is estimated that changes in yields may range between a reduction of 30 % for Mexico and an increase of 5 % in Argentina, refer to the pronounced heterogeneity of this macro-region in terms of climate, ecosystems, culture and population distribution. The cyclical phenomenon in the south known as the Southern Oscillation or El Niño is the climatic variable with greatest socio-economic impact on agriculture in Latin America, especially considering that agriculture represents about 10 % of gross domestic product in the region (Magrin et al. 2007).

As part of the research agenda concerning climate change for Latin America, it has been emphasized that policies and financial resources such as perception, action, as well as organizational and institutional efforts have been primarily directed towards mitigation. This is due to the fact that they may contribute directly to reducing the emission of greenhouse gases or likewise increase their capture from the atmosphere. It is assumed that mitigation reduces the likelihood of adverse impacts due to climate variability in the very long term. Adaptation

Table 1.1 World risk report for selected Latin American countries (2012)

Rank	Country	World risk index[a] (%)	Vulnerability[b] (%)	Lack of coping capacities[c] (%)	Lack of adaptive capacities[d] (%)
7	Costa Rica	17.38	40.80	65.63	35.19
10	El Salvador	16.89	51.82	76.71	49.82
14	Nicaragua	15.36	56.43	82.68	48.21
19	Chile	12.26	39.60	57.84	40.01
20	Jamaica	12.15	47.06	72.49	42.21
21	Haiti	11.96	73.54	90.43	67.48
25	Dominican Republic	11.63	50.23	75.74	44.96
30	Honduras	11.02	55.09	81.68	47.40
58	Ecuador	7.94	49.19	76.93	43.85
64	Panama	7.69	46.74	68.89	41.86
77	Peru	7.18	49.84	74.93	43.77
81	Colombia	6.89	49.80	76.89	42.76
89	Cuba	6.55	37.54	58.95	34.48
94	Mexico	6.39	46.15	71.59	43.12
97	Venezuela	6.13	46.62	74.59	41.84
110	Bolivia	5.13	57.13	80.34	47.43
124	Brazil	4.30	45.18	68.39	41.83
125	Bahamas	4.17	38.99	57.10	42.59
126	Uruguay	4.12	37.06	51.31	39.19
129	Paraguay	3.84	54.56	79.63	51.14
133	Argentina	3.80	39.82	61.56	35.84

Source Based on Alliance Development Works (2012)
[a]The World Risk Index is calculated by combining the four individually calculated components of exposure, susceptibility, lack of coping capacities and lack of adaptive capacities
[b]Vulnerability comprises the components of susceptibility, coping capacities and adaptive capacities
[c]Coping capacities depending on governance, disaster preparedness and early warning, medical services and social and material coverage
[d]Adaptive capacities are relating to forthcoming natural events, to climate change and to other challenges

requires devoting more attention in terms of international and national agendas towards instigating various actions that either anticipate and/or offset the potential adverse effects of climate change. The economic resources destined towards both processes, whether public or private investment funds are not mutually exclusive and in many cases may be complementary.

According to the World Risk Report (Alliance Development Works 2012), there are differences in vulnerability and coping and adaptive capacities among Latin American countries (see Table 1.1). In this region, it is needed to define agendas and list of spending priorities for adaptation processes and actions applying a cost analysis and defining financial needs, and also referring to economic and political principles implying cost-effectiveness (Chisari and Galiani 2010). Internationally and nationally, in order for adaptation to represent a key strategy in this geographic region, assessment and strengthening of institutional and government systems is important, in order to ensure the effectiveness of the initiatives and financial incentives designed to meet the challenges related to the consequences of climate impact in critical contexts such as ecosystems, water, territory, population and the rural economy.

Agriculture and Mitigation

Agriculture is an activity that provides human populations with food, raw materials for clothing, medicines and other products related to their welfare and vital ecosystem services, for example biodiversity, soil accumulation, regulation of water cycles and trapping of carbon, among others. Taking into account the fact that world population is expected to reach 9.1 billion by 2050, agricultural production needs to increase with growing demand, thus climate change represents a major challenge, especially as 75 % of the population living in rural areas of the world depend on agriculture, forestry and fisheries. Agriculture is not only affected by climate change, but is itself a leading emitter of pollutants, so climate change has profound implications in terms of the capacity of agriculture to meet demands for food and plant products, whilst needing to reduce the environmental impact it causes (Fedoroff et al. 2010).

Variations in demand for water (quantity and reliability of supply), resulting from the effects of climate change on agriculture; the largest consumer of this resource in the world, will thus threaten not only the welfare of million farmers and resource-poor farmers, but also affect the food supply for local consumers (Rosenzweig et al. 2004). The effects of climate change on agriculture have major impact on environmental, productive and economic performance. Agricultural production is sensitive to two main types of climate-induced effects: (1) direct effects caused by changes in temperature, precipitation and carbon dioxide concentrations, and (2) indirect effects that occur as a result of changes in ground moisture and the distribution and frequency of pests and diseases.

Globally, it is estimated that the potential impact of climate change may reduce yields from irrigated wheat and rice by 15–30 %, respectively (Nelson et al. 2009). Maize production in Africa and Latin America for the year 2050, will be reduced by 10 %, equating to 2 billion U.S. dollars per year (Jones and Thornton 2003). The latter aggregate figure does not reflect the great regional variability and cropping systems involved, so it is necessary to evaluate the type of impact according to site-specific environment and production system (Liu et al. 2004). Methodological tools will be required to test the response of agricultural systems (at specific sites) to environmental factors (i.e. temperature increase and change in hydrological regimes) and the provision of inputs (i.e. nitrogen fertilization and biofertilization) as part of construction of models to indicate functional change at the regional level (Schneider and Smith 2009).

In early 2010, the project "Mitigation of Climate Change in Agriculture" (MICCA), sponsored by the Organization for Food and Agriculture of the United Nations (FAO), was initiated to promote efforts to mitigate agriculture climate change in developing countries, to the effect that agricultural practices are geared towards a reduction of GHG emissions with biomass and soils aiding to carbon sequestration, creating resilience and increasing productivity in agricultural systems. By promoting adaptation and mitigation, this project encourages decision making on the part of farmers themselves, who are influenced by the possibility of obtaining a fringe benefit in the form of increased agricultural productivity, as a result of choosing appropriate technology to manage local conditions and weather circumstances (i.e. the use of organic fertilizers to reduce pollutants and increase the organic matter content in the soil of each agro-ecological zone).

The options when seeking mitigation through agriculture should promote adaptation to climate change, food security and rural development, in order to promote long-term sustainability. Of those projects so far evaluated by the FAO, 74 % have been required to incorporate agricultural practices such as conservation agriculture, compost production, organic farming, agro-forestry, natural resource management, reforestation, forest conservation and bio-energy (Niggli et al. 2009). The main benefits have included payment for environmental services, conservation of watershed and biodiversity and increased productivity. Other benefits relate to environmental education, increased awareness of the value of ecosystem services and greater financial discernment. Carbon permits have also been distributed, aiming to improve the climate and carbon credit schemes are being planned and managed by central governments (Seeberg-Elverfeldt and Tapio-Bistrom 2010).

As climate change has significant effect on the factors governing the absorption and accumulation of carbon in ecosystems and hence plays a key role in their future capacity for carbon capture, then it is clearly necessary to maintain large tracts of land for agricultural use, with considerable potential for mitigating carbon emissions, as these have been estimated to increase globally to 0.6 Gt CO_2e per year by 2030. If agriculture widely adopts better management practices for carbon, it is estimated that 5.5–6 Gt CO_2e could be captured annually by 2030; comparable to the level of emissions from this sector. About 90 % of this could potentially be achieved by enhancing carbon sinks and approximately 10 % by reducing emissions. Although most of this (70 %) can be carried out in developing countries, it is evident that the greatest possibility for mitigation lies in managing cropland and pasture, and in rehabilitating cultivated organic soils and degraded land (Trumper et al. 2009).

Ecosystem Services Approach for Adapting to Climate Change

In areas with high levels of agricultural production there is correspondingly poor diversity in terms of ecosystems and limited multifunctional socio-ecological systems. Thus, areas with great geographical landscape value in terms of environmental potential for facing climate change (regulation: provision and culture, e.g. nutrient recycling, food production and tourism, respectively) offer better options for the future, including diversified and multifunctional agriculture (Raudsepp-Hearne et al. 2010). This will require the development of socio-ecological systems that provide many faceted environmental services, by conserving and restoring the ecosystem, together with an explicit declared intention on the part of management and social advocacy policies, aimed at reducing vulnerability and increasing agricultural resilience itself through time and space.

Because agro-ecosystems are at risk from practices that enhance production beyond ecological limits (causing for example soil degradation), one would expect that by restoring the ecological functions of agricultural systems, rural producers might contribute to reducing GHG emissions. It is essential to recognize that the focus of Ecosystem-based adaptation can form part of an overall adaptation strategy that will help people contend with the adverse effects of climate change (IUCN 2009). Investment in strengthening or maintaining these services, by means of conservation, restoration and sustainable use of natural resources may help to maintain climates and reinforce the resilience of agro-ecosystems, by promoting mitigation of emissions and strengthening ecosystem-based adaptation.

The focus of ecosystem-based adaptation has not been fully recognized in development policies in terms of food security, by financing and creating incentives for the consolidation of sustainable agriculture. Immediate economic advantages have impeded the implementation and success of policies that resist climate change, reflected in the lack of technical expertise applied to agricultural systems to combat GHG emissions, and likewise the lack of political support for defining an agenda aimed at adaptation and mitigation. By focusing on ecosystem services and considering ancillary benefits, both aspects could be aligned to international efforts and objectives currently being formulated in this context. It is evident that the adaptation approach based on ecosystems should form a fundamental component of social conscience, political thought and economic processes (Munang et al. 2013).

Overall, agro-environmental processes are faced with the challenge of providing food for a world population which will reach almost 9 billion by 2050, exerting constant pressure on natural resources, whilst climate change impacts multiply the risk to agricultural production. Production practices and resources based on ecosystem services, such as integrated management of pests, production systems for conservation and minimum tillage, water conservation and the use of organic waste in agricultural plots among others are approaches that together help to improve the survival and resilience of rural inhabitants and their territories. The main objectives constitute adaptation processes based on the ecosystem with community work focussed on reducing poverty and achieving sustainable development, while strengthening achieved environmental benefits, climate resilience and methods for lowering carbon output.

It has been reported that efforts are needed to integrate data, models and maps used by disparate groups for vulnerability and impact assessment. Mining of large data sets pertaining to agronomy, human geography, climate and other factors will be critical for identifying vulnerabilities. Important research includes comparison studies of vulnerability assessments to identify best practices, development of protocols for future assessments, use of agreed-upon protocols across groups, sectors and places and improved documentation and systems for recording, sharing and evaluating results (UNEP 2013). In this sense, in this paper some agro-environmental agendas in Latin America facing regional vulnerability to climate change are presented in Table 1.2.

Table 1.2 Agro-environmental agendas in Latin America facing regional vulnerability to climate change

Priority	Actions	Issues	Country	Reference
Agriculture	Agro-ecological practices, soil conservation, sustainable harvesting techniques, delivering organic crops without polluting. Water harvesting techniques, wetlands recovery, participatory watershed management, irrigation systems and breakwaters (DRP 2012; PACC-MAE 2013)	Water pollution	• Brasil	• DRP (2012)
		Water management	• Ecuador	• PACC-MAE (2013)
			• Mexico	• CONAGUA, (2012), Eakin et al. (2007)
			• Costa Rica	• MIDEPLAN (2012)
			• Argentina	• Urcola et al. (2010)
		Environmentally-friendly soil management practices	• Brasil	• World Bank (2012)
			• Central America	• Tiessen et al. (2003)
			• Mexico	• Méndez et al. (2010)
		Crop and grain production	• Brasil, Argentina	• World Bank (2012)
			• Mexico	• Eakin (2005), Eakin and Luers (2006)
	Compare moisture conditions between months and seasons (Gamble et al. 2010).		• Latin America	• CEPAL (2012)
			• Nicaragua, Colombia	• Vermeulen et al. (2013)
	Establish trees on agricultural land in order to regulate soil, water and air quality (Smith et al. 2012)	Agroforestry	• Brasil, Ecuador, Colombia, Mexico, Costa Rica, El Salvador, Nicaragua	• Bhagwat et al. (2008)
	Improve regional and global agriculture trade systems		• Puerto Rico, Costa Rica	• Smith et al. (2012)
			• Honduras	• Ayarza et al. (2010)
		Coastal crop vulnerability	• Bahamas, Islas Caimán, Suriname, Guatemala, Brazil, México, Argentina, Venezuela, Cuba, Puerto Rico	• CEPAL (2012)

(continued)

Table 1.2 continued

Priority	Actions	Issues	Country	Reference
		Agroecological practices	• Brasil	• Rival (2012)
			• Mexico	• Schroth et al. (2009), Hausermann and Eakin (2008)
		Pests	• Puerto Rico, Costa Rica	• Smith et al. (2012)
		Droughts	• Ecuador, Nicaragua	• PACC-MAE (2013)
			• Jamaica	• Gamble et al. (2010)
			• Brasil	• Rival (2012)
		Floods	• Nicaragua	• PACC-MAE (2013)
			• Mexico	• Eakin and Appendini (2008)
		Improved crop varieties	• Brasil	• World Bank (2012)
		Agroforestry payments	• Mexico	• OCDE (2013)
			• Honduras	• Ayarza et al. (2010)
		Agriculture's carbon footprint	• Brasil	• World Bank (2012)
			• Mexico	• Eakin et al. (2009b)
Emission of greenhouse gases (GHG)	Quantify carbon emissions	Reducing emissions from deforestation	• Brasil	• Rival (2012)
	Avoid deforestation	Solar thermal technology	• Guatemala	• CDKN (2013)
	Use of renewable energy sources (Rival 2012)	Biomass power plant	• Belize	• CDKN (2013)
		Solar electrical energy	• Nicaragua	• CDKN (2013)
		Avoid oil pollution	• Ecuador	• Rival (2012)
	Design sustainable livestock systems (World Bank 2012)		• Mexico	• Masera et al. (2005)
		Sustainable cattle ranching	• Colombia	• World Bank (2012)

(continued)

Table 1.2 continued

Priority	Actions	Issues	Country	Reference
Local environmental risk		Reduction of gross domestic product (agriculture, fisheries, electricity and infrastructure)	• Bolivia, Colombia, Ecuador, Perú	• DRP (2012)
		Coastal flood vulnerability	• México, Brasil, Cuba, Bahamas, Argentina	• CEPAL (2012)
		Water vulnerability	• Mexico	• CONAGUA (2012), Eakin et al. (2007)
		Habitat degradation and loss	• Brasil, Ecuador, Colombia, Costa Rica, El Salvador, Nicaragua	• Bhagwat et al. (2008)
			• Mexico	• Eakin and Bojórquez-Tapia (2008)
		Rainfalls	• Ecuador	• PACC-MAE (2013)
			• Costa Rica	• MIDEPLAN (2012)
			• Argentina	• Podestá et al. (2009)
		Floods	• Nicaragua	• PACC-MAE (2013)
			• Mexico	• Schroth et al. (2009), Eakin and Appendini (2008)
		Soil preservation	• Mexico	• CONAGUA (2012)
		Sustainable environment management	• Ecuador	• Rival (2012)

(continued)

Table 1.2 continued

Priority	Actions	Issues	Country	Reference
Biodiversity	Management of the conservation units through monitoring, control and scientific development	Species conservation	• Colombia	• PACC-MAE (2013)
			• Brasil, Peru, Bolivia, Colombia, Ecuador, French Guyana, Guyana, Surinam, Venezuela	• Schulman et al. (2007)
		Deforestation	• Mexico	• OCDE (2013)
		Forest degradation	• Brasil	• Rival (2012)
		Forest regeneration		
		Migration	• Costa Rica	• Harvey and González-Villalobos (2007)
		Floods	• Colombia	• PACC-MAE (2013)
		Agroforestry	• Puerto Rico, Costa Rica	• Harvey and González-Villalobos (2007)
			• Honduras	• Ayarza et al. (2010)
		Depletion of natural resources	• Mexico	• OCDE (2013)
			• Brazil	• World Bank (2012)
		Payments for forest conservation	• Brasil, Costa Rica	• World Bank (2012)
			• Mexico	• Schroth et al. (2009)
Ecosystems	Reduce climate vulnerability and build climate resilience	Conservation of area-sensitive species	• Colombia	• PACC-MAE (2013)
			• Ecuador	• Rival (2012)
			• Peru, Bolivia	• Paniagua-Zambrana et al. (2007)
		Coastal management	• Uruguay	• PACC-MAE (2013)
		Water recharges	• Mexico	• Schroth et al. (2009), OCDE (2013), Eakin et al. (2007)
		Renewable water resources	• Latin America	• World Bank (2012)

(continued)

Table 1.2 continued

Priority	Actions	Issues	Country	Reference
		Adaptive land planning	• Uruguay	• PACC-MAE (2013)
			• Mexico	• Schroth et al. (2009)
			• Argentina	• Urcola et al. (2010)
			• Nicaragua, Colombia	• Vermeulen et al. (2013)
		Efficient nutrient cycling	• Mexico, Argentina	• Eakin and Wehbe (2009)
		Global warming	• Ecuador	• Rival (2012)
			• Brasil, Peru, Bolivia, Colombia, Ecuador, French Guyana, Guyana, Surinam, Venezuela	• Schulman et al. (2007)

Conclusions

Current agricultural practices that are being promoted as appropriate tools for adaptation to climate change are either already considered standard 'best practices' for reasons other than climate change, or else are not currently widely adoptable due to the natural resources and the ecosystems services on which they depend when implemented at large scale. In terms of public policy for adaptation to climate change in the Latin American region, the greatest benefits are likely to be generated by research and development on regional agro-environmental processes. It is highlighted the need for assessing the local climate risks, agricultural vulnerability and adaptive capacities. There are also likely to be benefits from research to improve understanding and prediction of climate change and its impacts, including the local changes in hazard exposure caused by climate change, the geographic and spatial distribution of vulnerability, as well as underlying socioeconomic factors.

The main policy guidelines for meeting Latin American agendas of adaptation and institutional systems are: (1) Governments should consider the options of ecosystem-based adaptation as an integral component of risk reduction and strategies for adaptation to climate change, principally as part of local, regional and national development planning processes; (2) Projects related to climate change must take into account the local environmental conditions and ecosystems involved, as well as identifying opportunities for maximizing ecosystem services for risk reduction and adaptation to climate variability, also implying the conservation of biodiversity; (3) Communities and locally participating sectors, without losing their right of access to their resources, must become involved in all processes of adaptation to ensure the proper design of projects that are directed towards sustainable development; (4) All ecosystem services should be recognized when carrying out assessments of cost-effectiveness relating to different adaptation options; (5) The resilience of social and ecological systems to natural hazards; those caused by humans and individual impacts of climate change will be facilitated through improved ecosystem management and sustainable use of resources, and (6) The ecosystem-based adaptation is not the only solution, but represents a cost-effective approach to long term problems that can be used in conjunction with other measures of adaptation to climate change and disaster management, with the purpose of reducing the vulnerability of agro-environmental processes as well as rural populations and their territories.

Latin American countries need to identify their current vulnerability and adaptation needs at national, regional and local levels with an urgent obligation to promoting sustainable approach to ecosystems in terms of the design and development of policies of multi-sectoral adaptation and planning within their territories. This must meet the requirement to fortify institutional capacity and scientific research related to the implementation of relevant policies, regarding particular ecosystem and sectoral adaptation options.

References

Alliance Development Works (2012) Focus: Environmental degradation and disasters. Alliance Development Works, United Nation University and the Nature Conservancy, Berlin

Ayarza M, Huber-Sannwald E, Herrick J, Reynolds J, García-Barrios L, Welchez L, Lentes P, Pavón J, Morales J, Alvarado A, Pinedo M, Baquera N, Zelaya S, Pineda R, Amézquita E, Trejo M (2010) Changing human–ecological relationships and drivers using the Quesungual agroforestry system in western Honduras. Renew Agric Food Syst 25(3):219–227

Bellarby J, Foereid B, Hastings A, Smith P (2008) Cool farming: climate impacts of agriculture and mitigation potential. Greenpeace, The Netherlands

Bhagwat S, Willis K, Birks H, Whittaker R (2008) Agroforestry: a refuge for tropical biodiversity? Trends Ecol Evol 23(5):261–267

Chisari O, Galiani S (2010) Climate change: a research agenda for Latin America and the Caribbean. Technical Notes No. IDB-TN-164. Inter-American Development Bank. Available at: http://www.iadb.org

Climate and Development Knowledge Network (CDKN) (2013) Renewable energy pathway for Central America. Clim Dev Outlook 6:1–4

Comisión Económica para América Latina y el Caribe (CEPAL) (2012) Efectos del cambio climático en la costa de América Latina y el Caribe. Vulnerabilidad y Exposición. Instituto de Hidráulica Ambiental de la Universidad de Cantabria, CEPAL y Programa Iberoamericano de Adaptación al Cambio Climático. Naciones Unidas, Santiago de Chile

Comisión Nacional del Agua (CONAGUA) (2012) Agua y cambio climático en México, 2007–2012. CONAGUA, México

Diálogo Regional de Política de Agua y Adaptación al Cambio Climático en las Américas (DRP) (2012) Agua y Adaptación al Cambio Climático en las Américas; Soluciones del Diálogo Regional de Política (DRP). Available at: http://www.cna.gob.mx/Contenido.aspx?n1=4&n2=42&n3=73&n4=73

Eakin H (2005) Institutional change, climate risk, and rural vulnerability: cases from Central Mexico. World Dev 33(11):1923–1938

Eakin H, Appendini K (2008) Livelihood change, farming, and managing flood risk in the Lerma Valley, Mexico. Agric Hum Values 25:555–566

Eakin H, Bojórquez-Tapia L (2008) Insights into the composition of household vulnerability from multicriteria decision analysis. Glob Environ Change 18:112–127

Eakin H, Luers A (2006) Assessing the vulnerability of social-environmental systems. Annu Rev Environ Resour 31:365–394

Eakin H, Wehbe M (2009) Linking local vulnerability to system sustainability in a resilience framework: two cases from Latin America. Clim Change 93:355–377

Eakin H, Magaña V, Smith J, Moreno J, Martinez J, Landavazo O (2007) A stakeholder driven process to reduce vulnerability to climate change in Hermosillo, Sonora, Mexico. Mitig Adapt Strat Glob Change 12:935–955

Eakin H, Jebe M, Avila C, Sánchez G, Bojórquez-Tapia L (2009a) Social vulnerability of farmers in Mexico and Argentina. In: Leary N, Conde C, Kulkarni J, Nyong A, Pulhin J (eds) Climate change and vulnerability. Earthscan, London

Eakin H, Winkels A, Sendzimir J (2009b) Nested vulnerability: exploring cross-scale linkages and vulnerability teleconnections in Mexican and Vietnamese coffee systems. Environ Sci Policy 12:398–412

Fedoroff N, Battisti D, Beachy R, Cooper P, Fischoff D, Hodges C, Knausf V, Lobell D, Mazur B, Molden D, Reynolds M, Ronald P, Rosegrand M, Sánchez P, Vonshak A, Zhu J (2010) Radically rethinking agriculture for the 21st century. Science 327:833–834

Feng S, Krueger A, Oppenheimer M (2010) Linkages among climate change, crop yields and México-US cross-border migration. PNAS 107(32):14257–14262

Gamble D, Campbell D, Allen T, Barker D, Curtis S, McGregor D, Popke J (2010) Climate change, drought, and Jamaican agriculture: local knowledge and the climate record. Ann Assoc Am Geogr 100(4):880–893

Harvey C, González-Villalobos J (2007) Agroforestry systems conserve species-rich but modified assemblages of tropical birds and bats. Biodivers Conserv 16:2257–2292

Hausermann H, Eakin H (2008) Producing "viable" landscapes and livelihoods in Central Veracruz, Mexico: institutional and producer responses to the coffee commodity crisis. J Latin Am Geogr 7(1):109–131

Instituto Nacional de Ecología (INE) (2009) y Secretaría de Medio Ambiente y Recursos Naturales (Semarnat). México. Cuarta comunicación nacional ante la convención marco de las naciones unidas sobre el cambio climático. INE-Semarnat. México

Inter-American Institute for Cooperation on Agriculture (IICA) (2012) The food security situation in the Americas: document to facilitate the dialogue scheduled to take place at the 42nd general assembly of the organization of American States. IICA, Costa Rica

International Union for Conservation of Nature (IUCN) (2009) *Ecosystem-based adaptation*. Position paper for UNFCCC Climate Talks September 28 to October 9, Bangkok. Available at: http://cmsdata.iucn.org/downloads/iucn_position_paper_eba_september_09.pdf

International Union for Conservation of Nature (IUCN) y World Business Council for Sustainable Development (WBCSD) (2008) Agricultural ecosystems. Facts and trends. WBCSD-IUCN, Switzerland

Jones P, Thornton P (2003) The potential impacts of climate change on maize production in Africa and Latin America in 2055. Glob Environ Change 13:51–59

Kok M, Tyler S, Prins A, Pintér L, Baumüller H, Bernstein J, Tsioumani E, Venema H, Grosshans R (2010) Prospects for mainstreaming ecosystem goods and services in international policies. Netherlands Environmental Assessment Agency e International Institute for Sustainable Development, La Hague

Liu H, Li X, Fischer G, Sun L (2004) Study on the impacts of climate change on China's agriculture. Clim Change 65:125–148

Lobelland D, Ortiz-Monasterio I (2007) Impacts of day versus night temperatures on spring wheat yields. A comparison of empirical and CERES model predictions in three locations. Agron J 99:469–477

Magrin G, Gay C, Cruz D, Giménez J, Moreno A, Nagy G, Nobre C, Villamizar A (2007) Latin America. In: Parry M, Canziani O, Palutikof J, van der Linden P, Hanson C (eds) Climate change 2007: impacts, adaptation and vulnerability. Contribution of working group II to the 4th assessment report of the intergovernmental on climate change. Cambridge University Press, UK, pp 581–615

Masera O, Díaz R, Berruela V (2005) From cookstoves to cooking systems: the integrated program on sustainable household energy use in Mexico. Energy Sustain Dev 11(1):25–36

Méndez E, Bacon C, Olson M, Petchers S, Herrador D, Carranza C, Trujillo L, Guadarrama-Zugasti C, Cordón A, Mendoza A (2010) Effects of fair trade and organic certifications on small-scale coffee farmer households in Central America and Mexico. Renew Agric Food Syst 25(3):236–251

MIDEPLAN (2012) La agricultura tropical frente al cambio climático: Costa Rica carbono neutral 2012. Ministerio de Planificación Nacional y Política Económica, Costa Rica

Munang R, Thiaw I, Alverson K, Goumandakoye M, Mebratu D, Liu J (2013) Using ecosystem-based adaptation actions to tackle food insecurity. Environ Sci Policy Sustain Dev 55(1):29–35

Nelson G, Rosegrant MW, Koo J, Robertson R, Sulser T, Zhu T, Ringler C, Msangi S, Palazzo A, Batka M, Magalhaes M, Valmonte-Santos R, Ewing M, Lee D (2009) Climate change: impact on agriculture and costs of adaptation. IFPRI. Available at: www.ifpri.org/sites/default/files/publications/pr21.pdf

Niggli U, Fließbach A, Hepperly P, Scialabba N (2009) Low greenhouse gas agriculture: mitigation and adaptation potential of sustainable farming systems. FAO, Rome

Organización para la Cooperación y el Desarrollo Económicos (OCDE) (2013) Evaluaciones sobre el desempeño ambiental: México 2013. OCDE, Paris, pp 1–8

Ostrom E (2009) A general framework for analyzing sustainability of socio-ecological systems. Science 325:419–425

PACC-MAE Team (2013) Supporting climate change adaptation in Latin America. Clim Change Adapt Bull 12:1–7

Paniagua-Zambrana N, Byg A, Svenning J, Moraes M, Grandez C, Balslev H (2007) Diversity of palm uses in the western Amazon. Biodivers Conserv 16:2771–2787

Podestá G, Rajagopalan B, Apipattanavis S, Laciana C, Weber E, Easterling W, Katz R, Letson D, Menendez A (2009) Decadal climate variability in the Argentine Pampas: regional impacts of plausible climate scenarios on agricultural systems. Clim Res 40:199–210

Raudsepp-Hearne C, Peterson G, Bennett E (2010) Ecosystem service bundles for analyzing tradeoffs in diverse landscapes. PNAS 107(11):5242–5247

Rival L (2012) Sustainable development through policy integration in Latin America. A comparative approach. Occasional paper seven. Social dimensions of green economy and sustainable development. United Nations Research Institute for Social Development and the Friedrich-Ebert-Stiftung, Geneva

Rosenzweig C, Strzepek K, Major D, Iglesias A, Yates D, McCluskey A, Hillel D (2004) Water resources for agriculture in a changing climate: international case studies. Glob Environ Change 14:345–360

Schneider U, Smith P (2009) Energy intensities and greenhouse gas emission mitigation in global agriculture. Energ Effi 2:195–206

Schroth G, Laderach P, Dempewolf J, Philpott S, Haggar J, Eakin H, Castillejos T, Garcia-Moreno J, Soto-Pinto L, Hernandez R, Eitzinger A, Ramirez-Villegas J (2009) Towards a climate change adaptation strategy for coffee communities and ecosystems in the Sierra Madre de Chiapas, Mexico. Mitig Adapt Strateg Glob Change 14:605–625

Schulman L, Ruokolainen K, Junikka L, Sääksjärvi IE, Salo M, Juvonen S-K, Salo J, Higgins M (2007) Amazonian biodiversity and protected areas: do they meet? Biodivers Conserv 16:3011–3051

Seeberg-Elverfeldt C, Tapio-Biström M (2010) Global survey of agricultural mitigation projects. Mitigation of Climate Change in Agriculture Series No. 1. FAO, Rome

Smith J, Pearce BD, Wolfe MS (2012) Reconciling productivity with protection of the environment: is temperate agroforestry the answer? Renew Agric Food Syst 28(1):80–92

Tiessen H, Menezes R, Salcedo I, Wick B (2003) Organic matter transformations and soil fertility in a treed pasture in semiarid NE Brazil. Plant Soil 252:195–205

Trumper K, Bertzky M, Dickson B, van der Heijden G, Jenkins M, Manning P (2009) The natural fix? The role of ecosystems in climate mitigation. A UNEP rapid response assessment. United Nations Environment Programme, UNEP-WCMC, Cambridge, United Kingdon

United Nations Environment Programme (UNEP) (2013) Research priorities on vulnerability, impacts and adaptation. Responding to the climate change challenge. Programme of Research on Climate Change Vulnerability, Impacts and Adaptation (PROVIA), UNEP, Nairobi

Urcola H, Elverdin J, Mosciaro M, Albaladejo C, Manchado J, Giussepucci J (2010) Climate change impacts to rural societies: stakeholders perceptions and adaptation strategies in Buenos Aires, Argentina. Innovation and Sustainable Development in Agriculture and Food. ISDA, France

Varming M, Seeberg-Elverfeldt C, Tapio-Biström M (2010) Agriculture, forestry and other land use mitigation project database. An assessment of the current status of land-based sectors in the carbon markets. Mitigation of Climate Change in Agriculture Series No. 2. FAO, Rome

Vermeulen S, Challinor A, Thornton P, Campbell B, Eriyagama N, Vervoort J, Kinyangi J, Jarvis A, Läderach P, Ramirez-Villegas J, Nicklin K, Hawkins E, Smith D (2013) Addressing uncertainty in adaptation planning for agriculture. PNAS 110(21):8357–8362

World Bank (2012) Inclusive green growth in Latin America and the Caribbean. World Bank's Sustainable Development Department of the Latin America and Caribbean Region (LCSSD). Available at: http://siteresources.worldbank.org/INTLAC/Resources/green_growth_full.pdf

Chapter 2
Risk Analysis in Transboundary Water of the Rivers Pilcomayo and Paraguay

Julian Báez, Roger Monte Domecq and Lisa Lugo

Abstract Analysis of Trans-boundary Risks in the Pilcomayo and the Paraguay River Basins could generate relevant technical information about the management of water resources in the proposed basins. The Pilcomayo River Commission (Paraguay, Bolivia and Argentina) has conducted important studies in the water resources field of the Pilcomayo Basin that are currently stored in the Commission's data base. Less information is available for the case of the Paraguay River Basin, especially in the area of study: the confluence area with the Paraná River. The study presents a border management tool for managing the risks associated with extreme events water in rivers shared between Argentina and Paraguay, in order to contribute to an efficient coordination work between local agencies that manage emergencies. This study also aims to identify the main risks that can occur in urban and rural environments of potential areas to be affected by floods, droughts and other natural and anthropogenic phenomena such as erosion, sedimentation and pollution from various sources. A web platform has been developed to collect all available information regarding these Rivers. Alpha 3 CapWEM project (Capacity development in Water Engineering and Environmental Management) in which eight Universities from European and Latin American countries collaborate

J. Báez (✉) · R. Monte Domecq
Unidad de Estudios Hidro-Ambientales, Centro de Tecnología Apropiada, Universidad Católica Nuestra Señora de la Asunción. Campus Santa Librada, Asunción, Paraguay
e-mail: julian_baez@uca.edu.py

R. Monte Domecq
e-mail: roger.montedomecq@gmail.com

L. Lugo
Centro de Tecnología Apropiada, Universidad Católica Nuestra Señora de la Asunción. Campus Santa Librada, Asunción, Paraguay
e-mail: lugo@uca.edu.py

W. Leal Filho et al. (eds.), *International Perspectives on Climate Change*,
Climate Change Management, DOI: 10.1007/978-3-319-04489-7_2,

on improving higher education in the fields of Water and Environmental Management has provided the framework for the present study.

Keywords Water risk • Transboundary analysis • Water resources management • Río Pilcomayo • Río Paraguay

Introduction

The Republic of Paraguay is located entirely in the La Plata River Basin, within are located the Paraguay River, a tributary of the Paraná River, and the Pilcomayo River, a tributary of the River Paraguay. Paraguay and Pilcomayo Rivers have significant cross-border sections with Brazil and Argentina. Particularly the Paraguay River divides the country into two very distinctive natural regions: the eastern region, which accounts for 97 % of the population, with rich soils for agriculture and the western region, representing 60 % of the territory but just with 3 % of the total population and is where major cattle and milk production of the country is concentrated.

The River of Paraguay and Pilcomayo produces cyclic flood affecting population, including those of the capital of the country, Asuncion. The most extreme flooding affected more than 100,000 people in 1983. This works describe the flooding process and develop a Web based information system for the Pilcomayo and Paraguay River, which pretends to become as a tool of flood management.

Paraguay and Pilcomayo River

The La Plata River Basin is the second biggest Basin in South America and the fifth largest in the world; it spreads over a 3.1 million km^2, Fig. 2.1 (Collischonn et al. 2001). The Paraguay River is a major tributary of the La Plata River, and its basin covers 1,095,000 km^2, through the countries of Brazil, Bolivia, Paraguay and Argentina. It has an average flow of 2,700 m^3/s.

In Asuncion, Paraguay, the average flow of Paraguay River is about 3,000 m^3/s, with a maximum of 4,000 m^3/s in June and minimum 2,500 m^3/sec in December, Fig. 2.2. Although during periods of flooding in the port of Asuncion, in coincidence with the warm phase of ENSO, the flow of Paraguay River can be as high as 11,000 m^3/s (Barros et al. 2004).

The Paraguay River overflows cyclically producing population displacement in the cities of Asunción, Alberdi and Pilar. The five major floods in the Paraguay River were in 1905, 1983, 1992, 1988 and 1982 (Barros et al. 2004). In 1983 occurred the highest rise in hydrometric level at the port of Asuncion, Paraguay, reaching 9.02 m.

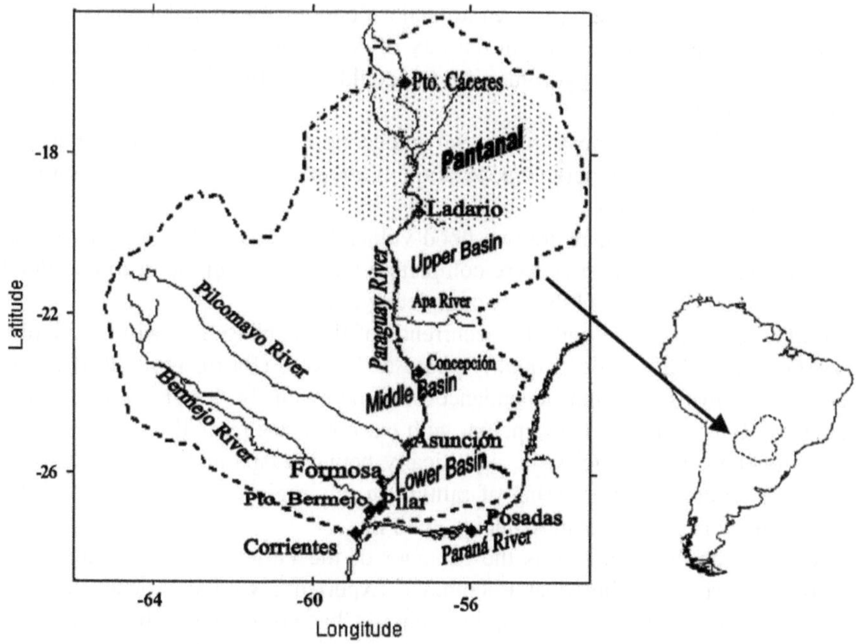

Fig. 2.1 La Plata Basin. Original from Barros et al. (2004)

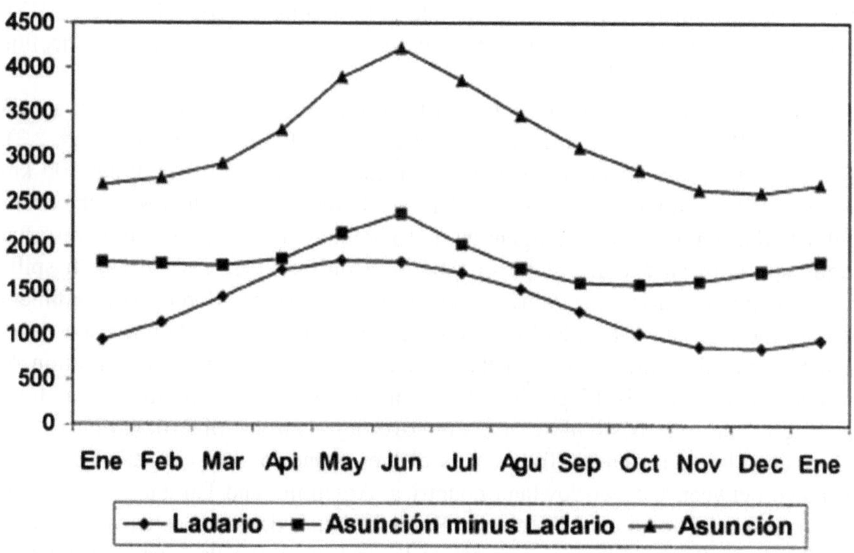

Fig. 2.2 Mean annual stream flow in Asuncion and Ladario. Original from Barros et al. (2004)

After these years there have been no major flooding, and population has grown in flood-prone areas, especially in the city of Asuncion. At present, a river level height of 5 m, referred to hydrometric cero, could cause the displacement of more than 1,500 inhabitants.

River flooding in Asuncion has social and economic consequences, as it causes the displacement of a vulnerable population of about 100,000 inhabitants at northern and southern marshes.

South of Asunción, there are two flood-vulnerable cities to the Paraguay River, Alberdi and Pilar. Both cities were completely under water in the last great flood of 1983. Figure 2.5 shows Pilar city completely flooded.

Large floods coincide with the occurrence of the El Niño-Southern Oscillation (ENSO), particularly with its quality phase or El Niño (Barros et al. 2004). This behavior confirms the high dependence of flooding in the lower reaches of the Paraguay River, south of Concepción, with extraordinary rainfall recorded during the summer in the middle basin, specifically between Apa and Pilcomayo rivers (Fig. 2.1). Therefore, monitoring of rainfall in this region allows forecasting the future behavior of the river in Asuncion and Pilar.

Another variable that affects the behavior of the Paraguay River hydrometric levels is its tributary, the River Pilcomayo. Experience with significant flooding rains due to high volume in 2012 in the Central Chaco of Paraguay, had significant impact on the Paraguay River in Asuncion during the month of May of that year.

Pilcomayo River has very particular geomorphological and hydrological characteristics that have result in the continued waters retreat of this river to its source, in Bolivia. The most significant feature is that the Pilcomayo begins in the Bolivian highlands at altitudes above 1,000 m, reaching a vast plain in the territories of Argentina and Paraguay. This situation generates an annual sediment transport of extraordinary magnitude, about 140 million tons (Martin-Vide et al. 2012). These sediments have to be removed each year in order to avoid the rapid decline of the river.

The Pilcomayo River flows south-eastwards from the Bolivian Andes, across the Chaco Plains, down, in principle, to the Paraguay River at Asunción (Fig. 2.3). The drainage basin covers the southern Andean ranges of Bolivia along 500 km of Main River with an average slope of 1 %, whereas most of the rest of the channel sets the border between Argentina and Paraguay along 835 km in a very flat landscape with an average slope of 0.04 %, (Martin-Vide et al. 2012). It spills over the plains during the rainy season from January to March. The sediment load of the Pilcomayo is one of the largest in the world: 140 million tons per year, which is mostly wash load from the upland Andes. The mean concentration of suspended sediment is 15 g/l. The maximum-recorded concentration is as high as 60 g/l. The river has built a large fan covering a surface of 210,000 km^2, with many abandoned channels. Today, it is a river prone to avulsion, raising border disputes between the two lowland countries, Argentina and Paraguay. Moreover, the very special feature of Pilcomayo River is that it does not actually flow into the Paraguay River. Very far upstream of the mouth in the Paraguay the channel blocks itself with sediment and wood debris forcing water and sediment to spread

Fig. 2.3 Location map of Pilcomayo River and its hydrographic network

across the plains. Moreover, the point of blockage has moved hundreds of kilometers upstream throughout the 20th century. Many environmental issues arise because of this collapse (channel discontinuity), not the least of them is the migration of fish. The future of the river concerns Bolivia and the two lowland countries.

The hydrological regime of the Pilcomayo described by (Martin-Vide et al. 2012) is quite predictable in its general trends. Every year a rainy season produces high discharges in the period of January, February and March. On the contrary, the dry season brings the discharge down from June to October. For the gauging station 1 (Fig. 2.3) located at the end of the mountainous upper part of the river, 1,000 km upstream of Asunción, where the catchment area is 82,000 km^2, the mean discharge is some 720 m^3/s in February but only 35 m^3/s in September (twenty times lower).

The Pilcomayo is a tributary featuring an annual 3-month "pulse" of water and sediment that cannot reach its main river (the Paraguay). Whereas water spills over the plains, feeds the marshes and contributes to the groundwater so that it is finally drained far downstream as clear water, sediment has no other choice except to raise the land of the plains and fill the river channel itself, contributing altogether to the river channel instability in the alluvial fan. The river essentially reaches a local base level as it crosses the Chaco Plains, so that it spreads across the floodplain and fan complex.

The last major flood of the river took place in April of 2012, affecting mainly the Paraguayan Chaco, and it was associated with the lack of sediment removal in Argentine territory in addition to extraordinary rainfall in the central Chaco in Paraguay. The volumes of precipitation exceeded 500 mm in 3 days, which represents 70 % of the annual rainfall in this region. The consequences were catastrophic, causing even the loss of human lives.

Both rivers have social and economic implications for Argentina and Paraguay. Paraguay River is a vital communication channel for trade in Paraguay and a threat in flood situations for coastal populations. In drought seasons Paraguay River level arises not only navigation problems but also scarcity for drinking water from both countries. Pilcomayo River, in contrast, is a source of drinking water in the province of Formosa as well as for use in irrigation systems in Argentina and Paraguay.

Uncertainties and the consequent related risks in water resources engineering design and operation are unavoidable. Water resources projects are always subject to a probability of failure in achieving their intended purposes. As an example, a flood control project may not protect an area from extreme floods. A water supply project may not deliver demanded water. This failure may be due to failure of the delivery system or may be due to lack of supply. A water distribution system may not deliver water meeting quality standards even though the source quality does. The rational in the selection of the design and operation parameters and the design and operation standards are continually questioned.

The EU ALFA III project "Capacity Development in Water Engineering and Environmental Management—CapWEM" has considered a pilot project named "Risk Analysis in Transboundary Rivers". One of the deliverables of the pilot project is a Web platform called "Information System for Transboundary Risk Management in Paraguay and Pilcomayo rivers". This system aims to organize the information of hydrological extremes, droughts and floods, border sections of the rivers Pilcomayo and Paraguay to serve as a support for decision-making.

Risk Analysis of Transboundary Rivers in the Context of Climate Variability and Change

Extreme flood events and the economic, social and environmental impacts and losses in human life they cause have significantly increased in recent years. Against this already serious background, enhanced climate variability and climate change are expected to increase the frequency and intensity of floods (UNECE 2009).

Floods are part of the water cycle and supply floodplains with sediment and nutrients, the main reason for early settlement in and development of floodplains. Both natural characteristics and human interventions and activities in river basins influence the amplitude, frequency, duration and impact of floods. Increasing climate variability and climate change have the potential to exacerbate flood problems in many regions around the world due to their effects on precipitation volume and timing. Population and economic growth are the dominant drivers behind observed increases in flood damage. Human behavior often reduces the resilience of the land and water resources in the system.

Flood risk is defined as "a combination of probability of a flood event and of the potential adverse consequences for human health, environment, cultural heritage and economic activity associated with a flood event" (UNECE 2009).

Another risk factor for the study area is drought, especially in the region of the Pilcomayo River, where the recurrence of drought threatens not only agricultural and livestock production, but also drinking water availability for the poorest inhabitants of the region.

The SINERGIA project (International System of Water Resources Studies and Management of Impacts due to Global Warming in the Paraguay River Basin 2011), developed a vulnerability map for the Paraguay River Basin, shown in Fig. 2.4. The survey area corresponds to the areas 16, 19, 20, 21 and 22, which are described below:

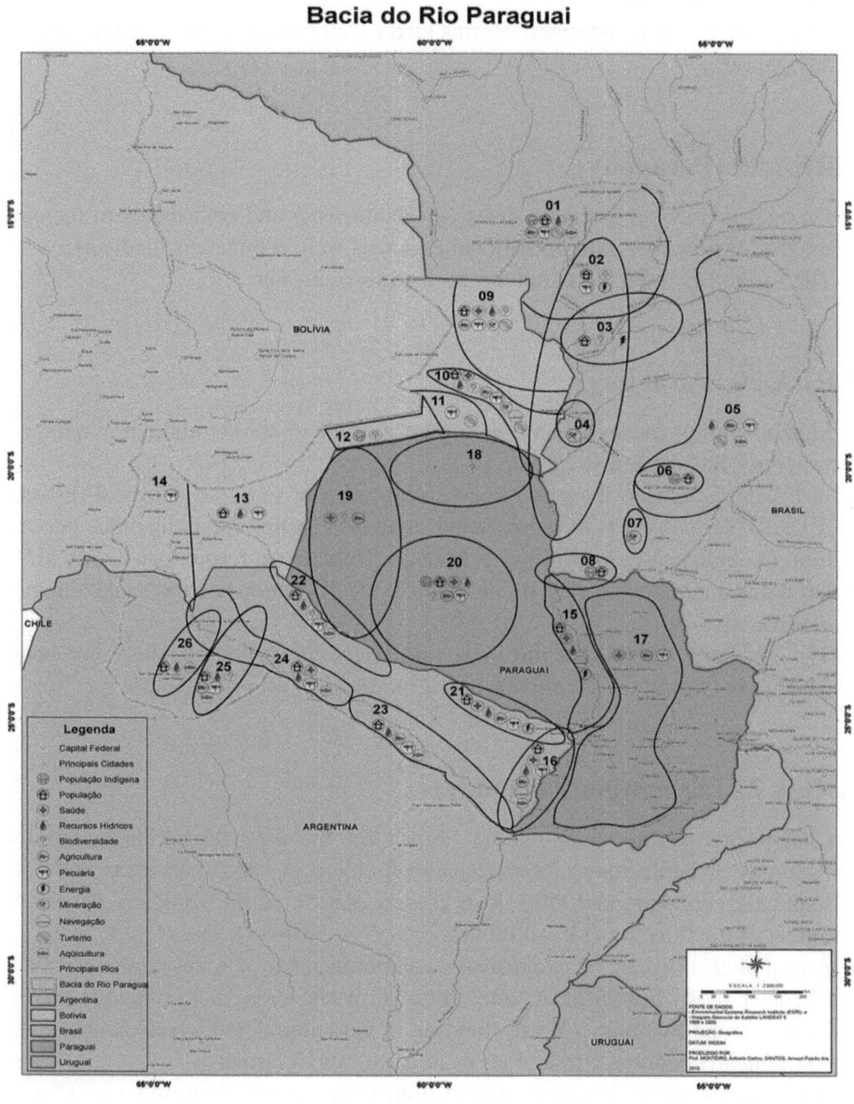

Fig. 2.4 Identified vulnerabilities by SINERGIA project as consequence of climatic factors

AREA 16 (Paraguay/Argentina)

- Any change in hydrological regime has impacts on the area of confluence of the Paraguay and Pilcomayo River, causing flooding in low urban areas and consequently economic losses, as well as coastal erosion of the Paraguay River and retreat of the mouth of the Pilcomayo River. In drought seasons there are water uptake problems and subsequent treatment for drinking water.
- Increased rain causes vector-borne diseases as dengue, yellow fever, and leptospirosis and because of the drought, occurs vector-borne diseases as Hantavirus and respiratory diseases.
- Worsening drought and severe storms affect water transport at critical points for navigation.

ÁREA 19 (Paraguay)

- Drought causes vector-borne diseases (Hantavirus) and contributes to increase the risk of desertification of National Parks, as well as threatens biodiversity.
- The variability in precipitation threats harvest rainwater.

ÁREA 20 (Paraguay)

- Intense rainfall causes urban flooding because there are not enough gutters and waterproofing asphalt streets.
- In Filadelfia, Chaco there is an increasing amount of acute diarrheal diseases (EDAS), serious cases of dehydration, mainly affecting the indigenous population. Drought causes vector-borne diseases (Hantavirus, respiratory diseases).
- Droughts increase the risk of desertification of National Parks, and threaten biodiversity.
- The variability in rainfall threatens rain water harvesting. In the other hand, drought in Central Chaco affects agricultural and forest productivity.

ÁREA 21 (Argentina/Paraguay)

- Changes on hydrological regime impact on the area of confluence of the Paraguay and Pilcomayo River, causing flooding in low urban areas and consequently economic and labor lost, coastal erosion of the Paraguay River and retreat of the Pilcomayo River mouth.
- Appearance of hydric related diseases and permanence of water related vectors.
- Changes in hydrological regime impacts in the area of confluence of the Paraguay and Pilcomayo River, flooding of agricultural and livestock areas, and consequently economic loss of crops affecting small farmers.
- Water consumption increases due to high temperatures (November to March) which generates a deficit to the timber industry and the cities located along Route 86 (Argentina).

Fig. 2.5 City of Pilar complete flooded, located south of Asuncion. May 1983

ÁREA 22 (Argentina/Paraguay)

- Intense rainfall causes flooding due to Pilcomayo River overflows and land loss due to coastal erosion.
- Long droughts cause forest fires, resulting in loss of biodiversity and ecosystems.
- Long droughts and floods result in loss of grazing (for goats, pigs, horses, sheep, and cattle). Droughts that occurs during winter season causes animal mortality for lack of food.
- The change in the hydrological regime resulting in high pressure on the isinglass sector of Pilcomayo River.

SINERGIA project identified major vulnerabilities in rivers Pilcomayo and Paraguay at their cross-border section. Water risk analysis in these rivers is completed with the knowledge of the behavior of droughts and floods in the region, considering that Risk is the product of the threat of occurrence of an extreme natural event by the vulnerability and exposure of the region where extreme event occurs (Fig. 2.5).

Information System for Risk Management at Paraguay and Pilcomayo Transboundary Rivers

The information system for risk management at Paraguay and Pilcomayo transboundary Rivers, was the result of three seminars developed between 2012 and 2013 in Asuncion, Paraguay and Resistencia, Chaco, Argentina. During these seminars, developed with the support of CapWEM project, it was defined the scope of the study and the facilities of the system itself.

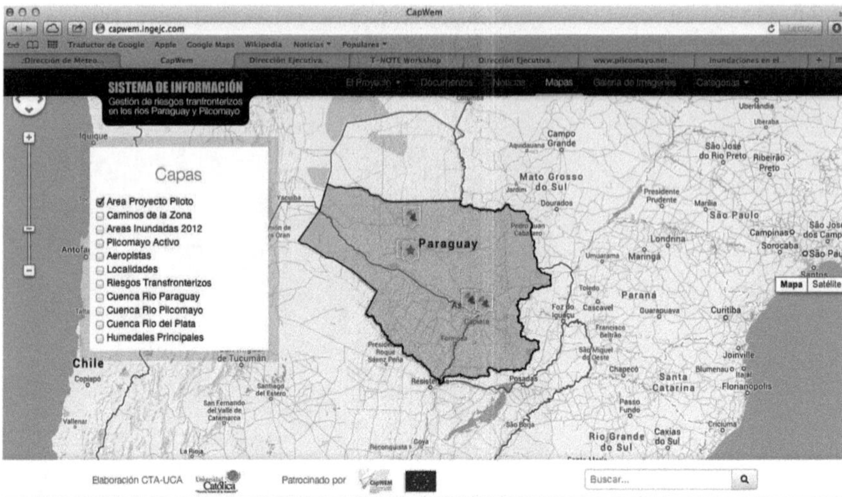

Fig. 2.6 Main screen of the system developed. Figure shows the project area

The general objective of the system is to implement a transboundary management tool for managing the risks associated with extreme hydrological events in rivers shared between Argentina and Paraguay, to establish coordination between local emergency Agencies. The specific objectives are:

- To identify and classify the main natural hydrologic threats in the study area, flood, drought, river coastal erosion, sedimentation and water quality, as well as, anthropogenic threats like urbanization.
- To identify the vulnerable areas in the study region.
- To evaluate the socioeconomic impact of extreme Hydro meteorological events which affect human life, urban and rural infrastructure, and impacts on the agricultural and cattle sector.

This study do not pretend to develop new specific knowledge in flood, drought, water quality or another things in the basin of Pilcomayo and Paraguay River, it pretends to gather published information, organize and store it in a data base, which can be visualized in the Web based plataform.

The system provides information related to hydrographic basins, scientific studies related to flooding and drought, meteorological and hydrological data from public institutions, as well as distribution of communities with their populations, many of them vulnerable.

Major main structure of the platform can be viwed Fig. 2.6 and the first results in—Fig. 2.7. If you want to test the system, go to http://capwem.ingejc.com.

The platform is still under development. However, there are important meteorological and hydrological information derived from several automatic weather stations from DINAC (Civil Aviation National Direction, Paraguay) and other institutions.

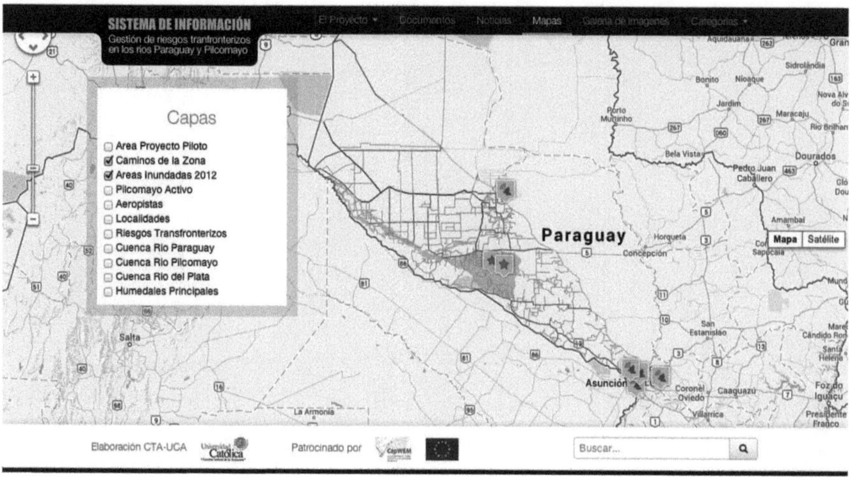

Fig. 2.7 Main screen of the system developed. Road (*brown line*) and flood impacts area of 2012

Conclusions

The Pilcomayo and Paraguay rivers are very important for the country. The Pilcomayo is under continuous threat, flood, drought, sediments and water quality. On the other hand, the Paraguay Rivers get importance because is the main communication way of the country and during low flow and high level of rivers will be a threat for both, companies and populations.

The platform under developing and continuous work will have a great chance to become as a tool where the users will have the ability to get information in order to take decisions in case of floods and drought. Our focus is in the rivers Pilcomayo and Paraguay, but the platform has capacity to incorporate information for all the country.

References

Barros V et al (2004) The major discharge events in the Paraguay River: magnitudes, source regions, and climate forcings. J Hydrometeorol 5:1161–1170

Collischonn W et al (2001) Further evidence of changes in the hydrological regime of the River Paraguay: part of a wider phenomenon of climate change? J Hydrol 245:218–238

Martín-Vide JP et al (2012) Collapse of the Pilcomayo River. Geomorphology. Elsevier. Available online 13 Dec 2012. http://dx.doi.org/10.1016/j.geomorph.2012.12.007

SINERGIA (2011) International system of studies on water resources and the impacts of global warming in the Pantanal Basin. http://www.portalsinergia.org.br/homepage.php

UNECE (2009) Transboundary flood risk management. Experiences from the UNECE region. http://www.unece.org/env/water/meetings/transboundary_flood_workshop.htm, ISBN: 978-92-I-II70II-5

Chapter 3
Analysis of Temporal Variability of Droughts in Southern Paraguay and Northern Argentina (1961–2011)

Maria Manuela Portela, Artur Tiago Silva, João Filipe dos Santos, Julián Baez Benitez, Carlos Frank and José Miguel Reichert

Abstract The study presents an analysis of droughts using monthly rainfall data, from January 1961 to December 2011, from 11 rain gauges located in Paraguay and Northern Argentina. The characterization of the drought events used the standardized precipitation index (SPI) applied at different time scales (3, 6, and 12 consecutive months). The temporal variability of the droughts in the study period was analyzed in terms of changes in their frequency—regardless of the severity, has the frequency of droughts increased or decreased?—and in their severity—are we experiencing more severe droughts or not? The results achieved, despite proving the suitability of the approaches applied, did not reveal any trend towards an increase or a decrease either in the frequency of the droughts or in their severity in the studied area.

Keywords Drought • Standardized precipitation index (SPI) • Kernel occurrence rate estimator (KORE) • Drought occurrence rate • Drought severity

M. M. Portela (✉) · A. T. Silva · J. F. dos Santos
Instituto Superior Técnico, IST. DECivil, Av. Rovisco Pais, 1, 1049-001 Lisbon, Portugal
e-mail: maria.manuela.portela@ist.utl.pt

A. T. Silva
e-mail: artur.tiago.silva@ist.utl.pt

J. F. dos Santos
e-mail: Joaof.santos@ipbeja.pt

J. B. Benitez
Dirección de Meteorología e Hidrología (DINAC), Francisco López
No. 1080 y de la Conquista, Asunción, Paraguay
e-mail: baez.julian@gmail.com

C. Frank
Universidad Tecnologica Nacional, Buenos Aires, Argentina
e-mail: carlosfrank@criba.edu.ar

J. M. Reichert
Department of Soil Science, Federal University of Santa Maria (UFSM), Santa Maria, Brazil
e-mail: reichert@smail.ufsm.br

W. Leal Filho et al. (eds.), *International Perspectives on Climate Change*,
Climate Change Management, DOI: 10.1007/978-3-319-04489-7_3,

Introduction

Droughts are generally associated with the persistence of low rainfall, soil moisture and water availability relative to the normal levels in a designated area. Although there is no universally accepted definition for drought, Tallaksen and Van Lanen (2004) defines it as "a sustained and regionally extensive occurrence of below average natural water availability". Unlike other extreme events, like floods and earthquakes, droughts remain a less visible natural risk, whose impacts are not systematically recorded. Droughts are among the most complex and least understood natural hazards, affecting more people than any other one. They are also recurrent hazards particularly in areas with pronounced natural climate temporal variability, as those under analysis.

The objective of the present study was to quantify the yearly drought occurrence rates in the northern Argentina and in Paraguay, using a new approach, the Kernel occurrence rate estimation method coupled with bootstrap confidence band and the standardized precipitation index, SPI, for drought recognition.

Some authors have studied drought occurrences in several countries. Recently, for Japan, Lee et al. (2012) have analyzed the spatiotemporal characteristics of drought occurrences over Japan for the period from 1902 to 2009 using an effective drought index (EDI). With hierarchical cluster analysis applied to drought characteristic data (such as duration, severity and onset and end dates) available at 50 observational stations, drought regions were identified and drought occurrences analyzed.

Using future climate scenarios, Sheffield and Wood (2008) analyzed changes in drought occurrence using soil moisture data. According to the same authors for the future projections, the models showed decreases in soil moisture globally for all scenarios with a corresponding doubling of the spatial extent of severe soil moisture deficits and frequency of short-term (4–6-month duration) droughts from the mid-20th to the end of the 21st centuries. Long-term droughts become three times more common. Regionally, the Mediterranean, west African, central Asian and central American regions show large increases most notably for long-term frequencies as do mid-latitude North American regions but with larger variation between scenarios.

For Argentina some related studies have also been pursued, namely the work of Capriolo and Scarpati (2012). Given the soil water balance obtained by the evapotranspiration formula of Penman–Monteith, soil water deficit and surplus were considered as triggers of extreme hydrologic events. The authors considered annual threshold values of 200 mm of soil water deficit and 300 mm of soil water surplus for drought and flood recognition, respectively. Using the Mann–Kendall statistical test the results have shown significance trends at level 0.1 for drought and for two periods, one of 20 years (1991–2010) and the other of 10 years (2001–2010). Ravelo (2000), have analyzed droughts for the period 1931–1999 in the plains region of Argentina. Meteorological drought indices were used in a time and space analysis to establish drought intensity, frequency, probability distribution and levels of probability for the occurrence of given drought intensities. More extensively Barrucand et al. (2007) studied the frequency and spatial distribution of droughts in different regions of Argentina during the 20th century. The behavior of the mean monthly atmospheric circulation associated with dry conditions in the Pampas during the second half of the century was analyzed.

Within the study area, Paraguay is particularly affected by droughts. The potential impacts of such hazards on the economy of the country can be significant, especially taking into account that Paraguay is the sixth largest producer of soybean in the world (Masuda and Goldsmith 2009a, b). As a consequence of one of the most severe drought periods, from November 2008 to March 2009, the Gross Domestic Product (GDP) fell by 4.2 % in the first trimester of 2009, and the yield of soybean suffered a reduction of 30 %. In Argentina, the impact event was worst.

Impact assessment of a specific drought requires knowing its causes and the spatial and temporal distribution of the rainfall anomalies. Grimm et al. (2000) and Grimm (2004), analyzed the influence of El Niño Southern Oscillation (ENSO) warm (El Niño) and cold (La Niña) phases in the rainfall patterns of southeastern region of South America, providing a comprehensive view of the anomalies of rainfall and atmospheric circulation associated with both ocean–atmosphere phase events. The La Niña phase coincides with a reduction of rainfall in northern Argentina, southern Brazil and southeastern Paraguay. Fraisse et al. (2008) analyzed soybean yields in Paraguay and rainfall amounts during different phases of crop phenological development, finding significant rainfall reductions during La Niña phases. Podestá et al. (2002) also demonstrated the influence of ENSO phases on agriculture in central and eastern Argentina.

To assess the drought occurrence in the region and to understand the historical and recent climatic variability, it is worthwhile to study the long-term time series of rainfall regarding their nonhomogeneous climatic and hydrological conditions. The present study was based on 51 years of rainfall data, from 1961 to 2011, in 11 rain gauges distributed over the study area (Argentina and in Paraguay). The yearly drought occurrence rates were quantified using a new approach, the Kernel occurrence rate estimation (KORE) method coupled with bootstrap confidence band. To recognize the drought occurrences the standardized precipitation index (SPI), developed McKee et al. (1993), was applied. This index was designed to quantify the precipitation deficit at multiple time scales, which reflect the impact of droughts on the different types of reservoirs of fresh water at the watershed level. Indeed, soil moisture conditions respond to precipitation anomalies on a relatively short scale, while streamflow, reservoir storage and groundwater reflect the longer-term precipitation anomalies.

The perception of the meaning of drought and of its impacts varies significantly (Vogt and Somma 2000) as, in fact, drought does not have a precise and universally accepted definition. According to Rossi (2003), most of drought definitions refer to one or more components of the hydrological cycle, and to the impacts of the water shortages on ecosystems or on specific water users, according to the respective branch of science or activity (agriculture, socioeconomics, health, etc.). Nevertheless, it should be stated that there is a consensus regarding the following different types of drought: meteorological, agricultural, hydrological and the socio-economic (Wilhite and Glantz 1985). Such types of drought can also be defined in straight connection with the SPI index.

Meteorological drought, also termed climatological drought, is caused by a precipitation deficit over an extended period of time. This deficit may be accumulated and expressed relative to a climate norm and to the duration of the dry period (Lloyd-Hughes 2002). The SPI is linked to this drought type when calculated at 1–3 months' time scale (Hayes et al. 1999).

The water soil deficiency is usually connected to agricultural drought and is caused by a deficit of fresh water relative to evapotranspiration losses. A drought exists when the water availability at the root-zone is insufficient to sustain crops and pasture between precipitation events (Tate and Gustard 2000). For agricultural drought Sims et al. (2002), reported a strong relationship between SPI over short time scales (3–6 months) and temporal variations of soil moisture.

A hydrological drought results directly from reduced precipitation, which originates reduced surface runoff and, indirectly, from reduced groundwater discharge to the river channel. Key indicators are reduced river flows and low water levels in lakes and reservoirs. According to Lloyd-Hughes (2002), hydrological droughts are the most visible and important in terms of human perception. The SPI at 12 month time scale is considered a hydrological drought index, having been tested for monitoring surface water resources, e.g., river flows and water levels in lakes (Szalai and Szinell 2000; Hayes et al. 1999).

At longer time scales of the SPI (24 or 36 months), droughts last longer, but are less frequent. They are used to monitor the impact of droughts on aquifers, which are systems that respond more slowly to changing conditions (Changnon and Easterling 1989).

In the applications carried out the time scales considered were 3, 6 and 12 months (SPI3, SPI6 and SPI12, respectively).

Study Region and Rainfall Data

The study region spreads from the southern half of Paraguay to the contiguous Northeastern states of Argentina, namely the states of Formosa, Chaco, Corrientes, and Misiones. The data consists of 11 series of monthly rainfall, from January 1961 to December 2011 (50 years), from rain gauges distributed on that region. Figure 3.1 shows the study region, along with the location of the rain gauges. The rain gauges were spatially grouped as further justified in Sect. Changes in Drought Occurrence Rate.

The rainfall data from Paraguay and Argentina comes from the National Meteorological Services of both countries. In the case of Paraguay, the institution is the Direction of Meteorology and Hydrology of the National Directorate of Civil Aeronautical (DINAC—Spanish acronym, www.meteorologia.gov.py) and in Argentina is the National Weather Service (www.smn.gov.ar) that depends on the Defense Ministry of Argentina.

The measurement of the rainfall is accomplished at stations equipped with standard pluviometers (Hellman type) and in some cases with automatic weather stations, in both cases following the World Meteorological Organization, WMO,

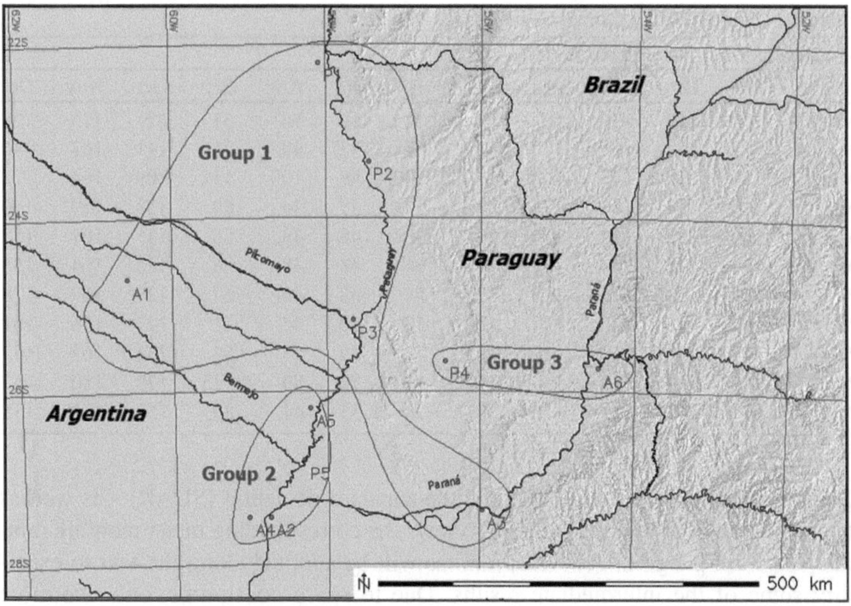

Fig. 3.1 Map of the study area. Location and grouping of the 11 rain gauges

Table 3.1 Name and location of the 11 rain gauges; mean annual rainfall (MAR); standard deviation of the annual rainfall (SDAR)

Sample code	Country	Location	Latitude	Longitude	MAR (mm)	SDAR (mm)
A1	Argentina	Las lomitas	−60.58	−24.7	911	189
A2		Corrientes	−58.77	−27.45	1,436	341
A3		Posadas	−55.97	−27.37	1,758	361
A4		Resistencia	−59.05	−27.45	1,379	317
A5		Formosa	−58.23	−26.2	1,413	276
A6		Iguazu	−54.47	−25.73	1,843	276
P1	Paraguay	Pto. Casado	−58.09	−22.21	1,245	250
P2		Concepcion	−57.45	−23.35	1,356	268
P3		Asuncion	−57.66	−25.18	1,392	299
P4		Villarica	−56.46	−25.65	1,668	327
P5		Pilar	−58.31	−26.8	1,397	297

standards and normative. The data, which is subjected to quality control (QC) is processed in the climatology department of both meteorological services. Normally the quality control search for integrity of daily data and missing data. Another QC applied is the verification of the rainfall occurrence between near meteorological stations, which represent the coherence with the meteorological system that produce the rainfall. The information of both meteorological services is available for research and exchangeability between national intuitions.

The specific rainfall records utilized in this study are identified in Table 3.1 which includes, for each rain gauge, some statistical characteristics of the annual

Table 3.2 Mean monthly rainfall at the 11 rain gauges

Sample	Mean monthly rainfall (mm)											
code	Jan	Feb	Mar	Apr	May	Jun	Jul	Aug	Sep	Oct	Nov	Dec
A1	127	117	139	90	52	20	13	16	31	62	115	128
A2	171	152	163	180	88	61	43	49	75	141	167	146
A3	159	152	142	162	127	130	98	100	142	209	164	173
A4	169	154	166	174	81	57	37	44	69	128	159	140
A5	169	142	148	155	99	66	48	48	86	135	164	155
A6	169	142	148	155	99	66	48	48	86	135	164	155
P1	141	128	126	107	83	54	32	40	57	128	169	178
P2	159	139	133	127	120	59	40	44	71	127	176	161
P3	147	134	128	157	115	71	49	58	84	134	162	153
P4	174	139	148	160	142	114	72	89	115	178	180	157
P5	165	142	151	173	84	58	53	44	80	157	155	135

rainfall series—the mean (MAR) and the standard deviation (SDAR)—as well as its geographic coordinates. Table 3.2 shows the corresponding mean monthly rainfalls. The rain gauge of A1—Las lomitas will be adopted along the text to exemplify some of the intermediate results. Due to space constraints, only the more relevant results from this study are summarized for the all rain gauges.

SPI and Drought Occurrences

The analysis of the temporal variability of droughts in the study region was assessed via the standardized precipitation index (SPI), one of the most popular and common drought indexes (Santos et al. 2010; Vicente-Serrano 2006).

The SPI, originally developed by McKee et al. (1993), remaps the rainfall records into a standardized probability distribution function so that an index of zero indicates the median rainfall amount, while a negative index stands for drought conditions and a positive one, for wet conditions, Santos et al. (2011). A comprehensive description of the calculation and of the advantages of the SPI index can be found in Edwards and McKee (1997), Guttman (1998, 1999), Hayes et al. (1999), Lloyd-Hughes and Saunders (2002) and Santos and Portela (2010).

As summarized by Santos et al. (2010), there are several advantages to the SPI, namely (1) its flexibility, as it can be applied at various time scales; (2) there is less complexity involved in its implementation, relative to other drought indices; (3) it is adaptable to other hydroclimatic variables besides precipitation, Santos et al. (2010); and (4) its suitability for spatial analysis, allowing comparison between sites in a given region as it is a normalized index.

Though originally the computation of the SPI index utilized a Gamma distribution function applied to the monthly rainfall series, for SPI1, and to cumulative rainfall series, for the other time scales (McKee et al. 1993), several authors tested different distributions based on different time scales and concluded that, compared

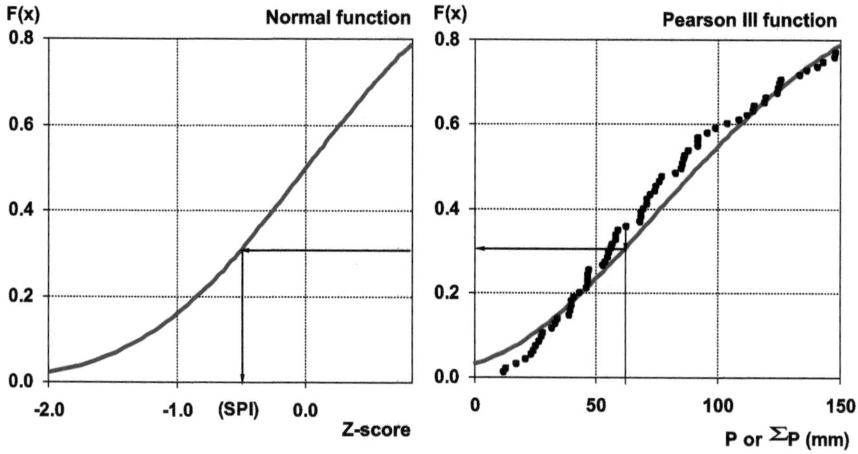

Fig. 3.2 Schematic representation of the SPI calculation procedure (reproduced from Santos et al. 2013)

with the Gama distribution, the Pearson type III distribution ensured the best fit due to its higher flexibility given by its three parameters (Guttman 1999; Ntale and Gan 2003; Vicente-Serrano 2006). The Pearson III probability distribution function is given by:

$$f(x) = \frac{1}{\alpha \Gamma(\beta)} \left(\frac{x-\gamma}{\alpha} \right)^{\beta-1} \exp\left(-\frac{x-\gamma}{\alpha} \right) \tag{3.1}$$

where γ, α, and β are the location, scale, and shape parameters, respectively. The parameters of the previous distribution should be estimated using the L-moments method.

The value of SPI assigned to each rainfall is the z-standard normal associated to the probability of non-exceedance of that rainfall, according to the Pearson type III distribution, as represented in Fig. 3.2.

The relationship among the SPI values, the probabilities of non-exceedance and the drought categories are shown in Table 3.3.

The SPI calculated at 1 month, SPI1, is the normalized monthly rainfall and essentially conveys meteorological drought identification. At the 3–6-month time scales it may be taken as an agricultural drought index, and at 6–12-month time scales it constitutes a hydrological drought index, useful for monitoring surface water resources, as previously specified.

Figure 3.3, exemplifies how the evolution of the SPI index for increasing time scales can be synthetically represented. The figure (concept taken from http://joewheatley.net/visualizing-drought/), which illustrated the SPI at the time scales from 1 to 12 months at rain gauge A1—Las lomitas, in Northern Argentina, clearly shows a pronounced recurrence of dry periods in the 1960s and 1970s

Table 3.3 Drought categories given by the SPI, after Agnew (2000)

Probability	SPI	Drought category
0.05	>1.65	Extremely wet
0.10	>1.28	Severely wet
0.20	>0.84	Moderately wet
0.60	>−0.84 and <0.84	Normal
0.20	<−0.84	Moderate drought
0.10	<−1.28	Severe drought
0.05	<−1.65	Extreme drought

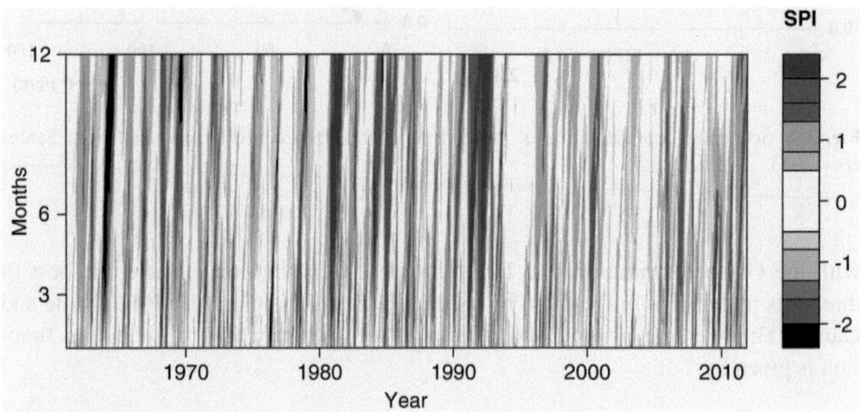

Fig. 3.3 SPI at rain gauge A1—Las lomitas at several time scales

(yellow/brown colors especially marked in the higher time scales of SPI), followed by a wetter period (green/blue colors) in the 1980s and early 1990s. Figure 3.3 also shows that as the SPI time scale increases towards the 12 months the number of drought events decreases, because the droughts at longer time scales tend to be longer. Thus, the number of droughts of longer duration increased with longer SPI time scales.

To study the spatial and temporal variability of different types of droughts in Paraguay and in the Northeastern states of Argentina, the SPI was used at three time scales: 3 (SPI3), 6 (SPI6) and 12 (SPI12) months, mainly to be focused on agricultural and hydrological drought monitoring, in respect to two of the most important South American resources.

Figure 3.4, included in the next page, shows the time series of SPI3, SPI6, and SPI12 at rain gauge A1—Las lomitas along with the threshold at −1.28, which is used to identify severe droughts (see Table 3.3). The red tick marks along the time axis indicate the months with droughts. The figure clearly shows the higher occurrence of drought events in the first decades of the study period, as already suggested by Fig. 3.3 for the same rain gauge.

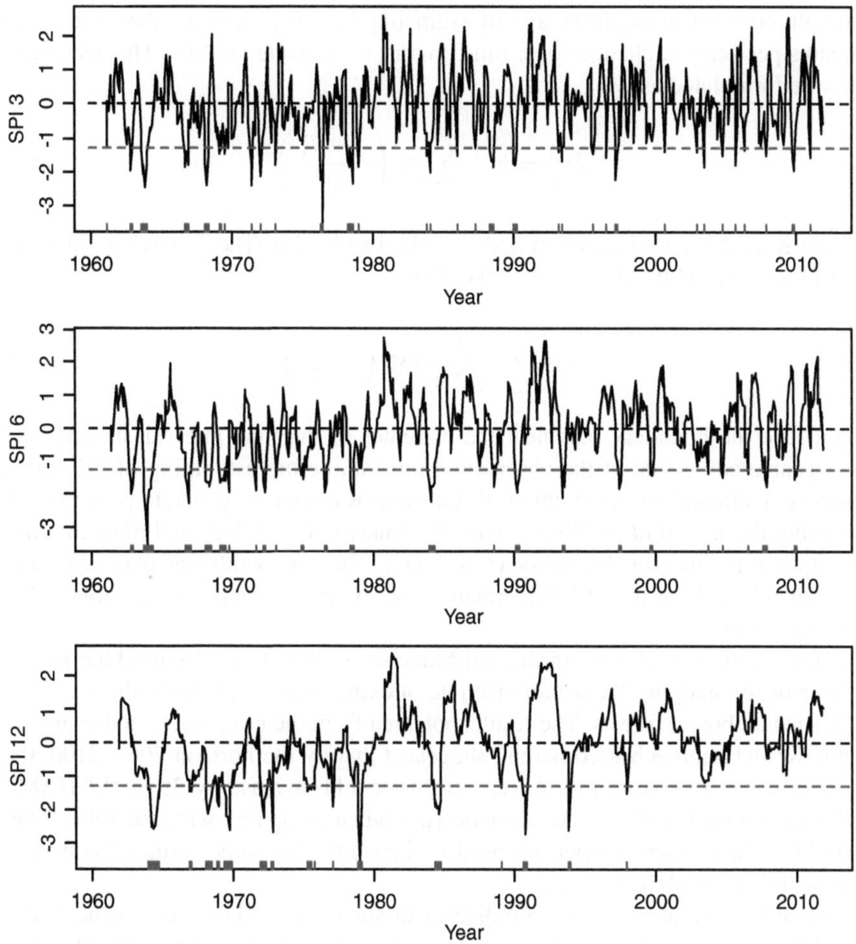

Fig. 3.4 SPI3 (*top*), SPI6 (*middle*) and SPI12 (*bottom*) at rain gauge A1—Las lomitas (Northern Argentina). The *red tick marks* indicate the occurrence of droughts (attributed to the last month of the interval), as the SPI drops below -1.28

Temporal Variability of Droughts

Changes in Drought Occurrence Rate

Analysis of changes in temporal occurrence attempt to answer the question: regardless of the severity of the drought, that is to say, of the rainfall deficit, has the occurrence of droughts increased or decreased? Hence the analyzed variable is not related to the SPI themselves but to the temporal distribution of the occurrence of droughts.

To address the previous question, a kernel occurrence rate estimator, KORE (Mudelsee et al. 2003; Silva et al. 2012), may be applied to a historical series of drought occurrences with the aim of estimating how the mean number of drought months per year, λ, changes over time, that is, to characterize λ(t). This technique may be formulated as:

$$\widehat{\lambda}(t) = h^{-1} \sum_{i=1}^{m} K\left(\frac{t - T_i}{h}\right) \tag{3.2}$$

where K is the kernel function and h is the bandwidth. The following Gaussian kernel was used (Mudelsee et al. 2004, 2006):

$$K(y) = \frac{1}{\sqrt{2\pi}} \exp\left(-\frac{y^2}{2}\right) \tag{3}$$

To reduce the boundary bias near the extremes of the time interval, pseudo data was generated outside of the observation interval, before estimating λ(t). For that purpose a straightforward method of reflection was used to generate pseudo data, covering the amplitude of three times the bandwidth h before and after the limits of the time interval (Mudelsee et al. 2004). The bandwidth was obtained using Silverman's 'rule of thumb', Silverman (1986, p. 48). Its values range from 1.387 to 7.513 years.

Other authors, such as Girardin and Mudelsee (2008), have also used the Kernel estimator for studying the occurrence rate of extreme events, namely the fire years in Canadian boreal forests. The results obtained from their approach, and using the same Kernel approach used herein, suggested that by the horizon 2061–2100, the median number of large forest fires per year could increase by 39 % (ECHAM4 B2 scenario run) to 61 % (A2 scenario run) when compared with the 1901–1940 and 1781–1820 reference periods used in the study. The same results if considering the full 1999–2100 horizon.

Regarding the use of drought indices, Christie et al. (2011) finds for the Andes Cordillera, that severe and extreme drought events reveal unprecedented increments on the respective occurrence rates during the last century when compared with the previous six centuries. For that purpose the previous authors considered the Palmer Drought Severity Index (PDSI) to account with regional moisture and the same advanced statistical tool as in the present study, the Kernel estimation method, to analyze the occurrence rates of droughts.

Figure 3.5 summarizes the application of the KORE estimator to the SPI at rain gauge A2—Las lomitas. The analysis focused on the occurrence of severe droughts, that is, the occurrence of SPI values lower than −1.28. To account for the uncertainty of the KORE estimates, a point wise 90 % confidence band was constructed around λ(t), by means of a bootstrap simulations (Cowling et al. 1996, Mudelsee 2011), according to the methodology described in Silva et al. (2012). The Kernel occurrence rate estimation method coupled with bootstrap confidence band construction was first introduced into the analysis of climate extremes, by Mudelsee et al. (2003); with detailed description given by Mudelsee et al. (2004).

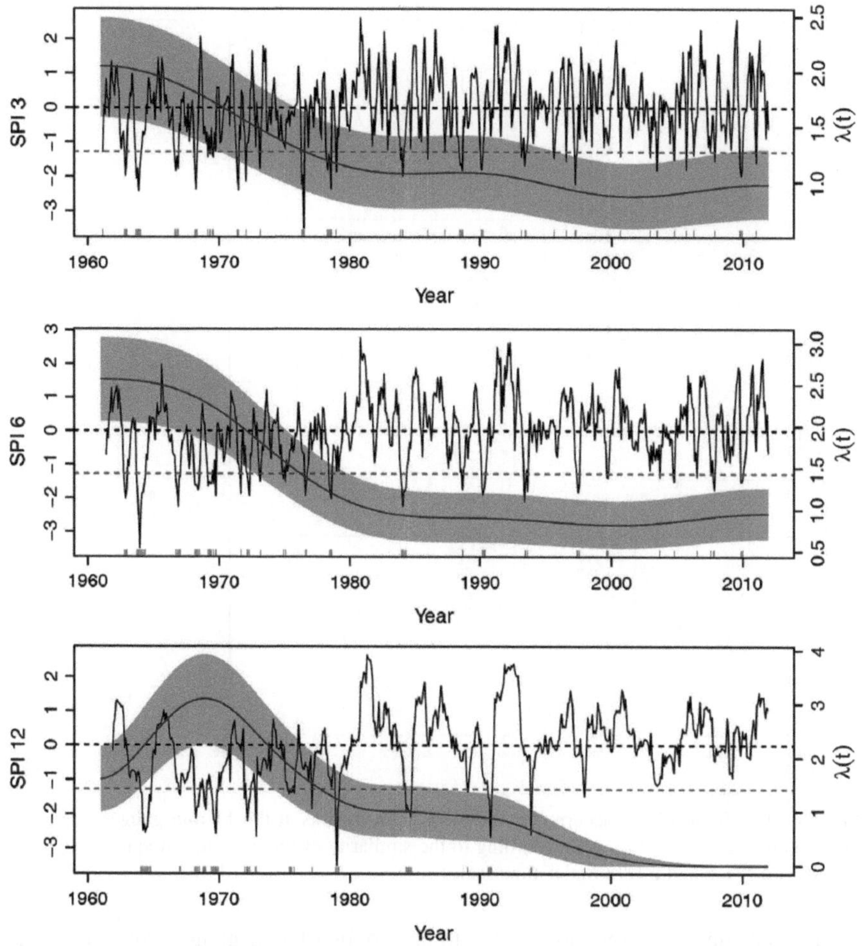

Fig. 3.5 SPI3, 6 and 12 at rain gauge A1—Las lomitas (*left axis*); occurrence of droughts (*red tick marks* along the time axis), and time-dependent occurrence rate of droughts, λ(t), and corresponding 90 % bootstrap confidence intervals

Figure 3.5 shows, that, for all the studied SPIs (3, 6 and 12) the occurrence rate of drought months in A1—Las lomitas is lower in recent years than it was in the past. The width of the confidence bands attest to the significance of the changes in the observed occurrence rate.

Similar figures were obtained for all 11 historical monthly rainfall series but are not shown due to space constraints. Significant changes in occurrence rate of severe drought months were detected at all the studied sites. However, those changes were not homogeneous in space: in some samples there is a visible increase in the occurrence rates of the droughts towards the present, while other samples show a decrease in the same temporal direction.

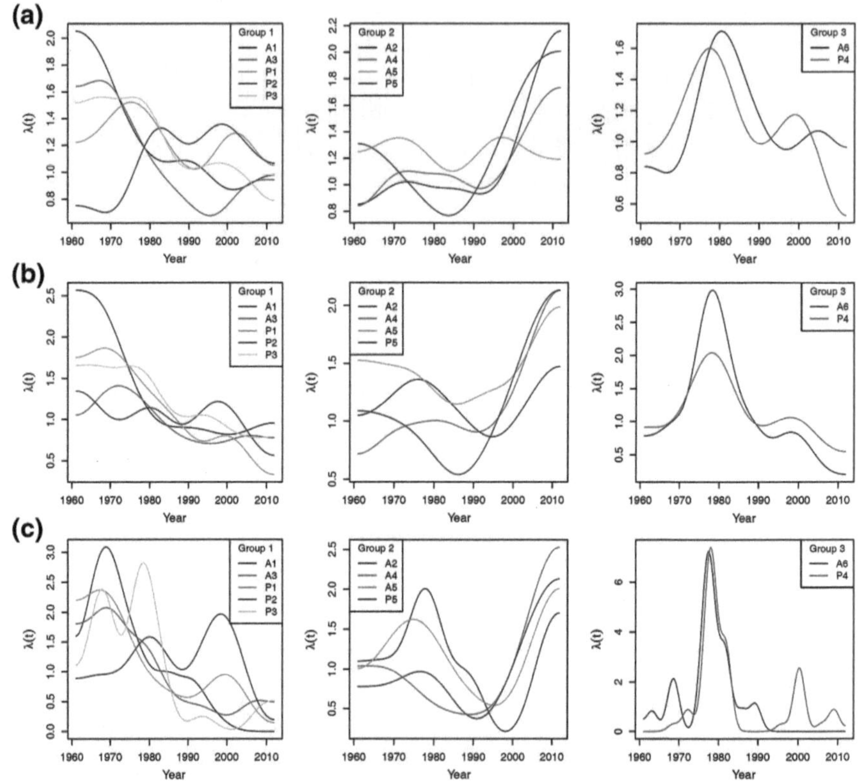

Fig. 3.6 Time-dependents occurrence rate of drought months at the 11 rain gauges, organized into the three groups of Fig. 3.1, according to the similarity of the exhibited frequency changes. Results at **a** 3, **b** 6 and **c** 12-months' time scales

By analyzing the previous results, three groups of rain gauges were identified where the temporal evolution of the KORE estimates behave similarly, at least for most of the time scales of SPI. Except for the rain gauge of A3—Posadas of group 1, the areas that encompass the rain gauges of the three groups do not overlap, and are spatially apart, as represented in Fig. 3.1.

Figure 3.6 shows the KORE estimates (without the bootstrap confidence bands) for all the studied samples, grouped by similarity in the behavior of the frequency changes over time.

The results of Fig. 3.6 show that, generally, the sites in group 1 have seen a recent decrease in the occurrence of severe drought months, while the sites belonging to group 2 have seen a recent increase in all of the studied time scales. The figure shows that the rain gauges of group 3 have verified a clustering of drought months at the end of the 1970s.

It should be mentioned that the behavior detected for group 1 is in agreement with the one reported by Barrucand et al. (2007) and Penalba and Vargas (2001) that find a reduction in the number of dry cases during the second half of the 20th century. In their study the drought index was calculated as the percentage of

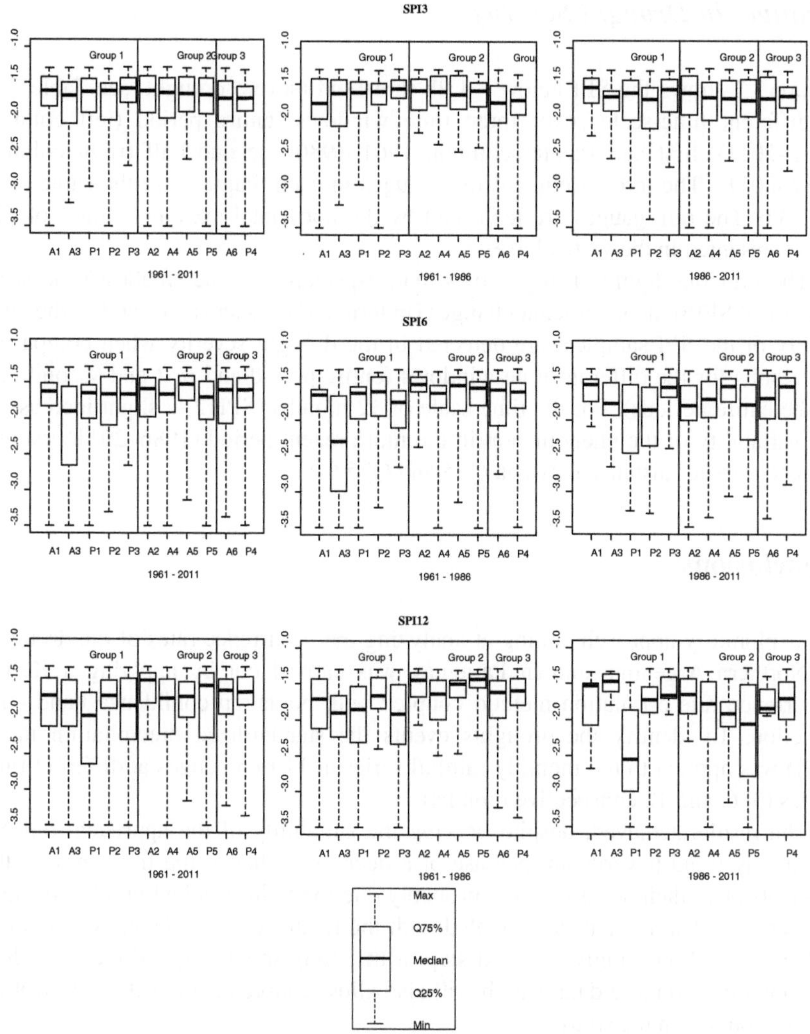

Fig. 3.7 Boxplots of the SPI of drought months (SPI <−1.28) at the 11 rain gauges over three time windows: the entire period (*left column*, 1961–2011); the first half (*center column*, 1961–1986); and the second half (*right column*, 1987–2011)

stations with rainfall lower than the median in a given region and dry months were determined as those with an index over 0.8

Changes in Drought Severity

Figure 3.7 shows the SPI below −1.28 threshold (severe drought, see Table 3.3) at different time-scales over three time windows: entire period (left column, 1961–2011); first half (center column, 1961–1986); second half (right column, 1987–2011). The three groups of rain gauges in each diagram are the same as in Fig. 3.6. The rain gauges are identified by the horizontal axes and values of SPI are represented in the vertical axes.

The previous figure clearly shows that, especially for the smallest time scale (SPI3 and SPI6), no significant changes (in terms of median values and of the variability of the SPI samples) are detected in the drought severity when comparing the results for the different time windows (entire period and first and second half of the entire periods). For the time scale of 12 months (SPI12) a slight increase in the variability of the phenomenon in the more recent period (1986–2011) may be noticed (greater distance between Q25 and Q75 %).

Conclusions

An exploratory approach aiming at analyzing the occurrence rates of the droughts in Northern Argentina and in Paraguay was carried out by applying a Kernel occurrence rate estimation method coupled with bootstrap confidence band construction. To identify the droughts events the standardized precipitation index (SPI) was applied to the monthly rainfall series at 11 rain gauges at different time scales (3, 6, and 12 consecutive months).

The results achieved, despite proving the suitability of the approach, did not reveal any trend towards an increase or a decrease either in the frequency of the droughts or in their severity, most probably due to the insufficient number of rainfall series that were analysed coupled with the relatively small length of the rainfall samples. Accordingly, the next step of the study should expand the procedure by applying it to more data, which will also allow a more detailed characterization of the droughts in the study area.

Acknowledgments The study was supported by the project CapWEM [Capacity development in Water Engineering and Environmental Management DCI-ALA/19.09.01/10/21526/254-922/ ALFA III (2010)55], financed by the European Commission (ALFA III Programme) in which eight partner institutions collaborate: six from Latin American countries (Argentina, Brazil, Chile, Costa Rica, El Salvador, and Paraguay) and the remaining two from European countries (Germany and Portugal). The objectives of the project concern capacity building of Higher Education Institutes in the scope of Environmental and Water Resources Engineering, including the aspect of hydrological risk analysis.

References

Agnew CT (2000) Using the SPI to identify drought. Drought Netw News 12:6–12

Barrucand MG, Vargas WM, Rusticucci MM (2007) Dry conditions over Argentina and the related monthly circulation patterns. Meteorol Atmos Phys. doi:10.1007/s00703-006-0232-5

Capriolo AD, Scarpati OE (2012) Extreme hydrologic events in north area of Buenos Aires Province (Argentina). ISRN Meteorol, vol 2012. Article ID 415081, 9 p. doi:10.5402/2012/415081

Changnon SA, Easterling WE (1989) Measuring drought impacts: the Illinois case. Water Resour Bull 25:27–42

Christie DA, Boninsegna JA, Cleaveland MK, Lara A, Le Quesne C, Morales MS, Mudelsee M (2011) Aridity changes in the temperate–mediterranean transition of the Andes since AD 1346 reconstructed from tree-rings. Clim Dyn, Springer 36:1505–1521

Cowling A, Hall P, Phillips M (1996) Bootstrap confidence regions for the intensity of a Poisson point process. J Am Stat Assoc 91:1516–1524

Edwards DC, McKee TB (1997) Characteristics of 20th century drought in the United States at multiple time scales. Climatol. Report 97–2, Department of Atmospheric Science, Colorado State University, Fort Collins, Colorado

Fraisse CW, Cabrera VE, Breuer NE, Baez J, Quispe J, Matos E (2008) El Niño—southern oscillation influences on soybean yields in Eastern Paraguay. Int J Climatol 28:1399–1407

Girardin MP, Mudelsee M (2008) Past and future changes in Canadian boreal wildfire activity. Ecol Appl, Ecol Soc Am 18(2):391–406

Grimm AM (2004) How do La Niña events disturb the summer monsoon system in Brazil? Clim Dyn 22:123–138

Grimm AM, Barros VR, Doyle ME (2000) Climate variability in southern South America associated with El Niño and La Niña events. J Clim 13:35–58

Guttman NB (1998) Comparing the Palmer Drought Index and the standardized precipitation index. J Am Water Resour Assoc (JAWRA) 34:113–121

Guttman NB (1999) Accepting the standardized precipitation index: a calculation algorithm. J Am Water Resour Assoc (JAWRA) 35(2):311–322

Hayes M, Wilhite DA, Svoboda M, Vanyarkho O (1999) Monitoring the 1996 drought using the standardized precipitation index. Bull Am Meteorol Soc 80:429–438

Lee SM, Byun HR, Tanaka HL (2012) Spatiotemporal characteristics of drought occurrences over Japan. J Appl Meteorol Climatol 51:1087–1098

Lloyd-Hughes B (2002) The long-range predictability of European drought. Thesis submitted for the degree of Doctor of Philosophy. University of London, England

Lloyd-Hughes B, Saunders MA (2002) European drought climatology and prediction using the Standardized Precipitation Index (SPI). 8.11 In 13th conference on applied meteorology

Masuda T, Goldsmith P (2009a) Produção mundial. Área plantada, produtividade e projeções em longo prazo, *Boletim de Pesquisa de Soja 2009* – Fundação MT, Editora Central de Texto Carrion ET Carracedo Editores Associados, pág. 45–57

Masuda T, Goldsmith P (2009b) World soybean production: area harvested, yield, and long-term projections. Int Food Agribus Manage Assoc (IAMA). Int Food Agribus Manage Rev 12(4):143

McKee TB, Doesken NJ, Kleist J (1993) The relationship of drought frequency and duration to time scales. In: Proceedings of the 8th conference on applied climatology, American Meteorology Society, pp 179–184

Mudelsee M (2011) The bootstrap in climate risk analysis. In: Kropp JP, Schellnhuber HJ (eds) Extremis: disruptive events trends in climate and hydrology. Springer, Berlin, pp 45–58

Mudelsee M, Borngen M, Tetzlaff G, Grunewald U (2003) No upward trends in the occurrence of extreme floods in central Europe. Nature 425:166–169

Mudelsee M, Borngen M, Tetzlaff G, Grunewald U (2004) Extreme floods in central Europe over the past 500 years: role of cyclone pathway Zugstrasse Vb. J Geophys Res 109(D23):D23101. doi:10.1029/2004JD005034

Mudelsee M, Deutsch M, Borngen M, Tetzlaff G (2006) Trends in flood risk of the River Werra (Germany) over the past 500 years. Hydrol Sci J 818–833

Ntale HK, Gan T (2003) Drought indices and their application to East Africa. Int J Climatol 23:1335–1357. doi:10.1002/joc.931

Penalba O, Vargas W (2001) Propiedades de déficits y excesos de precipitación en zonas agropecuarias. Meteorologica 26:39–55

Podestá G, Letson D, Messina C, Royce F, Ferreyra RA, Jones J, Hansen J, Llovet I, Grondona M, O'Brien JJ (2002) Use of ENSO-related climate information in agricultural decision making in Argentina: a pilot experience. Agric Syst 74:371–392

Ravelo AC (2000) Agroclimatic characterization of extreme droughts in the Pampas region of Argentina. Revista de la Facultad de Agronomía (Universidad de Buenos Aires) 20(2):187–192. ISSN 0325-9250

Rossi G (2003) Requisites for a drought watch system. In: Rossi G et al (eds) Tools for drought mitigation in Mediterranean regions. Kluwer Academic Publishing, Dordrecht, pp 147–157

Santos JF, Portela MM (2010) Caracterização de secas em bacias hidrográficas de Portugal Continental: aplicação do índice de precipitação padronizada, SPI, a séries de precipitação e de escoamento. *10º Congresso da Água*, 18 pp., Associação Portuguesa dos Recursos Hídricos (APRH), Alvor

Santos JF, Pulido-Calvo I, Portela MM (2010) Spatial and temporal variability of droughts in Portugal. Water Resour Res 46(3)

Santos JF, Portela MM, Pulido-Calvo I (2011) Regional frequency analysis of droughts in Portugal. Water Resour Manage 25:3537–3558. doi:10.1007/s11269-011-9869-z

Santos JF, Portela MM, Naghettini M, Matos JP, Silva AT (2013) Precipitation thresholds for drought recognition: a complementary use of the SPI. In: 7th International conference on river basin management including all aspects of hydrology, ecology, environmental management, flood plains and wetlands, RBM13, Wessex Institute, New Forest, UK

Sheffield J, Wood EF (2008) Projected changes in drought occurrence under future global warming from multi-model, multi-scenario, IPCC AR4 simulations. Clim Dyn. 31:79–105. doi:10.1007/s00382-007-0340-z

Silva AT, Portela MM, Naghettini M (2012) Nonstationarities in the occurrence rates of flood events in Portuguese watersheds. Hydrol Earth Syst Sci 16(1):241–254

Silverman BW (1986) Density estimation for statistics and data analysis. In: Monographs on statistics and applied probability, vol 26. Chapman & Hall/CRC

Sims AP, Nigoyi DS, Raman S (2002) Adopting indices for estimating soil moisture: a North Carolina case study. Geophys Res Lett 29:1183. doi:10.1029/2001GL013343

Szalai S, Szinell C (2000) Comparison of two drought indices for drought monitoring in Hungary—a case study. In: Vogt JV, Somma F (eds) Drought and drought mitigation in Europe. Kluwer Academic Publishers, Dordrecht, pp 161–166, 325 pp

Tallaksen LM, Van Lanen HA (2004) Hydrological drought: processes and estimation methods for streamflow and groundwater, vol 48. Elsevier Science Limited

Tate EL, Gustard A (2000) Drought definition: a hydrological perspective. In: Vogt JV, Somma F (eds) Drought and drought mitigation in Europe. Kluwer Academic Publishers, Dordrecht, pp 23–48, 325 pp

Vicente-Serrano SM (2006) Spatial and temporal analysis of droughts in the Iberian Peninsula (1910–2000). Hydrol Sci J 51(1):83–97. doi:10.1623/hysj.51.1.83

Vogt JV, Somma F (2000) Drought and drought mitigation in Europe. Kluwer Academic Publishers, Dordrecht, 325 pp

Wilhite DA, Glantz MH (1985) Understanding the drought phenomenon: the role of definitions. Water Int 10(3):111–120

Chapter 4
Sedimentation and Life Expectancy of Lake Amatitlán, Guatemala: Increased Vulnerability Under Future Climate Change

Astrid Michels, Michael Lloyd, Roger Woeller, Jean-Francois Dionne, Lorena Grande, Mario Sorto, Ana-Beatriz Suarez and Manuel Basterrechea

Abstract In the last few decades, Lake Amatitlán a highland lake (surface area 15 km^2, 1,188 masl) located south of Guatemala City, has been severely impacted by human activities. The western basin of the lake (Lago Oeste) receives high sediment loads and huge amounts of untreated industrial and domestic wastewater from the capital. In 2012, a bathymetric survey was conducted in Lake Amatitlán and sediment cores were retrieved to better understand the sediment input and to estimate the lake's loss in storage capacity over time. The life span of Lake Amatitlán's western basin, Lago Oeste, was estimated at ~114 years, from 2012, assuming constant annual sediment accumulation rates. A tephra layer in sediment cores was used as a temporal marker to assess the impact of extreme climate events on sedimentation. Our results indicate that in 2010, during the tropical cyclone Agatha, two days after the explosion of Pacaya volcano, sediment loads increased two-fold. It is likely, that future climate change will increase the frequency and intensity of extreme climate events such

A. Michels (✉) · M. Lloyd · R. Woeller · J.-F. Dionne
BluMetric, PO Box 430-3108, Carp Road, Ottawa, ON K0A 1L0, Canada
e-mail: astrid.michels@outlook.com

L. Grande · M. Sorto
BluMetric, 7 Calle Poniente Bis, Pasaje 9, casa No. 7, Colonia Escalon, San Salvador, El Salvador, Centro América
e-mail: sansalvador@blumetric.ca

A.-B. Suarez
Autoridad para el Manejo Sustentable de la Cuenca y del Lago de Amatitlán (AMSA), Km. 22 Ruta al Pacífico, Bárcenas, Villa Nueva, Guatemala
e-mail: asuarez@amsa.gob.gt

M. Basterrechea
Asesoria Manuel Basterrechea, Avenida La Reforma 1-50, zona 9, Edificio el Reformador, Guatemala, Guatemala
e-mail: asebaste@gmail.com

W. Leal Filho et al. (eds.), *International Perspectives on Climate Change*, Climate Change Management, DOI: 10.1007/978-3-319-04489-7_4,

as tropical storms, heavy rains, and landslides and will consequently impact and enhance sedimentation of Lake Amatitlán. Climate change impacts will be exacerbated by increasing demographic pressures and unregulated settlements in the Lake Amatitlán watershed. Therefore, immediate management actions and adaptation strategies are urgently needed to protect the sedimentation and the disappearance of Lake Amatitlán.

Keywords Climate change • Central America • Amatitlán • Life span • Sedimentation • Watershed management

Introduction

Countries located within Central America are expected to be especially affected by climate change and variability, even though they contribute the least to global greenhouse gas emissions (Agrifor Consultants 2009; Galindo et al. 2011). Guatemala is located on the Central-American Isthmus, a region with a high level of risk from natural phenomena such as volcanic eruptions and earthquakes and water-related disasters including extreme rainfall, landslides, and floods (Valladares Cerezo 2003). Climate scenarios for Central America predict an increased variability in precipitation patterns and an increase in the intensity of extreme climatic events such as droughts, tropical storms, and floods (IPCC 2007a, b). This is partly due to the unique geographical features including a complex topographic relief on a narrow land strip surrounded by two oceans that shape Central America's climate and make its countries subject to greater frequency and intensity of extreme climatic events (IPCC 2007a). The number of tropical storms in Central America alone has increased 12-fold between 2000 and 2009 as compared to the period between 1970 and 1979 (ESI 2001). Guatemala has abundant freshwater resources that currently face a variety of challenges such as sedimentation, water contamination, and eutrophication (Perez et al. 2011). Climate change manifests itself in heavy precipitation events, altered runoffs, and increased soil erosion, and is expected to exacerbate water-related issues in this region (IPCC 2007b). In addition to climate change impacts, pressures such as population growth, poverty, unregulated urban growth on steep slopes make Guatemala extremely vulnerable with wide-ranging social, economical and ecological consequences (Calderon 2010; Valladares Cerezo 2003). Of particular interest in Guatemala, is Lake Amatitlán, the country's fifth largest lake located in one of the most contaminated watersheds of Central America (Calderon 2010).

Lake Amatitlán has a unique landscape with surrounding valleys, mountains and volcanoes and despite its severe water quality degradation the lake still remains a popular site for recreation. Paleolimnological studies reveal the presence of Mayan settlements in the Lake Amatitlán catchment area in the past 2,700 years (Velez et al. 2011). During the Mayan classic era (AD 1300–BC 300) the lake was the main water supply for indigenous populations and considered a sacred place (AMSA 2003). The move of Guatemala's capital in 1776 from Antigua to the present location of Guatemala City marked the beginning of the environmental degradation of Lake Amatitlán. Since 1950, the lake's watershed has been severely affected by increases in population, deforestation, intensified land use and the discharge of huge amounts of untreated industrial and domestic wastewater. In 2010, the Lake Amatitlán watershed had more than 2.2 million inhabitants and population is annually increasing by 3.5 % (Gil 2011). The lake receives approximately 50 % of Guatemala's City sewage and industrial effluents that are transported via Río Villalobos River into Lake Amatitlán (US ACE 2000).

In 1996 the Guatemalan Authority for Sustainable Management of Lake Amatitlán (AMSA) was founded to support the restoration of the lake and to develop effective management measurements. AMSA has implemented a number of activities such as soil conservation practises, reforestation activities, and environmental education. In addition, several studies have been initiated including a bathymetrical survey in 2001 (ESI 2001), sediment and contaminant investigations as well as the hydrological characterization of the Lake Amatitlán watershed (AMSA 2003; Basterrechea 1988; CYM-ESI 2000, 2002; García 2002; de Castillo 2008).

However, considerable amounts of eroded soils continue to be deposited in Lake Amatitlán reducing the lake's storage capacity and life time. The relationship between catchment attributes and hydrological response, especially the magnitude of sediment accumulation and its impact on the life expectancy of Lake Amatitlán are not well understood. In 2001 for example, estimates regarding the life span of Lake Amatitlán ranged between 20 to 400 years (ESI 2001).

Lakes can be viewed as sentinels of their catchment because they integrate historical events and allow the interpretation of hydrological processes of the watershed in which they are embedded over time and space (Schindler 2009). In the absence of appropriate monitoring data on continuous sediment load, especially during peak flows, sediment deposits on the bottom of Lake Amatitlán can be used to estimate annual sediment yields and to evaluate the impact of extreme climatic events such as tropical storms on lake sedimentation.

The main objectives of this study are to:

Perform a bathymetric survey to accurately assess the current state, trend and spatial variability of sediment deposition in Lake Amatitlán;

Analyze sediment startigraphy and main sediment properties including grain size, texture and bulk density in lake sediments;

To estimate the life span of Lake Amatitlán and evaluate its vulnerability to future climate extremes.

Bathymetric studies are cost-efficient and accurate methods for determining long-term sediment deposition patterns, if accurate data from a prior survey is available. Two bathymetrical surveys were completed for Lake Amatitlán, the first one in 1971 and the second in 2001 (ESI 2001).

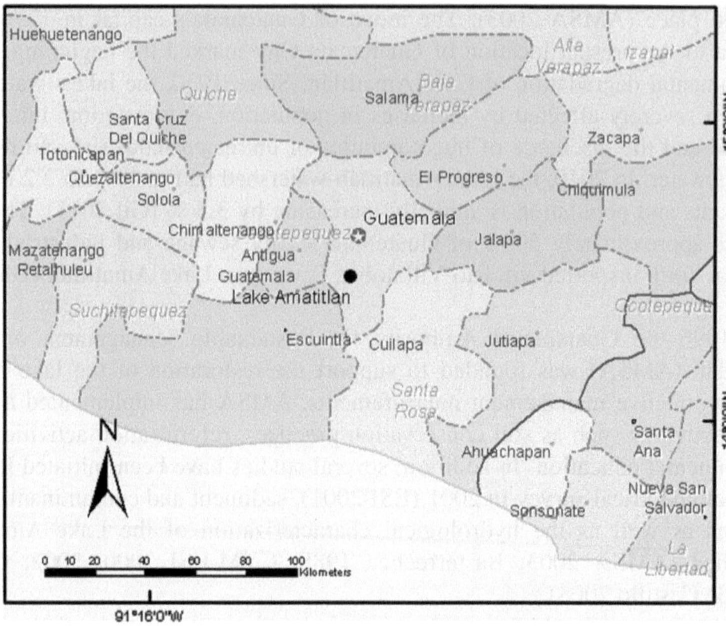

Fig. 4.1 Location map of Guatemala and Lake Amatitlán (ESRI 2010)

Methods

Study Site

Lake Amatitlán (14° 27′ 23″ N, 90° 33′ 58″ W), a highland lake in Guatemala, lies at an altitude of 1,188 m above sea level (masl) ~16 km southwest of Guatemala City (Fig. 4.1). The lake is a moderate-size elongated caldera of volcano-tectonic origin that formed approximately 40,000 years ago at the southern end of the Guatemala City graben (Gil 2011; Velez et al. 2011; Wunderman 1984). Four active volcanoes (Agua, Acatenango, Fuego and Pacaya) are within the Lake Amatitlán watershed. Mean direct annual rainfall on the lake is 1,220 mm (García 2002). The lake stretches north–west to south–east and a railroad bridge constructed in 1881 on the embankment at the narrowest point of the lakeshore divides it into two sub-basins with different physical, chemical and biological characteristics: the western basin or Lago Oeste and the eastern basin or Lago Este (Fig. 4.2). The main hydrologic input into the western basin (Lago Oeste) is Río Villalobos, which drains a watershed of 381 km^2 and transports substantial sediment loads and untreated industrial and wastewater from Guatemala City (de Castillo 2008). The outflow of Lake Amatitlán is Río Michatoya, which is used for hydroelectric power generation.

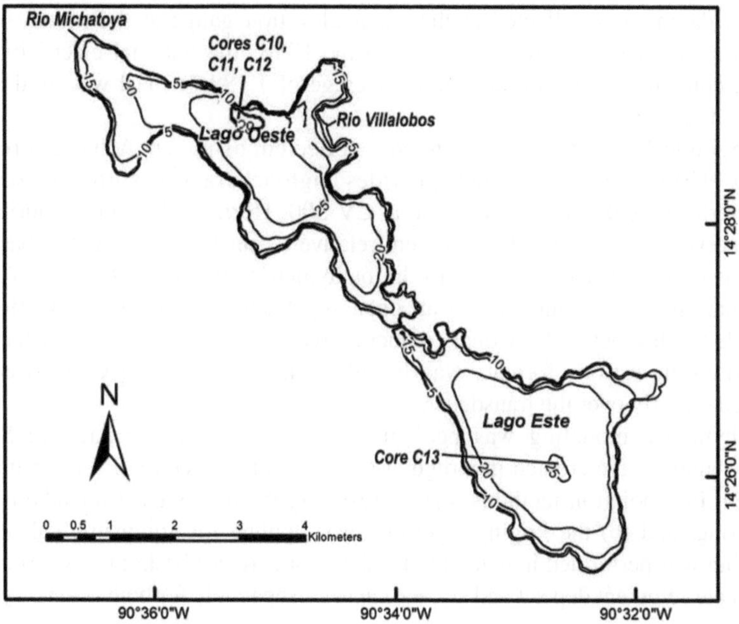

Fig. 4.2 Bathymetric map of *Lago Oeste* and *Lago Este* showing sediment core locations (C10–C13)

Bathymetric Survey

Bathymetric data was collected as a series of cross-section lines perpendicular to the north–south axis of the east and west basin of Lake Amatitlán. The data was collected using an Odom Echotrac CV300 echosounding system with a dual-frequency single-beam transducer (200–24 kHz) and an expected accuracy ranging between 5 to 10 cm. Odom eChart was used to control the echo sounder parameters, positioning and navigation with respect to lines.

The hydrographic software package HYPACK® MAX 2010 was used to design the single beam survey, collect, edit and post-process the data creating contours, cross-sections, volumes. A Digibar Pro Sound velocimeter was used to calibrate the acoustic system. Water column sound velocity were measured using the Digibar Pro Sound velocimeter three times a day (morning, mid day and night) to correct for differences related to diurnal temperature changes that affect the speed of soundings.

Similar to the 2001 bathymetry, horizontal data was collected on World Geographic System 1984 datum Zone 15 north. Vertical datum was established at water levels, the transducer was lowered in the water to an elevation of 0.9 m during the complete survey and this value was used to adjust the depths in post-processing in HYPACK to the official daily value from the official water elevation.

Vertical datum was collected at the water elevation gauge at the mouth of Río Michatoya managed by the EEGSA—Grupo EPM, the national electricity generating company. The four-day survey average of 1,188.71 masl was used as the bathymetric survey datum.

Longitude-latitude information was collected using an Ashtech Promark Proflex™500 GPS Base, which provides high-accuracy positioning (<5 cm) and interfaces with the Odom Echotrac CV 300. Horizontal accuracy during the entire survey was in the order of <5 cm relative to our base station. The base station was deployed one day prior to the bathymetric survey on the roof of Hotel Rocarena, Town of Amatitlán, in an open sky location. For base correction, an Ashtech Proflex 500 GNSS dual frequency receiver configured to transfer RTK corrections in various formats was installed on the boat. The GPS rover was installed directly over the transducer.

Bathymetric modeling was performed using the software package Surfer. Contour maps were created in a regular spaced grid surface (5×5 m) using two different interpolation methods: (a) nearest neighbour was used for lake bottom contouring, and (b) the kriging algorithm was applied for volumetric calculation. Modeling was performed individually for the west and east basin of Lake Amatitlán to estimate sediment deposit and storage capacity separately for both basins.

Sediment Collection

A total of thirteen sediment cores were collected in January 2012 in Lake Amatitlán (Fig. 4.2). An UWITEC hammer corer equipped with a 3 m long PVC tube (diameter 60 mm) was used to retrieve the cores. The four sediment cores discussed in this paper were collected from the deepest area of the lake, three cores were retrieved at 29 m depth in Lago Oeste and one core was collected at 25 m depth in the eastern basin (Lago Este). Sediment cores were carefully extruded from the tubes and the sediment–water interface identified. Cores were sliced into 5 cm pieces and the sediment stratigraphy described. Selected sediment samples were analyzed for particle size distribution, loss of ignition and specific mass density.

Results and Discussions

Bathymetry Comparison 2001–2012

The 2012 bathymetry survey was conducted between 19 and 23 January, 2012. Depth profiles in Lago Oeste were obtained from 107 transects (50–75 m separation) and from 35 transects (150 m separation) in Lago Este of Lake Amatitlán.

Table 4.1 Comparison of morphological data between 2001 and 2012 for Lago Oeste and Lago Este, water elevation 1,188.36 m

	Lago Oeste		Lago Este	
Year	2001	2012	2001	2012
Surface area (km^2)	7,938,600	7,705,030	7,149,300	7,173,579
Storage volume (m^3)	162,746,122	148,298,021	133,125,032	133,588,663
Average depth (m)	20.5	19.2	18.6	18.6
Max. depth (m)	31.9	29.7	25.0	25.0
		2011–2012 (%)		**2011–2012** (%)
Area loss (km^2)/year		0.3		0.0
Volume loss (m^3)/year		0.9		0.0

Figure 4.2 illustrates the 2012 bathymetry with the 5 m contour lines for Lago Oeste and Lago Este.

The 2012 survey was compared to the 2001 bathymetric data (ESI 2001) provided by AMSA to estimate changes in lake surface and storage volume over the 11-year period. For this purpose, the contours of the 2001 bathymetric map were digitized and the x, y and z elevations determined. Morphological parameters such as lake surface area and storage volume for 2001 were calculated and 2012 results were adjusted to the 2001 water elevation of 1,188.36 m. Surface area and storage volume estimated from the 2001 digitized map were compared to the values reported in the 2001 bathymetry report (ESI 2001) and indicated differences less than 1.5 %. This provides evidence that the bathymetry can be accurately reproduced.

Changes in surface area and storage volume over the 11-year period (2001–2012) are summarized in Table 4.1. Maximum depth in Lago Oeste in 2012 (water level elevation 1,188.36 m) was determined at 29.7 m and is 1.3 m less than the maximum depth of 31.9 m reported in 2001 and adjusted to the 2012 same water elevation. Lago Oeste shows significant sedimentation impacts during the 2001–2012 period with high amounts of sediments deposited in the delta of Río Villalobos.

Storage volume of Lake Oeste in 2012 has been estimated as 148,298,021 m^3 with a maximum of 50 % stored at 0–10 m depth, 40 % at 10–20 m and 10 % at more than 20 m. The storage volume of Lake Oeste decreases annually by 0.9 %.

Figure 4.3 illustrates the sedimentary accumulation in the 2001–2012 period for Lago Oeste. The morphology of the Río Villalobos delta on the northeast end of Lago Oeste has been severely impacted by sedimentation and sediment deposits up to 20 m thick have accumulated in the past 11 years. Significant sedimentation also occurred in the central and deepest area of Lago Oeste. The average depth of Lago Oeste has decreased by 2.2 m in the 11-year period. In May of 2010, the Lake Amatitlán watershed was strongly impacted by the tropical storm Agatha which followed the eruption of Volcan Pacaya. Heavy rainfall caused river levels to rise and high sediment loads were transported to the west basin of Lake Amatitlán. Sediment settling basins that were built in the mouth of the Río Villalobos to control and reduce the sediment inflow into Lake Amatitlán were completely destroyed during the tropical storm. The Guatemalan Meteorological

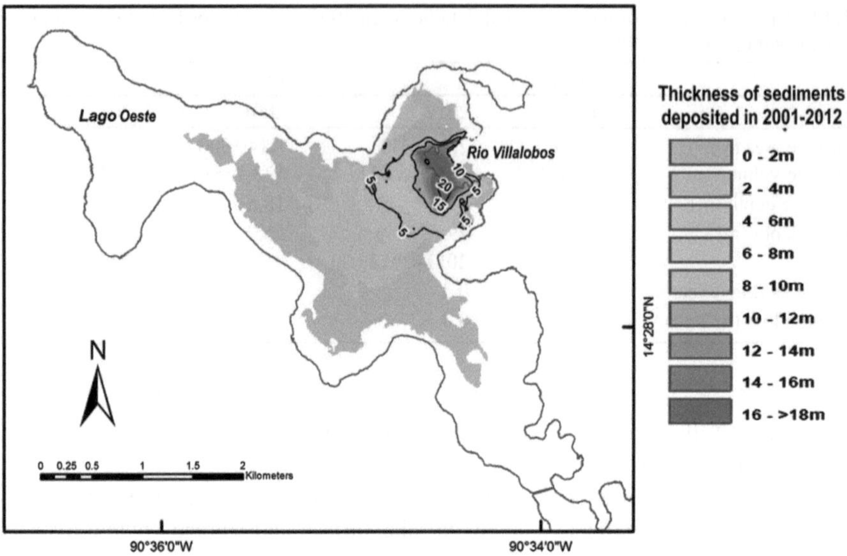

Fig. 4.3 Sediment accumulation in Lago Oeste (Lake Amatitlán) between 2001 and 2012

Institute registered the highest amount of rainfall since 1949 with precipitation reaching 200–359 mm between 25 and 28 May of 2010.

In contrast to Lago Oeste, Lago Este did not show any signs of sedimentation in the 2001–2012 period and 2012 depth profiles were always within 5 cm of the 2001 profiles.

Sediment Cores

Four sediment cores ranging between ranging from 1.8 to 2.7 m length were collected in the deepest portion of the western (C10, C11, C12) and the eastern basin (C13) of Lake Amatitlán (Fig. 4.2). All sediment cores were continuous from the water–sediment interface to their bottoms. Sediment samples were submitted to the lab for analysis of grain size distribution and other sediment properties, results are summarized in Table 4.2.

Lake Amatitlán sediments consist of coarse to medium coarse sands with a small percentage of clay and silt (Table 4.2). Grain size data was used to estimate dry mass density of sediment samples (Julien 2010), density values ranged between 1,230 and 1,380 kg/m^3. Sediment stratigraphy showed the presence of a volcanic ash layer that occurred consistently at 90 cm depth in three cores retrieved from Lago Oeste and at 5 cm depth in the core collected in Lago Este. This tephra layer was deposited on May 27 of 2010 during the eruption of Volcan Pacaya one day prior to the tropical storm Agatha. The unique coincidence of

Table 4.2 Grain size in sediment cores collected in Lago Oeste (Lake Amatitlán)

Particle	Particle diameter (mm)	%
Very coarse sand	>0.5	44
Coarse sand	0.7–0.5	20
Medium sand	0.5–0.3	22
Fine sand	0.25–0.105	11
Clay and silt	<0.05	1

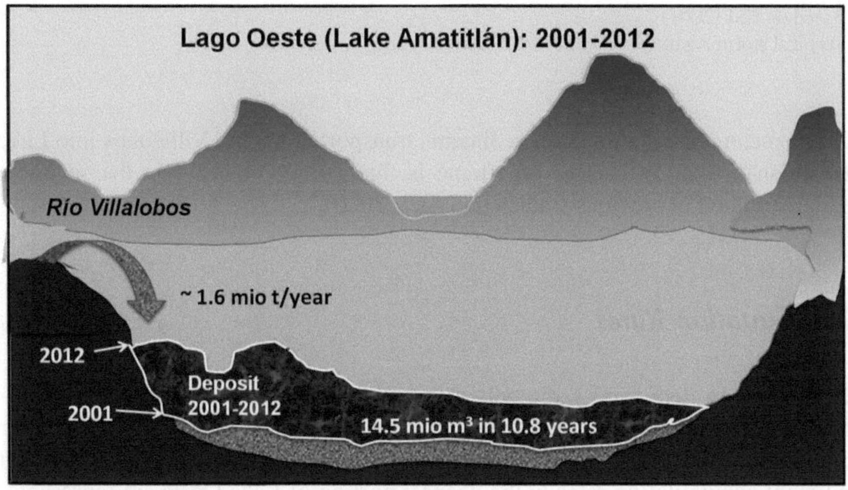

Fig. 4.4 Sediment deposition in Lago Oeste (2001–2002)

hurricane activity and volcanic eruption in May of 2010 provides valuable information on the impact caused by an extreme event such as tropical cyclones on the storage volume of the lake.

Sediment Loads in Rio Villalobos

Annual sediment loads transported by Río Villalobos can be estimated based on the 2012–2001 volumetric comparison. In the absence of any sediment outflow data, it was assumed that all sediments were trapped in Lake Amatitlán. The 2001–2012 volumetric difference (Table 4.1) corresponds to a sediment deposit of 14,408,101 m^3. Sediment volumes were converted to sediment mass using a dry specific mass of 1,250 kg/m^3. The mean annual sediment load transported by Río Villalobos was calculated at $\sim$1.6 million tons per year (Fig. 4.4). This information is very important for the effective sediment management of the Lake Amatitlán watershed, because it presents the baseline information against which the effectiveness of sediment management strategies should be compared.

Table 4.3 Estimation of sedimentation rates and sediment accumulation rates (m^3/yr)

Time period	Sedimentation rate (m/year)	Sediment accumulation rate (m^3/year)
1971–2001[a]	0.08	7.72E + 05
2001–2012	0.22	1.44E + 06
1971–2012[a]	0.11	9.49E + 05
2001–2010	0.18	1.38E + 06
Sedimentation rate during extreme climatic events (tropical storm Agatha on May 27 2010)		
2010–2012[b]	0.45	3.07E + 06

[a] Based on ESI (2001)
[b] Tropical storm Agatha

The mean annual amount of sediments transported by Río Villalobos into Lake Amatitlán in the 2001–2012 timeframe is three times higher than the sediment loads estimated in previous studies (US ACE 2000).

Sedimentation Rates

Differences in deposited sediments at maximum lake depth were used to estimate mean sedimentation rates (m/yr) for the 1971–2001–2012 time periods for Lago Oeste by dividing sediment thickness by sediment age. It was assumed that sediment accumulation rates in the deepest portion of Lake Oeste are representative of the mean annual sedimentation rates. The volcanic layer encountered at 90 cm depth and corresponding to May 20 of 2010 was used to estimate sedimentation rates for extreme climatic events. Sedimentation rates are summarized in Table 4.3.

Mean annual sedimentation rates for the 2001–2012 period were estimated at ~0.2 m/year. Sedimentation rate increased more than two-fold to 0.4 m/year during an extreme weather event such as the tropical storm Agatha. The comparison of current sedimentation trends in Lago Oeste for the 2001–2012 period to pre-colombian rates spanning the 1300 AD–300 BC period (Velez et al. 2011) indicates a 100 fold increase and illustrates the lake's accelerated sedimentation.

Volumetric Comparison 1971–2001–2012

Volumetric comparisons for the 1971–2010–2012 periods are summarized in Table 4.4. Annual sediment deposition for the 2001–2012 period is almost two times larger than for the 1971–2001 timeframe. These results are likely related to an increase in sediment transport from intensified anthropogenic activities, but it is also possible that lower spatial resolution of the 1971 bathymetric survey may have led to underestimation of the lake's storage volumes.

Table 4.4 Volumetric comparison for the time periods 1971–2001–2012

Lago Oeste, at 1187.66 masl	1971	2001	2012
Surface area (km^2)	8,313,524	7,787,824	7,677,474
Storage volume (m^3)	176,644,385	156,576,572	142,971,441
Max. depth (m)	33.7	31.14	28.94
Average depth (m)	21.2	20.1	18.6
		1972–2001 (%)	**2001–2012** (%)
Area loss (km^2)/year		6.3	1.4
Volume loss (m^3)/year		11.4	8.7
Annual area loss		0.2	0.1
Annual volume loss		0.4	0.8

Table 4.5 Lake life expectancy estimated by present study in comparison to earlier studies

Period selected for study	Method of estimation	Predicted life for Lake Amatitlán (year)	Predicted life for Lago Oeste, Lake Amatitlán (year)	Investigators
1975	Sediment transport model	53		Morataya (1975)
1999	Sediment transport model	300		CYM-ESI (2000); (2002)
1971–2012	Bathymetry	330	173	ESI (2001)
1971–2001	Bathymetry	423	213	2012 data
2001–2012	Bathymetry	220	114	2012 data

Life Expectancy of Lake Amatitlán

The life expectancy of Lake Amatitlán corresponds to the time at which the lake is completely filled with sediments. Life expectancy for Lake Amatitlán and its western basin Lago Oeste are summarized in Table 4.5 and illustrated in Fig. 4.5 for (Lago Oeste).

Based on the mean annual sediment deposit of 1.16 km^3 the life span of Lake Amatitlán (Lago Oeste and Este) is about 220 years indicating that in 2232 AD the lake will be filled. It is expected that the storage capacity of Lago Oeste will be depleted first. The life expectancy for Lago Oeste was estimated to be ~114 years in 2012 (water elevation 1,187.66 masl). This result are in notable contrast to previous investigations (CYM-ESI 2002).

Conclusions

The results of this study indicate that sedimentation in Lake Amatitlán was underestimated in previous investigations and the incoming annual sediment load of ~1.64 million tons is two to three times higher than estimated in previous

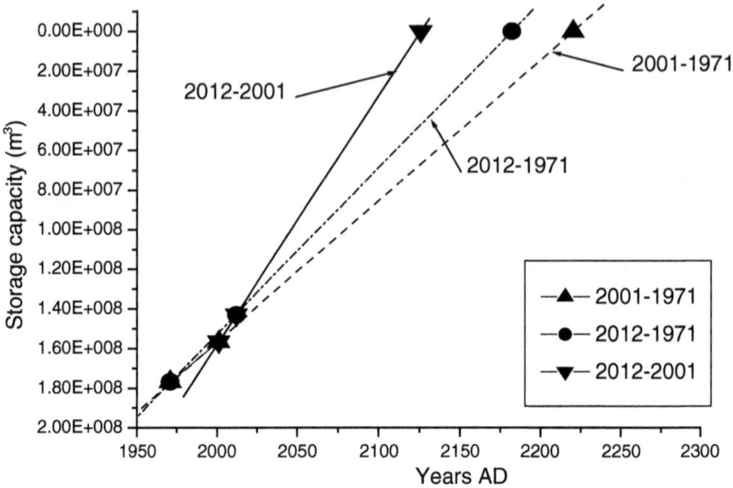

Fig. 4.5 Life expectancy of Lago Oeste (Lake Amatitlán) using recent and historic sediment yields

assessment (CYM-ESI 2002; US ACE 2000). The present study predicts that Lago Oeste, the western basin of Lake Amatitlán, has a life span of ~114 years if mean annual sediment loads from the 2001–2012 period are projected to the future. During extreme events, the mean annual sediment load increases twofold. The likelihood of the risk of one or several extreme weather events and their impact on the life expectancy of Lake Amatitlán needs to be considered in future management plans. The analysis of such extreme events is especially important given the short life span of the lake.

Immediate short-term solutions with long-term effects are needed to stop the sedimentation of Lake Amatitlán. Erosion control management measures need to consider prevention programs and address source control to reduce sediment production in vulnerable sub-watersheds. Source control measures should be combined with soil conservation practices and re-vegetation of riparian zones. In addition, sediment structures such as settling basins, sediment retarding structures, and channel stabilization need to be engineered to trap the eroded sediments before they reach the lake.

Several key vulnerability criteria demonstrate that the Lake Amatitlán area is a hotspot for vulnerability under future extreme climatic events: the magnitude that the disappearance of Lake Amatitlán and its transformation into a floodplain would have on nearby and downstream communities, the threat to public and private infrastructure, the high degree of social, ecological and economic damage, the timing within the next century, and the irreversibility of the impact. It is very likely that future climate change will increase the frequency and intensity of extreme climate phenomena in Central America (Agrifor Consultants 2009; IPCC 2007a, b) and will consequently accelerate sedimentation of Lake

Amatitlán. Additional non-climate related pressures such as rapid demographic growth (6.5 % in 2010 based on Gil 2011), will further shorten the lake's life expectancy. As sediments continue to accumulate in the lake and reduce its storage capacity, water-related problems such as flooding and water-borne diseases will increase in severity.

The consequences of failing to manage sedimentation and the restoration of the Amatitlán watershed within the next future would cause severe impacts and widespread damage to Guatemala's socioeconomic development. Given that the costs for failing to adapt are huge versus the costs of adaptation, coordinated actions and continued efforts at different scales should focus on engaging policy-makers and strengthening institutional capacities to develop effective sediment management and climate change adaptation strategies to cope with risks of landslides, floods and sedimentation that may affect the lifespan of Lake Amatitlán and consequently the livelihood of the people living in its surrounding communities. Timely and effective strategies embedded in policies are an imperative response to the complex issues of the Lake Amatitlán watershed to reduce vulnerability and increase resilience to future climate-related impacts.

Acknowledgments The authors thank the direction of AMSA (Autoridad para el Manejo Sostentable del Lago de Amatitlán) and the Interamerican Bank for Development (BID) for the financial support to complete this investigation.

References

Agrifor Consultants (2009) Climate Change in Latin America. Develop Cooperation in Latin America. Head of Unit EuropeAid/B/2, p 119

AMSA (2003) Plan Maestro de Manejo Integrado de la Cuenca y del Lago de Amatitlán, PLANDEAMAT

Basterrechea M (1988) Caracterización limnológica preliminar de 32 lagunas de Guatemala. Rev Biol Trop 36:115–122

Calderon IM (2010) Green movement against green water: using green infrastructure strategies in the lake Amatitlán watershed to promote environmental education in Guatemala Honors, Cornell University, p 88

CYM-ESI (2000) Análisis de los datos de caudales y sedimentación del año hidrológico 1998–1999, Reporte Final

CYM-ESI (2002) Estudio hidrológico de la cuenca del Río Villalobos, Informe final

de Castillo RNEG (2008) Interpretación quimiométrica de la calidad del agua de los ríos que conforman la microcuenca del Río Villalobos, principal tributario del Lago de Amatitlán, durante los años 1996 a 2006, Universidad de San Carlos de Guatemala, Tesis, p 143

ESI (Estudios, Servicios e Inversiones S.A.) (2001) Informe Final Batimetría 2001 Lago Amatitlán, Guatemala

ESRI (2010) ArcGIS desktop: release 10. Environmental Systems Research Institute, Redlands

Galindo LM, Miguel C, de Ferrer J (2011) Vital climate change graphics for Latin America and the Caribbean. Special edition. for the COP 16/CMP 6, Mexico. UNEP/GRID-Arendal, p 44

García HO, (2002) Cuantificación de la cuenca de la calidad del agua del Río Villalobos en época seca y lluviosa en un periodo de 24 horas 3 veces al mes en un punto previo a la entrada al Lago de Amatitlán. Tesis de Maestría. Escuela de Ingeniería Sanitaria y Recursos Hidráulicos, Universidad de San Carlos, p 46

Gil OF (2011) Tierra de amatles: monografía del Municipio de Amatitlán Impresiones Guatemala, p 326

IPCC (2007a) Climate Change 2007: The physical science basis. Contribution of working group Ii to the 4th assessment report of the intergovernmental panel on climate change, Solomon S, Qin D, Manning M, Chen Z, Marquis M, Averyt KB, Tignor M Miller HL (eds) Cambridge University Press, Cambridge, and New York, p 996

IPCC (2007b) Climate Change 2007: impacts, adaptation and vulnerability. Contribution of working group II to the 4th assessment report of the intergovernmental panel on climate change. Parry ML, Canziani OF, Palutikof JP, van der Linden PJ, Hanson CE (eds) Cambridge University Press, Cambridge, p 976

Julien PE (2010) Erosion and sedimentation, Colorado State University Cambridge University Press, Cambridge, p 371

Morataya JC (1975) Estudio sobre el transporte de sedimentos en la Cuenca del Río Villalobos. Universidad de San Carlos de Guatemala

Perez L, Bugja R, Lorenschat J, Brenner M, Cyrtis J, Hoelzmann P (2011) Aquatic ecosystems of the Yucatan Peninsula (Mexico), Belize and Guatemala. Hydrobiologia 661:407–433

Schindler DW (2009) Lakes as sentinels and integrators for the effects of climate change on watersheds, airsheds, and landscapes. Limnol Oceanogr 54:2349–2358

US ACE (US Army Corps of Engineers) (2000) Water resources assessment of Guatemala, p 56

Valladares Cerezo CE (2003) Urban slums report: the case of Guatemala City, Guatemala. Case studies for the global report on human settlements 2003. p 46

Velez MI, Curtis JH, de Hatch MP (2011) Environmental and cultural changes in highland Guatemala inferred from Lake Amatitlán. Geoarcheology 26(3):346–364

Wunderman RL, Rose IR (1984) Amatitlán an actively resurging cauldron 10 km south of Guatemala city. J Geophys Res Solid Earth 89:8525–8539

Chapter 5
Towards Valuing Climate Change Impacts on the Ecosystem Services of a Uruguayan Coastal Lagoon

Andrew L. Fanning

Abstract It has been well established that coastal zones in Latin America are particularly vulnerable to anthropogenic pressures including climate change impacts (e.g. sea-level rise, increased storm intensity, altered hydrological regimes, etc.). The twin challenge of maintaining human security (often via protection using dikes) and sustaining coastal ecosystems and the services they provide has been identified as a major issue for coastal management (Nicholls et al. 2010). This study was motivated by the dual observation that there are very few estimates of the local costs of climate change in developing countries and that the few studies that do exist rarely take into account the non-market value of ecosystem services. Using a case study of a coastal lagoon ecosystem in Uruguay, a preliminary interdisciplinary analysis that isolates changes in the economic value of ecosystem services was undertaken, which can be associated with historical climatic changes. The Economics of Ecosystems and Biodiversity (TEEB) framework is adapted to identify lagoon ecosystem services and three valuation methodologies are implemented to estimate non-market monetary values of climate change impacts on the artisanal shrimp fishery, carbon sequestration and habitat services. The results suggest that climate change is already affecting the economic value of the coastal lagoon ecosystem. Implications for local management and lessons learned from the case study are discussed.

Keywords Climate change • Monetary valuation • Ecosystem services • Coastal zones • Lagoons • Latin America • Uruguay

A. L. Fanning (✉)
International Development Research Centre, Latin America and Caribbean Regional Office, Climate Change & Water Program, Avenida Brasil 2655, 11300 Montevideo, Uruguay
e-mail: afanning42@gmail.com

W. Leal Filho et al. (eds.), *International Perspectives on Climate Change*, Climate Change Management, DOI: 10.1007/978-3-319-04489-7_5,

Introduction

There have been many calls from policy-makers and relevant stakeholders for estimates of the local impacts of climate change, especially in monetary terms (Agrawala and Fankhauser 2008; Parry et al. 2009; World Bank 2010). In particular, non-market impacts associated with climate change are estimated to make up a very significant share, if not the majority, of the total expected impacts (Tol et al. 2000; Rothman et al. 2003). It can be very difficult to estimate the non-market impacts of climate change with precision at any scale (local, national, regional, global) given available methodologies and inherent uncertainties, although there have been some promising attempts in Latin America at national and regional scales (ECLAC 2010; Vergara et al. 2013). By taking into account the impacts of climate change that take place outside the market, monetary estimates have the potential to at least be imprecisely correct, instead of being precisely wrong (Seo 2012).

The purpose of this chapter is to describe the results of an assessment seeking to value the local impacts of climate change on the ecosystem services of a coastal lagoon ecosystem in Uruguay. The novel aspect of the chapter lies in its focus on adapting and implementing monetary valuation methodologies to isolate climate impacts on ecosystem services that are often ignored in conventional economic analyses. By focusing on non-market impacts of climate change at the local level, this chapter is particularly relevant for protected area managers and stakeholders interested in communicating the monetary benefits of healthy ecosystems in a changing climate.

The remainder of the chapter is structured as follows. The first section briefly describes the observed climate changes in South East South America (SESA) with an emphasis on the Atlantic coast of Uruguay. Next, the case study site is characterized and ecosystem services are identified using the internationally recognized ecosystem assessment methodology of The Economics of Ecosystems and Biodiversity (TEEB) project. The following sections present the methodologies and results of valuing climate change impacts on three lagoon ecosystem services: (i) shrimp provision; (ii) carbon sequestration services; and (iii) habitat maintenance/cultural services. Finally, the chapter closes with a critical discussion of the valuation results and some concluding remarks.

Trends in Climate Affecting Uruguay

The South American Monsoon System (SAMS) affects most of the South American climate and, due to the importance of monsoon rains for human wellbeing, especially agriculture and hydroelectricity generation, its life cycle and variability has been the focus of considerable research and synthesis (Nogués-Paegle et al. 2002; Vera et al. 2006; Marengo et al. 2010 and references therein). Table 5.1 summarizes trends in sea level rise, temperature, precipitation and winds observed

Table 5.1 Summary of observed climate trends in Rocha province, Uruguay

Variable	Period	Trend
Sea level rise		
	1957–2008	2.94 mm/year—increase of 15 cm in the last fifty years[a]
Temperatures		
Average	1961–2010	0.2 °C/10 year—significant increase of 1 °C in the last 47 years[b]
Maximum	1950–2002	No significant change in warm days; decrease in cold days[c]
Minimum	1950–2002	Increase in warm nights; decrease in cold nights[c]
Precipitation		
Average	1961–2010	6.7 mm/year—increase of more than 300 mm in 49 years[b]
Heavy rain days	1961–2000	2–3 days/10 year—Increasing days with rainfall over 10 mm[d]
Dry days	1961–2000	Slight decrease in dry days[d]
Winds		
Velocity (m/s)	1961–2008	• Nearby stations on the Río de la Plata observe a decrease in wind velocities[e] • Global models estimate no change in average wind velocities on the Atlantic coast of Uruguay[e]
Direction	1979–2008	• Nearby stations observe increase in annual winds with an easterly influence and decrease in northerly winds[e] • Regional models estimate increase in 'summer' circulation patterns with more easterly winds at the expense of the 'winter' pattern[f]

[a]Verocai (2009); [b]DNM data; [c]Rusticucci and Renom (2008); [d]Rusticucci et al. (2009); [e]Bidegain et al. (2009); [f]Barros et al. (2005)

from climate change data (obtained from Uruguay's National Meteorology Authority) and literature as pertains to the Atlantic coast of Uruguay.

Verocai (2009) observes that average annual sea levels have risen more than 15 cm on the Atlantic coast of Uruguay (La Paloma, Rocha) between 1957 and 2008. Given that the coastline is considered micro-tidal (~40 cm) and that the Atlantic coast is only moderately vulnerable to changes in the tributary discharges of the Río de la Plata, changes in the prevailing winds (especially from the southeast) are highly relevant to explaining this substantial increase in annual sea levels (Re and Menendez 2007).

For temperature, several studies use observed data from Argentina, Uruguay and southern Brazil (among others) between 1900 and 2000 and the data suggests both average and extreme warming (e.g. less cold nights and more warm days) in SESA (Rusticucci and Barrucand 2004; Bidegain et al. 2005, Vincent et al. 2005; Marengo and Camargo 2007; Rusticucci and Renom 2008). The highest trends of annual warming in Uruguay have been recorded on the Atlantic coast having risen 1 °C between 1961 and 2008—two times the warming trend observed globally (Bidegain et al. 2009).

In SESA, an alarming trend of increasing average and extreme precipitation over the twentieth century, especially the second half, has been observed by numerous studies (Barros 2004; Haylock et al. 2006; Grimm 2010 and references therein). Table 5.1 reports an increasing linear trend in average annual

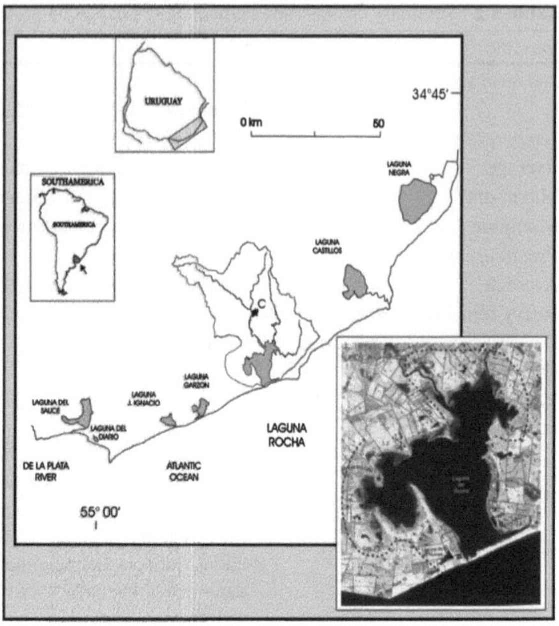

Fig. 5.1 Uruguayan coastal lagoons highlighting the Laguna de Rocha watershed. *Source* Department of Limnology, University of the Republic

accumulated rainfall of 6.7 mm per year between 1961 and 2010 in Rocha. Daily indices show a significant increase in extremes with 8–12 more days with heavy rainfall (e.g. more than 10 mm) between 1961 and 2000 (Rusticucci et al. 2009).

Case Study Site

The lagoon ecosystems along the Atlantic coast basin of Uruguay (9,266 km^2) (Fig. 5.1) provide food, water and the basis for local economic activities as well as protection from natural hazards and support for biodiversity of global importance (Conde et al. 2003).The Laguna de Rocha Protected Area (LRPA) (32° 40′ S, 54° 16′ W) is a 22,000 ha area located about halfway up the Atlantic coast of Uruguay. In 2010, the area received its current protected status and was incorporated into Uruguay's National Protected Area System.

The LRPA is a component of Uruguay's only UNESCO Biosphere Reserve (since 1976) which is also recognized under the Ramsar Convention's List of Wetlands of International Importance (since 1984). It is a designated Important Bird Area (IBA) site by Bird Life International for more than two hundred species of migratory and resident populations representing more than half of Uruguay's total 435 bird species (Aldabe et al. 2009). The lagoon ecosystem is also designated a 'Site of Regional Importance' by the Western Hemisphere Shorebird Reserve Network (WHSRN) recognizing an important wintering site for 24 species of shorebirds, some of whom migrate from the Arctic (WHSRN, 2011).

The LR can be considered 'choked' based upon its degree of exchange with the ocean by means of one single channel (Bonilla et al. 2005). This channel is frequently blocked by a sandbar that opens naturally under the combined forcing of: (i) continental run-off causing the depth of the lagoon to exceed a given level; and (ii) high wave and wind action on the seaward side (Conde et al. 2000). As a result, the physical and ecological structure of the lagoon is highly variable depending on the prevailing hydrological conditions and whether or not the sandbar is open or closed (Bonilla 2002). Comprehensive characterizations can be found in Bonilla et al. (2006) and Rodriguez-Gallego et al. (2008).

Primary economically productive activities within the LRPA include extensive livestock-raising (beef, mutton and milk), agriculture (soybean, potatoes, fodder) and tourism. The lagoon itself is a nursery and habitat for numerous crustacean and fish species of economic importance (Fabiano and Santana 2006). These provide the livelihoods of two artesanal fishing communities and sustain Uruguay's second most important inland fishery (Santana and Fabiano 1999; Rodriguez-Gallego et al. 2008). The protected area is mostly devoid of physical infrastructure which makes it an ideal case study site for assessing the local impacts of climate change on ecosystem services.

Selection of Ecosystem Services

Ecosystem services are defined by the Millennium Ecosystem Assessment (2005) simply as "the benefits humans derive from nature" arising through the pathways from ecosystem structure and processes to human decision-making and well-being. Despite the simplicity of the above definition, attempts to disentangle the myriad of ways that ecosystems support human life and well-being has led to much debate surrounding the appropriate means of classification (Costanza 2008; de Groot et al. 2010 and references therein). This study undertakes the first ecosystem service assessment ever conducted in the LRPA and follows the ecosystem service classification proposed by The Economics of Ecosystems and Biodiversity (TEEB) study which is itself adapted from MEA (2005) and other sources (de Groot et al. 2010). *S*pecific ecosystem services vulnerable to climate change in the LRPA are identified in Table 5.2.

Essentially, Table 5.2 divides eleven ecosystem services identified for the purposes of this study in the LRPA into four categories: (i) provisioning; (ii) regulating; (iii) habitat; and (iv) cultural. Within these ecosystem services, specific LRPA 'items' vulnerable to climate change impacts are identified along with the zone(s) where they can be found within the protected area. Justification for ecosystem service selection was made by borrowing heavily from the social and ecological characterizations provided in Rodriguez-Gallego et al. (2008) and integrating them with the climate trends summarized in the previous section.

The list of ecosystem services presented in Table 5.2 was used to assess the relevance and feasibility of various monetary valuation methodologies to estimate non-market impacts of climate change in the LRPA. The limitations

Table 5.2 Identification of LRPA ecosystem services vulnerable to climate change

Ecosystem services	LRPA item	Zone	Justification
Provisioning			
Food	Commercial fish and crustaceans	LR; ALZ; Marine zone	Subsistence/recreational use vulnerable to changes in water quality from precipitation extremes and sea level rise
Fuel and fibre	Natural grasslands and prairies	Coastal grassland; Medium and high plains	Fodder and agriculture vulnerable to precipitation extremes and soil/aquifer salinization
Regulating			
Flood and storm protection	Dunes; LR water body	ALZ; LR; coastal grassland	Laguna de Rocha, low-lying adjacent lands and littoral zone vulnerable to extreme precipitation, shift in prevailing winds and sea level rise
Pollution control	Emergent wetland plants; Riparian forest[a]	Northern wetland; Rocha River (and other tributaries)	Water purification services vulnerable to increased run-off from extreme precipitation and salinization
Erosion control	Dunes; emergent wetland plants	ALZ; northern wetland	Erosion control vulnerable to extreme precipitation, shifts in prevailing winds/waves and sea level rise
Carbon sequestration	Emergent wetland plants	Northern wetland; lagoon inlets	Vulnerable to precipitation extremes and salinization
Habitat			
Maintenance of life cycles of species	Migratory and resident species	All zones, especially the LR and sandbar (ALZ)	Ecological resilience vulnerable to increasing extreme events, sea level rise and wind changes
Cultural			
Recreational	Tourism	All zones, especially the LR and sandbar (ALZ)	Recreational uses vulnerable to changes in water quality, erosion, loss of fish and bird habitat from all climate change stressors
Educational	Formal and informal education/training	All zones	Educational uses vulnerable to extreme events and sea level rise
Aesthetic	Protected Area	All zones, especially LR and ALZ	Aesthetic appeal vulnerable to precipitation extremes in particular
Spiritual and inspirational	Protected area, especially fishing communities	All zones, especially LR and sandbar of the ALZ	Sense of place and informal local traditions vulnerable to changes in water quality, erosion, fish and bird habitat from all climate change stressors

LR Laguna de Rocha, *ALZ* Active Littoral Zone, *Sources* Typology adapted from MEA (2005) and de Groot et al. (2010). [a]not included in the currently delimitated protected area

of classification are recognized due to the multiple services provided by one ecosystem component (e.g. catching fish provides food *and* social identity to artisanal fishers in the Laguna de Rocha; emergent wetland plants provide water purification *and* erosion control). As well, the selection of specific ecosystem services and their justification for climate change vulnerability are debatable, given their qualitative nature. However, a consistent classification scheme of ecosystem uses is needed in order to demonstrate the applicability of methodologies for valuing climate change impacts on the LRPA in monetary terms.

Valuation of Climate Change Impacts on Three Ecosystem Services

When it comes to the monetary valuation of ecosystems, it's important to note that different empirical valuation methodologies are more or less valid in neoclassical economic theory based upon the different measures of welfare calculated for each. Furthermore, there is a heated conceptual debate surrounding the neoclassical economic notion of value as well as the information lost from reducing the multiple uses of ecosystems and the services they provide to a single (monetary) metric (Vatn and Bromley 1994; Goulder and Kennedy 1997). The approach taken here is to accept the conceptual arguments for using monetary valuations as one of many indicators of the value of ecosystems to society in order to explore some of their methodological challenges and limitations in developing countries for addressing the newer phenomenon of climate change.

Based on an assessment of both relevance and feasibility given limited resources, three ecosystem services were selected for monetary valuation of local impacts of climate change in the Laguna de Rocha. First, the impact of rainfall changes on production is estimated using a change in productivity method for the shrimp fishery. Next, the value of carbon sequestration by emergent wetland plants in the lagoon is estimated using the market price method from the carbon market. Third, investments in habitat maintenance and research are used to value the damage costs avoided by maintaining the ecosystem and increasing knowledge.

Changes in Productivity: Precipitation as an 'Input' to the Shrimp Fishery

The productivity method—or production function approach—estimates ecosystem service value as an input to the production of goods commercially sold in markets. The idea is that changes in the quantity or quality of an ecosystem service used as a factor of production will cause changes in total production costs and/or the productivity of other inputs (Barbier et al. 1997). A number of researchers have successfully implemented the Production Function (PF) approach to value the 'indirect' breeding/nursery function of wetland areas as inputs to near-shore fisheries production in diverse settings from Chesapeake Bay to Southern Thailand (Ellis and Fisher 1987; Sathirathai 1997; Barbier and Strand 1998; Barbier 2003, 2007).

The novel aspect of using the PF approach for this study is to argue that climate, especially precipitation, also provides an indirect regulating 'service' that influences the capture of shrimp (and other economically productive activities) in wetland areas.

Effects of Precipitation on the LRPA Shrimp Fishery

Shrimp fishing in the Laguna de Rocha generally begins at the end of austral summer (end of February) and runs until the end of autumn (sometime in May) (Norbis 2000). The target species is juvenile Sao Paulo Shrimp (*Penaeus Paulensis*) that, dependent upon environmental conditions, enter the lagoon in a post-larval phase to mature in an area relatively free of predators before migrating back to sea to reproduce as adults (ibid). Assuming that the post-larval shrimp are present in Uruguay's coastal waters, at least two hydrological conditions must take place in the dynamic lagoon ecosystem for the Sao Paulo shrimp to thrive. The first is that the sandbar is open to the ocean during austral spring so that the post-larval shrimp are able to enter the lagoon (Santana and Fabiano 1999). The second condition is that, while the juvenile shrimp can tolerate changes in salinity, they prefer salty water as an estuarine/marine species (Norbis 2000).

In particular, the salinity of the shallow lagoon is sensitive to changes in rainfall/run-off (Bonilla et al. 2006). As such, we could hypothesize a negative relationship between shrimp harvests and rainfall accumulated the previous year of harvest. In fact, for the years in which data is available, a negative correlation (−0.59) is observed between shrimp harvest and average annual rainfall of the year preceding harvest.

Valuing Precipitation as an Input to Shrimp Harvest

A 'static' one-period model adapted from Barbier (2007) was used to value how a change in precipitation affects the harvest, prices and consumer surplus of the open-access Sao Paulo shrimp market in the Laguna de Rocha. The management of the lagoon shrimp fishery is considered open access meaning that any profits will attract new fishers until all profits disappear and, following Freeman (1991), the welfare change in the environmental service will impact consumer surplus only.

Figure 5.2 is a stylized diagram of the welfare impact of a change in rainfall accumulation on the Sao Paulo shrimp fishery in the Laguna de Rocha. The diagram describes a change in precipitation that impacts the hydrological conditions of the lagoon causing a shift in the average cost curve, AC, of the open-access

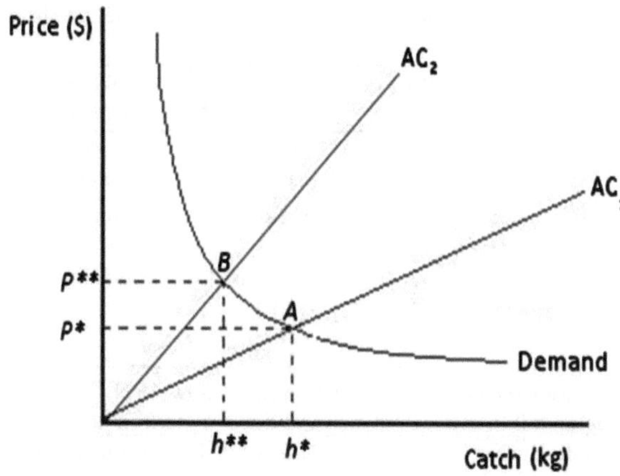

Fig. 5.2 Economic value effects of increased precipitation on the LR shrimp fishery. An increase in precipitation shifts the average cost curve (AC) upwards creating a new equilibrium price (P**) and shrimp harvest (h**) relative to the previous equilibrium price (P*) and harvest (h*). The decrease in consumer surplus after the change is given by the trapezoid area under the demand curve between the price differential (P**BAP*). *Source* Adapted from Barbier (2007)

shrimp fishery. The resulting welfare impact is the change in consumer surplus area, P**BAP* (Freeman 1991; Barbier 2007).

To conduct the one-period production function analysis of precipitation-shrimp linkages for the Laguna de Rocha, the methodology described in Barbier (2007) is adapted to model shrimp capture and effort in the lagoon using average annual precipitation data for Rocha from 1991 to 1999 as another 'input' to production. Due to space limitations, the formal methodology and econometric results are not included in this chapter but they are available along with the data from the author upon request.

The shrimp capture and effort data was the limiting factor in the analysis because of: (i) resource constraints with respect to the collection of primary data; and (ii) fishing in coastal lagoons is considered 'terrestrial' so there is minimal regulation or registration of secondary data by the Uruguayan authorities (Norbis, personal communication, August 18, 2011). The estimates of harvest/effort over the 9-year period were conducted by Uruguay's *Instituto Nacional de Pesca* (INAPE, National Fisheries Institute) as reported in Santana and Fabiano (1999). Harvest is estimated in kilograms of shrimp per season while the only time-series estimate of effort available was number of shrimp fishers. Precipitation data was obtained from Uruguay's *Dirección Nacional de Meteorología* (DNM, National Meteorology Authority).

Table 5.3 reports the equilibrium prices, harvests and resulting welfare loss estimated by the model from the 195.5 mm increase in rainfall observed from 1961 to 1998 in the Laguna de Rocha shrimp fishery. The ranges of values indicated in parentheses are the 95 % confidence bounds from the standard errors of the estimated parameters. The total welfare loss to the flow of shrimp capture in

Table 5.3 Equilibrium prices, quantities and valuation of precipitation-shrimp model for the Laguna de Rocha, 1961–1998

	1961 (Before rainfall increase of 195.5 mm)		1998 (After rainfall increase of 195.5 mm)	
Demand Function	P_0 (US$/kg)	h_0 (kg)	P_1 (US$/kg)	h_1 (kg)
$P = 43.05\ h^{-0.2}$	9.52	1,887	10.38	1,225
Value of rainfall as input to the Laguna de Rocha shrimp fishery (as measured by change in net welfare)				
Total welfare loss (US$)		Welfare loss per mm of rainfall (US$)		
1335		**6.83**		
(258–234,732)		(1.32–1,200.68)		

Notes Price *P* and harvest *h* data obtained from a 9-year dataset (1991–1999) reported in Santana and Fabiano (1999). Rainfall increase estimated as a linear trend from the *Dirección Nacional de Meteorología* rainfall data for Rocha from 1961 to 1998. Point estimates in bold. Figures in parentheses represent upper and lower bound welfare estimates based on the standard error of the estimated parameters. The statistically significant parameters for the iso-elastic demand function were estimated by ordinary least squares (OLS). Preferred production function model is groupwise heteroskedastic cross sectional time series generalized least squares (GLS)
Source Author's calculations

the Laguna de Rocha due to the linear increase in rainfall observed between 1961 and 1998 were estimated to be US $1,335 ($258 to 234,732 with 95 % confidence) or US $6.83/mm ($1.32 to 1,200.68 with 95 % confidence). In this particular case study, the model suggests that increases in average rainfall have had a significant negative impact on the welfare of shrimp fishers in the Laguna de Rocha.

While the point estimate values of precipitation impacts on the LRPA shrimp fishery were reasonable given the small size of the fishery (US $1,335), the range of welfare values estimated at 95 % confidence was very large (US $258–234,732). This large range could be related to the considerable variability among the small number of observations in the 9-year dataset. Another cause could be the 'static' approach modelled that Barbier (2007) suggests has a larger degree of uncertainty than dynamic analyses that incorporate changes in the stocks of the target species. Notwithstanding the considerable scope for improvement of the above model and, it is hoped that the results of this case study using the PF approach to value rainfall impacts on a coastal lagoon fishery can contribute to the discussion on the non-market valuation of climate as an input to production.

Valuing Carbon Sequestration Services of LRPA Emergent Wetland Plants

The second LRPA non-market valuation exercise presented in this chapter was driven by the broad recognition of wetlands for their global importance as carbon sinks; occupying only 4–6 % of the planet's surface yet storing 20–25 % of the world's soil carbon (Sampson and Scholes 2000). Coastal wetlands are particularly important for mitigation of GHG emissions because, unlike freshwater wetlands, they sequester carbon without emitting significant amounts of methane (Magenheimer et al. 1996; Choi and Wang 2004). As such, while there is a lack of quantitative valuation studies (Engle 2011), researchers note that coastal wetlands could be more valuable carbon sinks per unit area than many other ecosystems in a warming world (Choi and Wang 2004).

Estimating Carbon Stored by LRPA Wetland Soils

After consulting local experts and literature, specific soil carbon densities and accumulation rates for the Laguna de Rocha could not be found and presumably do not exist. As such, a proxy was obtained by calculating the average accumulation rate of 12 sites in the USA with fairly similar climatic regimes (±1.9 °C average annual temperature) found in the 154-site sample of tidal saline wetlands compiled by Chmura et al. (2003). This average carbon accumulation rate of 174 g m^{-2} $year^{-1}$ was then converted to annual metric tons per hectare (t ha^{-1} $year^{-1}$) and multiplied by the 2,910 ha area of the LRPA attributed to 'emergent wetland plants' and 'lagoon inlets' zones reported in Rodriguez-Gallego et al. (2008) to yield a proxy estimate of total annual carbon stored by Laguna de Rocha wetlands of just over 5,000 tons of carbon per year (5,063.4 t C $year^{-1}$).

Estimating the Market Value for the LRPA Carbon Sequestration Service

Carbon market data was obtained from the Bluenext environmental trading exchange platform on the daily trade of Certified Emission Reductions (CRUs) generated from projects under the Clean Development Mechanism of the Kyoto Protocol between 2008 and 2011. Given the large volumes of carbon traded on the global CRU carbon market (e.g. more than 46 million tons of carbon in 2010) relative to the Laguna de Rocha 'supply', the market demand curve was found to be essentially flat at the scale of changes in the coastal lagoon carbon sequestration volume. On the other hand, changes in total carbon volume explain next to none of the volatility in prices using linear and iso-elastic demand estimations ($R^2 = 0.009$ and 0.014, respectively). Based on the above, factors other than supply appear to play the dominant role in determining prices and demand for carbon. As a result, it's argued that the most relevant monetary estimate for the *local* carbon sequestration service of the LRPA is a financial assessment of the change in potential annual revenue flows to the protected area due to exogenous price shocks. Table 5.4 presents the results of the financial assessment used to calculate average annual revenue (in current euros) from carbon sequestration in the Laguna de Rocha using 2008–2011 prices for CERs.

Table 5.4 Potential annual revenue from Laguna de Rocha carbon sequestration service

Year	Carbon capture (t C/year)	Price (€/t CO_2e)			Annual revenue in current euros (€)		
		Min	Average	Max	Min	Max	Change
2008	5,063.4	12.83	16.84	20.65	238,416	383,732	145,317
2009	5,063.4	7.6	11.91	13.9	141,228	258,299	117,071
2010	5,063.4	10.97	12.53	14.59	203,852	271,121	67,269
2011	5,063.4	7.02	10.76	13.4	130,450	249,008	118,557
Annual average		**9.61**	**13.01**	**15.64**	**178,487**	**290,540**	**112,054**

Notes Prices are from BlueNext in current euros per ton of carbon dioxide equivalent (€/tCO_2e) traded in the global Certified Emission Reductions (CER) market. Based on atomic weights, for every one ton of carbon stored, 3.67 tons of carbon dioxide are pulled from the atmosphere.
Source Author's calculations

Given the significant price volatility of the carbon market, a range of sequestration values were calculated using the minimum (Pmin) and maximum (Pmax) prices observed for each year. The results indicate that average annual revenue from carbon sequestration in the Laguna de Rocha between 2008 and 2011 was between €178,487 and 290,540 current euros per year. In order to trade, a stream of these revenues could be discounted over the time horizon specified for a given transaction.

The objective of this valuation exercise was to provide a rough first indication of the potential value of the carbon sequestration service of emergent wetland plants in the Laguna de Rocha. That being said, a number of further steps beyond the scope of this chapter became apparent. The first action required is an in-depth analysis of carbon market regulations to determine how carbon sequestration services from protecting coastal wetlands can be certified (or not). The next step would be to determine if it's worthwhile to: (i) determine the specific carbon accumulation rates of Uruguay's saline wetlands; and (ii) go through the certification process in order to trade in carbon markets. The significant annual revenues reported in Table 5.4 indicate that it may be in Uruguay's (or any other countries') interest to investigate this option further as a means of financing the management of coastal wetland protected areas.

Valuing Climate Impacts on 'Habitat Maintenance' and 'Cultural' Ecosystem Services

The rationale for the final non-market valuation exercise presented here arises from two observations of the reality in the LRPA. First, there is very little physical or 'hard' infrastructure in the protected area that can be damaged by climate changes. Small, but important, exceptions are the fishing communities and other residences located on the lagoon sandbar and at the mouth of the Rocha River. The

Fig. 5.3 Climate change damage costs avoided through habitat conservation. *Notes* See preceding paragraph for explanation. *Source* Loosely adapted from Barbier (2007)

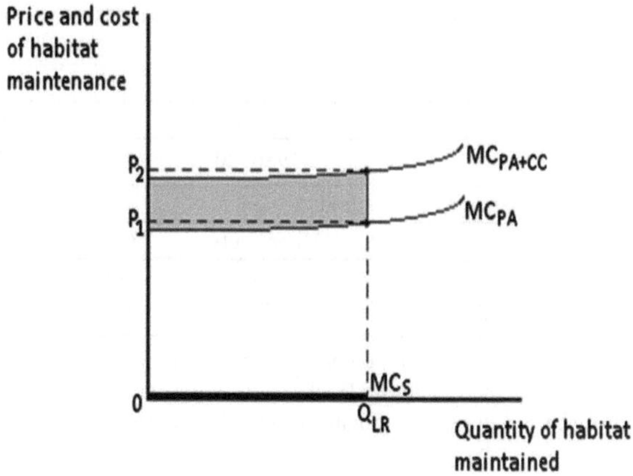

second observation is that there have been significant investments in 'soft' projects focused on conserving the relatively pristine state of the area through knowledge generation, capacity development, and policy planning. These projects have been implemented over the years by a number of actors including national and local government, academics, local producers and/or NGO's. In fact, the area has been the centre of so many projects in the region, often funded in part by international development organizations, that a feeling of 'project fatigue' was expressed by several individuals consulted throughout the course of this study.

Conventional economic analyses miss this significant mobilization of funds. A cost-based valuation methodology based on 'soft' project budgets can offer a rough indication of the economic value of conserving ecosystem services (that otherwise lack market or surrogate market indicators). The 'damage costs avoided' method simply takes the difference between the total costs of all projects from the costs of projects that do not explicitly address climate change to infer a monetary value of climate change impacts. This type of method is often considered a last resort by economists because it is not measuring economic willingness to pay. Instead, the assumption is that, if people incur costs to avoid damages from lost ecosystem services, then the value of those services must be at least what those people pay to avoid losing them (King and Mazzotta 2000). This assumption is debatable but it was felt that there was benefit in opening a space for discussion by demonstrating a cost-based methodology for valuing climate impacts on conservation efforts.

Figure 5.3 illustrates the effects of climate change on the costs of avoiding damages from lost habitat maintenance. Without people around, the cost of the habitat maintenance services provided by the lagoon ecosystem are 'free' as shown by the MC_S line along the horizontal axis to the quantity of habitat maintenance service provided by the Laguna de Rocha, Q_{LR}. However, in a world of scarcity, people incur costs to protect habitat maintenance services (and the cultural services upon which they depend), MC_{PA}, from being damaged or destroyed

Table 5.5 Summary statistics and results of damage cost method for valuing habitat maintenance and cultural services

Summary statistics			
Number of projects	Period	Mean	Standard deviation
46	1986–2012	49,326	112,690
Results of damage cost method			
	All projects	Cost of CC projects	% increase from CC
Total budget	2,269,000	300,000	**+13.2**
Average annual budget	84,037	11,111	

Notes Results based on total budgets of 46 conservation and research projects implemented/in execution in the Laguna de Rocha Protected Area between 1986 and 2012. Amounts are in current US dollars. *CC* climate change. *Source* Author's calculations

by other land-use alternatives. The assumption is that the value of the ecosystem service must be worth at least the amount that people pay to protect it, or the area under the MC_{PA} curve between 0 and Q_{LR}. Avoiding damages from climate change increase the costs of protecting the area, MC_{PA+CC}, thus resulting in an increase in the inferred value of the habitat maintenance service. The inferred economic value of climate change impacts on the habitat maintenance services of the LRPA is therefore at least the grey area between MC_{PA+CC} and MC_{PA}.

Table 5.5 reports the summary statistics and results of the damage costs avoided valuation of habitat maintenance and cultural services. Budgetary data dating back as far as 1986 was obtained for 46 projects (either in execution or completed) in the Laguna de Rocha Protected Area from various institutions.

The results of the damage costs avoided method suggest that an inferred value of climate change impacts in the protected area is US$ 300,000 over the 27-year period. This method infers that changes in climate have increased the economic value of the habitat maintenance and cultural ecosystem services provided by the LRPA by at least 13.2 %.

Conclusion

This chapter identified climate trends, selected ecosystem services and performed three monetary valuation methodologies for estimating a preliminary economic value of climate change impacts on ecosystem services in the Laguna de Rocha Protected Area. As with most empirical exercises, data availability, time and resources were constraining factors that largely determined the selection of methodologies that could be feasibly implemented in the study. That being said, a number of valuation results on the local non-market impacts of climate change were obtained with relatively minimal data requirements (and under specific assumptions).

The most general lesson learned from this study is that monetary valuation methodologies were found to be useful tools for linking the global issue of climate change with local impacts on ecosystem services as long as their limitations are kept in mind. Given that current efforts estimating non-market climate impacts on natural ecosystems are largely speculative (Seo 2012), there is ample scope for further research on isolating the monetary values of climate impacts on ecosystems. However, even as the research progresses it will be prudent for decision-making to be informed by multiple criteria as complex ecological-economic interactions and modelling of climate change scenarios far into the future push these methodologies to their limits (and beyond, in some cases).

References

Agrawala S, Fankhauser S (2008) Economic aspects of adaptation to climate change: costs, benefits and policy instruments. OECD, Paris

Aldabe J, Rocca P, Claramunt S (2009) Uruguay. In: Devenish C., Diaz Fernadez, Clay, Davidson & Yepez Zabala (eds), Important Bird Areas Americas: Priority sites for biodiversity conservation, BirdLife Conservation Series No. 16, BirdLife International, Quito, pp 383–392

Barbier E, Acreman M, Knowler D (1997) Economic valuation of wetlands: a guide for policy makers and planners. Ramsar Convention Bureau, Gland

Barbier EB, Strand I (1998) Valuing mangrove-fishery linkages: a case study of Campeche, Mexico. Environ Resour Econ 12(2):151–166

Barbier E (2003) Habitat-fishery linkages and mangrove loss in Thailand. Contemp Econ Policy 21(1):59–77

Barbier E (2007) Valuing ecosystem services as productive inputs. Econ Policy 22(49):177–229

Barros V (2004) Tendencias climáticas, Proyecto Agenda Ambiental Regional-Mejora de la Gobernabilidad para el Desarrollo Sustentable, UNDP ARG/03/001, Buenos Aires

Barros V et al (2005) Global climate change and the coastal areas of the Río de la Plata. Final report, AIACC project LA 26. Available at www.aiaccproject.org

Bidegain M et al (2005) Tendencias climáticas, hidrológicas y oceanográficas en el Rio de la Plata y costa uruguaya. In: Barros V, Menedez A, Nagy G (eds) El Cambio Climático en el Río de la Plata, Proyectos AIACC, Buenos Aires, pp 137–143

Bidegain M, Severova V, De los Santos B (2009) Climatología Regional de Base. Climatología de base de variables climáticas, nivel medio del mar y mapas de riesgo geológico, Informe No. 1: Proyecto URU/07/G32,Universidad de la República, Montevideo

Bonilla S (2002) Estructura y productividad de la comunidad de microalgas del ambiente pelágico en la Laguna de Rocha. Tesis de Doctorado PEDECIBA—Biología-Ecología, Universidad de la Republica, Montevideo, p 156

Bonilla S, Conde D, Aubriot L, Perez MC (2005) Influence of hydrology on phytoplankton species composition and life strategies in a subtropical coastal lagoon periodically connected with the Atlantic Ocean. Estuaries 28(6):884–895

Bonilla S et al (2006) Procesos estructuradores de las comunidades biológicas en lagunas costeras de Uruguay. In: Menafra R, Rodríguez-Gallego L, Scarabino F, Conde D (eds) Bases para la Conservación y el Manejo de la Costa Uruguaya. Vida Silvestre, Montevideo, pp 611–631

Chmura GL, Anisfeld SC, Cahoon DR, Lynch JC (2003) Global carbon sequestration in tidal, saline wetland soils. Global Biogeochem Cycles 17(4):1111. doi:10.1029/2002GB001917

Choi Y, Wang Y (2004) Dynamics of carbon sequestration in a coastal wetland using radiocarbon measurements. Global Biogeochem Cycles, GB4016. doi:10.1029/2004GB002261

Conde D, Aubriot L, Sommaruga R (2000) Changes in UV penetration associated with marine intrusions and freshwater discharge in a shallow coastal lagoon of the Southern Atlantic Ocean. Mar Ecol Prog Ser 207:19–31

Conde D, Rodríguez-Gallego L, Rodríguez-Graña M (2003) Análisis conceptual de las interacciones abióticas y biológicas entre el océano y las lagunas de la costa atlántica de Uruguay. Project No. PNUD/GEF/RLA/99/G31 (FREPLATA), Universidad de la República, Montevideo

Costanza R (2008) Ecosystem services: multiple classification systems are needed. Biol Conserv 141:350–352

de Groot R, Fisher B, Christie M (2010) Integrating the ecological and economic dimensions in biodiversity and ecosystem service valuation. The economics of ecosystems and biodiversity: ecological and economic foundations (Preliminary Draft), TEEB document available at www.teebweb.org

ECLAC (2010) Economics of climate change in Latin America and the Caribbean. Summary 2010, economic commission for Latin America and the Caribbean, Santiago

Ellis GM, Fisher AC (1987) Valuing the environment as input. J Environ Manage 25(2):149–156

Engle V (2011) Estimating the provision of ecosystem services by Gulf of Mexico coastal wetlands. Wetlands 31:179–193

Fabiano G, Santana O (2006) Las pesquerías en las lagunas costeras salobres de Uruguay. In: Menafra R, Rodríguez-Gallego L, Scarabino F, Conde D (eds) Bases para la Conservación y el Manejo de la Costa Uruguaya. Vida Silvestre, Montevideo, pp 557–565

Freeman AM (1991) Valuing the environmental resources under alternative management regimes. Ecol Econ 3:247–256

Goulder LH, Kennedy D (1997) Valuing ecosystem services: philosophical bases and empirical methods. In: Daily G (ed) Nature's services: societal dependence on natural ecosystems. Island Press, Washington DC, pp 23–48

Grimm AM (2010) Interannual climate variability in South America: impacts on seasonal precipitation, extreme events and possible effects of climate change. Stoch Env Res Risk Assess 25:537–554

Haylock M et al (2006) Trends in total and extreme South American rainfall in 1960–2000 and links with sea surface temperature. J Clim 19:1490–1512

King DM, Mazzotta MJ (2000) Dollar-based ecosystem valuation methods. U.S. Department of Commerce, NOAA. Available at http://www.ecosystemvaluation.org/dollar_based.htm. Accessed 10 April 2011

Magenheimer JF, Moore TR, Chmura GL, Daoust RJ (1996) Methane and carbon dioxide flux from a macrotidal salt marsh. Bay of Fundy, New Brunswick, Estuaries, vol 19. pp 139–145

Marengo JA, Camargo CC (2007) Surface air temperature trends in Southern Brazil for 1960–2007. Int J Climatol 28:893–904

Marengo JA et al (2010) Recent developments on the South American monsoon system Int J Climatol 32(1):1–21

Millennium Ecosystem Assessment (2005) Ecosystems and human well-being: wetlands and water synthesis. World Resources Institute, Washington, DC

Nicholls R, Brown S, Hanson S, Hinkel J (2010) Economics of coastal zone adaptation to climate change. Development and climate change discussion paper No. 10, The World Bank, Washington, pp 1–49

Nogués-Paegle J et al (2002) Progress in Pan-American CLIVAR research: understanding the South American monsoon. Meteorologica 27:3–30

Norbis W (2000) Estudios sobre la población de camarón rosado (Penaeus Paulensis) en las lagunas costeras de la reserva de biosfera bañados del este. Probides Documentos de Trabajo No. 28, Rocha

Parry M et al (2009) Assessing the costs of adaptation to climate change: a review of the UNFCCC and other recent estimates. International Institute for Environment and Development, London

Re M, Menendez AN (2007) Impacto del cambio climático en las costas del Río de la Plata. Revista Internacional de Desastres, Accidentes e Infraestructura Civil 7(1):25–34

Rodriguez-Gallego L et al (2008) Costos y beneficios socioeconómicos y ambientales del uso actual de la laguna de Rocha y su cuenca: Insumos para la gestión integrada de un área protegida costera. Programa de Desarrollo Tecnológico Proyecto 36-09, Universidad de la República, Montevideo
Rothman DS, Amelung B, Polomé P (2003) Estimating non-market impacts of climate change and climate policy. Working party on global and structural policies working paper ENV/EPOC/GSP(2003)12/FINAL, OECD, Paris
Rusticucci M, Barrucand M (2004) Observed trends and changes in temperature extremes in Argentina. J Clim 17:4099–4107
Rusticucci M, Renom M (2008) Variability and trends of extreme temperature indices in Uruguay. Int J Climatol 28:1083–1095
Rusticucci M et al (2009) An intercomparison of model-simulated in extreme rainfall and temperature events during the last half of the twentieth century—Part 1: mean values and variability. J Climatic Change 98:493–508
Sathirathai S (1997) Economic valuation of mangroves and the roles of local communities in the conservation of the resources: case study of Surat Thani, South of Thailand. Economy and Environment Program for Southeast Asia (EEPSEA), Singapore
Santana O, Fabiano G (1999) Medidas y mecanismos de administración de los recursos de las lagunas costeras del litoral atlántico del Uruguay (Lagunas José Ignacio Garzón de Rocha y de Castillos). Plan de Investigación Pesquera, INAPE-PNUD URU/92/003, Montevideo
Sampson RN, Scholes RJ (2000) Additional human-induced activities. In: Watson RT, Noble IR, Bolin B, Ravindranath NH, Verardo DJ, Dokken DJ (eds) IPCC special report on land use, land use change and forestry, Cambridge University Press, UK
Seo S (2012) Climate change impacts on non-market activities. Encyclopedia of Earth. Retrieved from http://www.eoearth.org/view/article/151246
Tol RSJ, Fankhauser RG, Richels G, Smith JB (2000) How much damage will climate change do? Recent estimates. World Econ 1(4):179–206
Vatn A, Bromley D (1994) Choices without prices without apologies. J Environ Econ Manage 26(2):129–148
Vera C et al (2006) Toward a unified view of the American monsoon systems. J Climate Spec Sect 19:4977–5000
Vergara W et al (2013) The climate and development challenge for Latin America and the Caribbean: options for climate-resilient, low-carbon development. Inter-American Development Bank, Washington DC
Verocai JE (2009) Base de datos del nivel del mar en la costa uruguaya. Climatología de base de variables climáticas, nivel medio del mar y mapas de riesgo geológico, Informe No. 1: Proyecto URU/07/G32, Universidad de la República, Montevideo
Vincent LA et al (2005) Observed trends in indices of daily temperature extremes in South America 1960–2000. J Clim 18:5011–5023
World Bank (2010) The costs to developing countries of adapting to climate change: new methods and estimates. Global report of the economics of adaptation to climate change study, Consultation Draft, Washington DC

Chapter 6
Integrating Climate Science, Monitoring, and Management in the Rio de la Plata Estuarine Front (Uruguay)

Gustavo J. Nagy, Nathalie Muñoz, José E. Verocai, Mario Bidegain, Bernardo de los Santos, Leonardo Seijo, Juan Martín García, Gabriella Feola, Beatriz Brena and Jimena Risso

Abstract The study presents an institutional arrangement for managing some risks associated with hydroclimatic variability in the Rio de la Plata River Estuarine Frontal System (EFS). The goals are to contribute to an efficient coordination work between local agencies that manage the use of ecosystem services, and to incorporate climate science and management options in coastal areas of Uruguay. We have identified and analysed climate forcing ENSO and the variables

G. J. Nagy (✉) · J. E. Verocai · B. de los Santos
Grupo de Cambio Ambiental Costero, IECA, Facultad de Ciencias, UdelaR, Montevideo, Uruguay
e-mail: gustavo.nagy56@gmail.com

J. E. Verocai
e-mail: otolito21@gmail.com

B. de los Santos
e-mail: bdelossantos@gmail.com

G. J. Nagy · N. Muñoz
GEF Project URU 07/G32 "Implementing Pilot Adaptation Measures to Climate Change in Coastal Areas of Uruguay", UCC-UNDP-DINAMA, Montevideo, Uruguay

J. E. Verocai
Oceanographic and Meterorological Service (SOHMA), Montevideo, Uruguay

M. Bidegain · B. de los Santos
Uruguayan Institute of Meteorology (INUMET), Montevideo, Uruguay
e-mail: bidegain.mario@gmail.com

B. de los Santos
e-mail: bdelossantos@gmail.com

L. Seijo
Oficina de Planeamiento y Presupuesto, Presidencia de la República, Montevideo, Uruguay
e-mail: leoseijo@gmail.com

J. M. García
Universitat Politècnica de Catalunya (UPC), Barcelona, Spain
e-mail: jmarting@catunesco.upc.edu

W. Leal Filho et al. (eds.), *International Perspectives on Climate Change*, Climate Change Management, DOI: 10.1007/978-3-319-04489-7_6,

river flow and winds, which govern the location and displacement of the frontal system, where the capital city Montevideo is located, and the fluctuations of salinity. The study analyses the austral spring–summer period (October–March) from 1997/1998 to 2012/2013. Fish catch and beach quality were selected as examples of environmental impacts of climate forcing which are presented as a causality loop diagram (CLD): ENSO → (River flow, wind, rainfall) → (EFS displacement, Salinity), and state/impact indicators → (Fish catch, Beach quality). Institutional cooperation has been developed to incorporate climate adaptation regarding the Frontal System. GEF Project "Implementing Pilot Adaptation Measures to Climate Change in coastal areas of Uruguay", in which the Directorate of the Environment, the University of the Republic, and the Government of Montevideo collaborate on improving climate science, monitoring, and management in the fields of Estuarine and Environmental Management has provided the framework for the present study.

Keywords ENSO variability • River flow • Environmental impacts • System dynamics • Adaptation

Introduction

The Republic of Uruguay is located in the La Plata River Basin (which spreads over a 3.1 million km^2), within are located the Paraná River and the Uruguay River (Fig. 6.1). The flows of both rivers mix with the Atlantic ocean within the Rio de la Plata, a large microtidal-river and estuary system (38,000 km^2, average depth <10 m). The mixing zone or estuarine frontal system—EFS may be defined by discontinuities of turbidity (Fig. 6.2) where fresh turbid water and marine green water prevail up and down-river the estuarine frontal system respectively (Nagy et al. 2008a).

J. M. García
Catedra UNESCO de Desarrollo Sostenible, UPC, Barcelona, Spain

J. M. García
Instituto Latinoamericano de Ciencias (ILC), Huancayo, Perú

G. Feola · B. Brena · J. Risso
Environmental Control and Quality Assessment, Intendencia de Montevideo, Montevideo, Uruguay
e-mail: gabriella.feola@imm.gub.uy

B. Brena
e-mail: bbrena@fq.edu.uy

J. Risso
e-mail: jimena.risso@imm.gub.uy

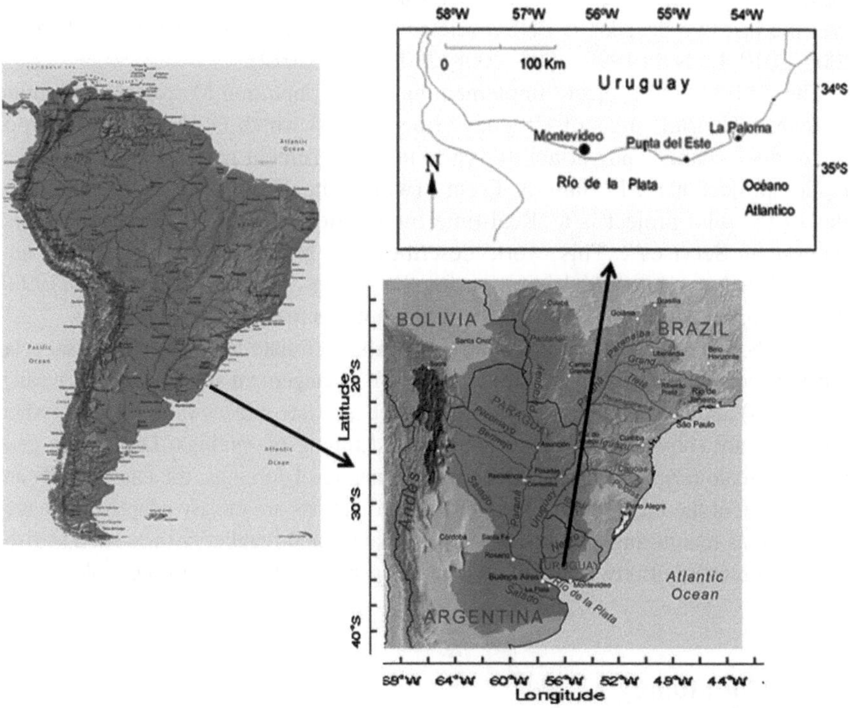

Fig. 6.1 Rio de la Plata basin and river estuary (Southeastern South America)

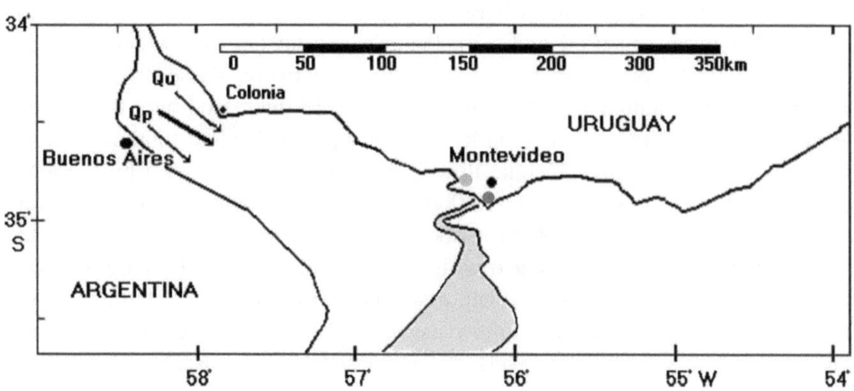

Fig. 6.2 Chart of Rio de la Plata, river flow corridors and schematic estuarine frontal system (*shadowed area*) location. Fishermen settlements (*light gray circle*) and beach area (*dark gray circle*). Modified from Nagy et al. (2008a) (Color figure online)

The cyclic variability of river flows displaces the EFS and changes salinity and turbidity on seasonal and interannual time-scales, which impact fish catch and beach waters quality at Montevideo, the Uruguay's capital city, particularly during spring and summer time (October–March). Most observed interannual variability over the last two decades is ENSO-related (i.e. El Niño: 1997–1998, 2002–2003, 2009–2010; La Niña 1999–2000, 2008–2009, 2011–2012).

The GEF-UNDP Project "Implementing Pilot Adaptation Measures to Climate Change in Coastal Areas of Uruguay" (the Project) which is devoted to incorporate climate change and variability issues into government policies has considered a pilot project named "Salinity Front" (within the EFS). One of the deliverables of the pilot project is a "Real-time Information System for Biodiversity and Ecosystem Services". This work describes the hydroclimatic variability and some related environmental impacts, for the spring–summer periods 1997/1998–2012/2013, and develops some management options to cope with them.

We follow a system dynamics approach to show climate forcings and causal relationships between climatic variables and the two management subjects: croaker catch (fish boxes) and beach quality (coliforms), which are state indicators or impacts (when thresholds are surpassed) of the magnitude and timing of forcings. Due to the geographical scale (La Plata Basin-LPB) and the nature of forcings, the causal roots are unavoidable at the local scale. Thus, the selected actions are mostly adaptive management such as monitoring, preparedness, surveillance, and regulatory ones such as fishing closed seasons or prevention of recreational use on spatial and/or time scales.

Hydroclimatology

The rivers Parana and Uruguay are major tributaries of the "Rio de la Plata". Total yearly river inflow (Q_{RP}) varies between 22,000 and 28,000 m^3/s during normal years and between <20,000 and >30,000 m^3/s during dry and wet years, often associated with La Niña and El Niño events, respectively. These flows and their seasonal and interannual variability have significantly increased over the last four decades. On the average the Paraná River (Q_P) supplies 75 % of total average discharge whereas the yearly average flow of Uruguay River shows a high interannual variability (from <3,000 to >10,000 m^3/s). Climatologically, maximum total outflow occurs in February–March associated with the peak of Paraná River flow (Q_P), during April–July, associated with the accumulated high discharges of both rivers ($Q_P + Q_U$) and during October, associated with the peak of Uruguay River flow (Q_U). The minimum total inflow usually occurs around January, associated with low and very low water levels of Paraná and Uruguay Rivers, respectively. Freshwater inflow trifurcates into three flow corridors, the flow of Uruguay River being the main contribution to the Uruguayan coast from April to October (high discharge period). This is coherent with the fact that monthly variability of salinity off Montevideo is well correlated with Q_U, which in turn is associated with ENSO SST 3.4 variability (Nagy et al. 2003, 2008a).

During the period 1997–2012 the yearly and spring–summer (October–March) average flow (Q_{RP}) were 23,000 and 23,500 m^3, respectively, 76 % of which from river Paraná. Rainfall anomalies from 100 (Montevideo) to 400 % (LPB) (Fig. 6.3)

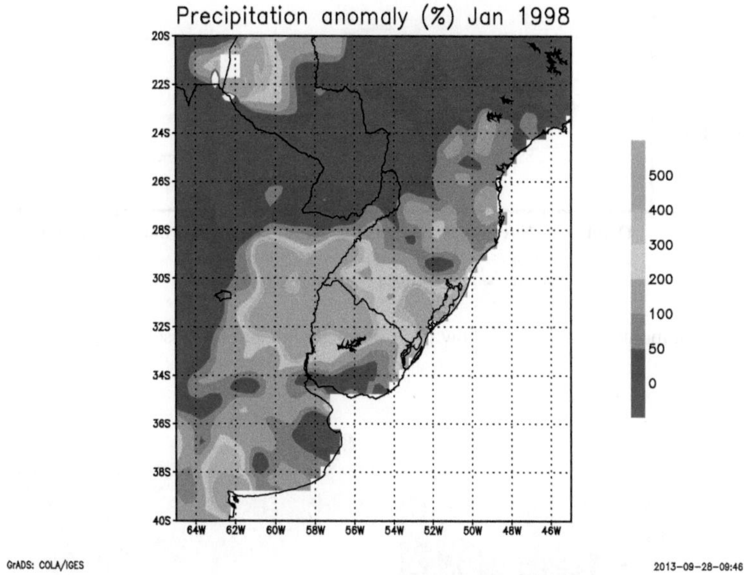

Fig. 6.3 January 1998 rainfall anomaly (%) over Rio de la Plata basin

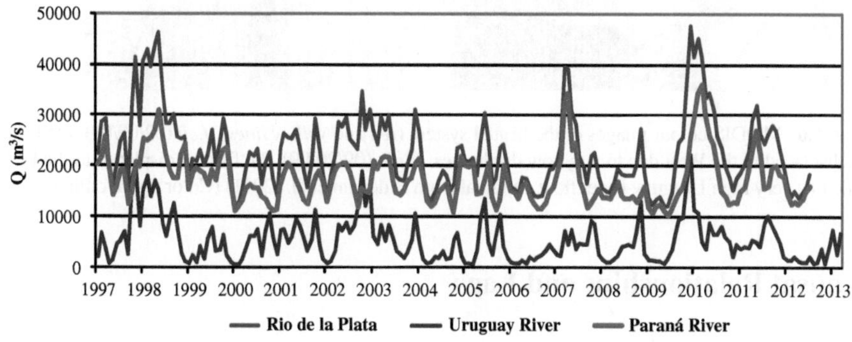

Fig. 6.4 Monthly river flow time-series for La Plata (Q_{RP}), Paraná (Q_P), and Uruguay (Q_U) rivers from January 1997 to July 2012 (Q_{RP} and Q_P) and March 2013 (Q_U)

and yearly extremes of 11,949 and 47,565 m^3/s occurred in coincidence with the cold and warm phases of ENSO respectively (Fig. 6.4).

Seven spring–summer flows are shown (Fig. 6.5) for El Niño (1997–1998, 2002–2003, 2009–2010) and La Niña years (1999–2000, 2008–2009, 2010–2011, 2011–2012).

During El Niño springs the overflow of Q_U strongly impacts the Uruguayan coast because of the preferential canalisation of its flow (Fig. 6.6).

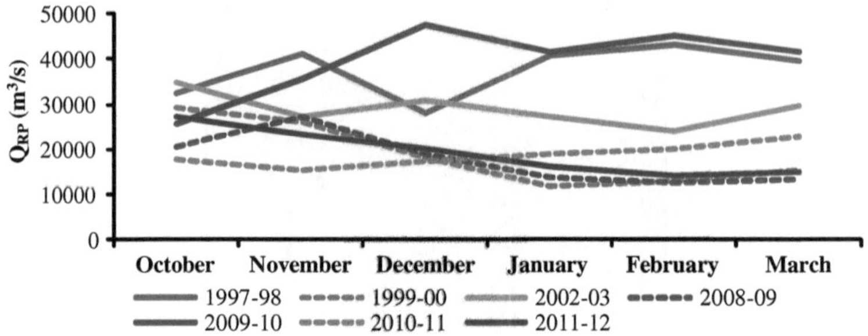

Fig. 6.5 Mean monthly total river inflow from selected high and low discharge periods

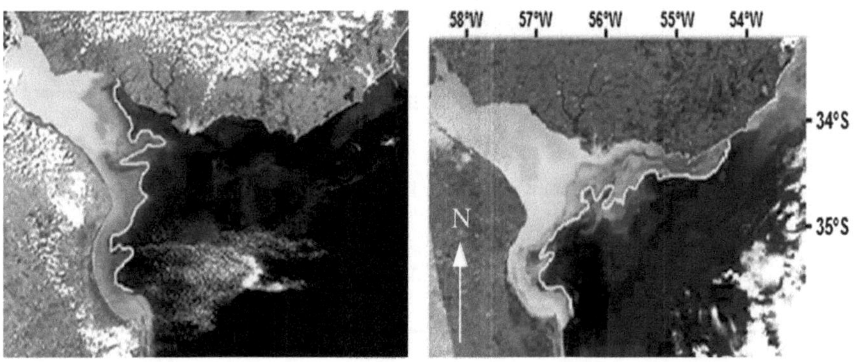

Fig. 6.6 MODIS colour images of the frontal system (*red* and *yellow lines*). *Left* (02/09/09): the front is displaced to the West due to very low discharges. *Right* (09/20/09): the front is displaced to the East due to a very high Uruguay River flow. Original from Bidegain et al. (2011) (Color figure online)

Causal Relationships and Lags

A causal loop diagram (CLD) shows the causal relationships (Fig. 6.7) between climatic variables and adaptation actions such as monitoring, modelling, forecasting, regulations, which pretends to become a tool of climate variability and change management. Even if both management issues share the climatic drivers, fishing season lasts from October to February, the peak of the catch being before December, whereas, the sun and beach season lasts from November to March, with a peak of beach attendants after December. Thus, as a consequence of El Niño events, when the River Uruguay peaks during spring, fish catch is strongly dependent on El Niño and Q_U. The beach quality is somewhat less dependent on them but on the overall freshwater inflow (Q_{RP}). Thus, salinity ranges and thresholds are slightly different, winds are a key specific variable for fish catch and local rainfall for beach quality (i.e. January 1998, Fig. 6.3), and timing differs in one month.

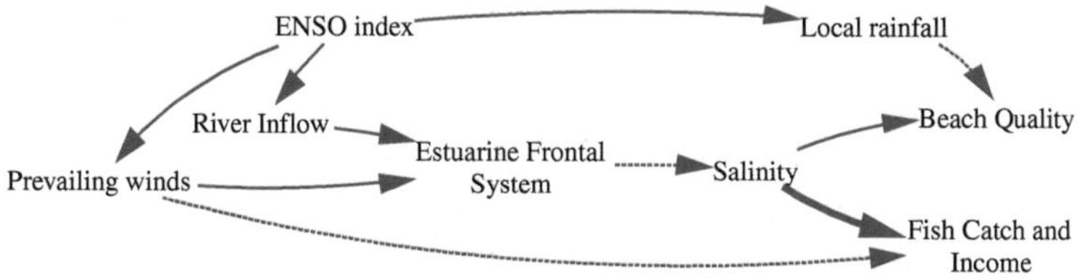

Fig. 6.7 Causal loop diagram for the Rio de la Plata estuarine frontal system. (+) feed-back = *continuous arrow*; (−) feed-back = *dashed arrow*; (+ or −) = *thick arrow* (Color figure online)

Table 6.1 Correlations (R_P or R_S) and lags (months) between ENSO index (SST 3.4) with river flow (Q_U and Q_{RP}), River flows and salinity, and Q_U and coliforms for the periods 1997–1998, 2009–2010 and 2010–2011

Year	Lag	SST-Q_U		SST-QRP		Q_U-Sal		Q_{RP}-Sal		Q_U-Col	
		R_P $p < 0.5$ $p < 0.1$						R_S $p < 0.5$ $p < 0.1$			
	m	A	M	A	M	A	M	A	M	A	M
1997–1998	0									−0.90	−0.90
	−1										**0.70**
	−2							−0.80	−0.60	**0.80**	
	−3						−0.60		−0.60		
	−4					−0.60		−0.80			
	−5					−0.80	−0.60	−0.80	−0.60		**0.70**
2009–2010	0	**−0.93**				**0.82**					
	−1				0.73	**0.97**	**0.66**				
	−2			0.66		**0.87**					
	−3	**−0.84**					−0.67	−0.67			**0.80**
	−4						−0.67	−0.67			**0.80**
	−5					−0.56	−0.67	−0.67		**0.90**	**0.80**
2010–2011	0			**0.86**				**1.0**			
	−1			**0.88**	**0.81**			**0.90**		0.80	
	−2			**0.91**							**0.90**
	−3		**−0.89**				**−0.90**		**−0.90**	**0.90**	
	−4					**−1.0**	**−0.90**		**−1.0**		
	−5	−0.77			−0.94	**−1.0**		**−0.90**	**−1.0**		

A accumulated months, *M* specific month lag

We have analysed the relationships and lags between monthly average ENSO index (Sea Surface Temperature anomaly at the Pacific Region 3.4—SST 3.4 and Multi ENSO Index—MEI), river flows (Q_{RP} and Q_U), salinity at Punta Brava (where the main submarine sewage outfall of the city is placed), and fecal coliforms concentration in two beaches (Ramírez and Pocitos) close to P. Brava. Both ENSO index showed good fit (Table 6.1) with river flows (Q_U and Q_{RP}). Significative ($p < 0.1$ or 0.5) correlations (Pearson: R_P or Spearman: R_S) are shown for lags zero to five months with the average of accumulated months (A)

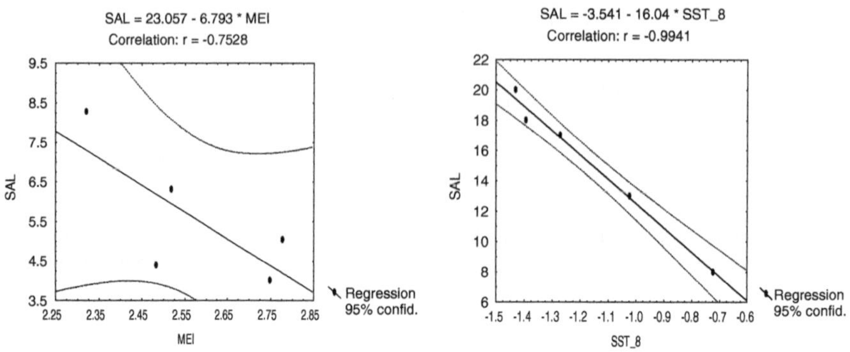

Fig. 6.8 Linear regression between ENSO and salinity (lags −5 to 1 month) during El Niño 1997–1998 (MEI vs. Salinity, *left*) and La Niña 2010–2011 (SST 3.4 vs. Salinty, *right*)

or specific ones (M). Examples of ENSO—salinity relationships are shown for El Niño and La Niña years (Fig. 6.8).

These correlations and lags suggest that monitoring of ENSO from July gives an input to early warning and that the best indicator is river flow(s) since September.

Institutional Arrangements to Identify and Implement Adaptation Measures

All available information was shared with experts from the academia and institutional stakeholders (Directorate of the Environment—DINAMA, Directorate of Aquatic Resources—DINARA, Naval Oceanographic Department—SOHMA, and the Government of Montevideo—IdeM) during a workshop held in March 2012. The Project experts presented research on climate threats and scenarios, identified vulnerabilities, and a list of best adaptations measures based on criteria drawn from international adaptation projects (Nagy et al. 2013). The goals of the workshop were to:

- Communicate scientific results to increase awareness with regard to climate change and variability.
- Present a list of adaptation actions identified by the Project's experts.
- Receive feedbacks from the attendants and adjust the expectative to the institutional needs and implementation possibilities.
- Involve those who wished to participate in the process.

The institutional involvement varied from making proposals (DINAMA, UdelaR), waiting for inputs without engagement (DINARA), or expressing interest to cooperate and receive inputs (IdeM). Project's experts redefined the decision criteria and adaptation actions based on the discussion held during the workshop. Six

Table 6.2 Adaptation research lines and prioritisation criteria for the Pilot Adaptation Area "Salinity Front"

Selected research lines proposals to become measures	Prioritisation criteria						
	R	A	V	C	N	U	Σ
1. To supply climate scenarios to assess impact thresholds for biodiversity	5	5	5	5	5	4	29
2. To elaborate a list of vulnerable and invasive species and their relationship with climatic changes	5	5	5	5	5	4	29
3. To supply climate information and scenarios to make a zonification of vulnerable habitats and ecosystem	5	5	5	5	5	4	29
4. To integrate real-time climatic variables into monitoring, modeling, forecasting, and early warning systems to manage the environment, biodiversity and resources	5	4	4	5	5	5	28
5. To inform fisheries agency about current and future climate scenarios to manage fishing close seasons	5	4	3	5	5	4	26

Criteria: *R* Relevance, *A* Acceptability, *V* Viability, *C* Cost-Benefit, *N* No-regret, *U* Urgency and uncertainty, Σ Overall. Original from UCC (2013)

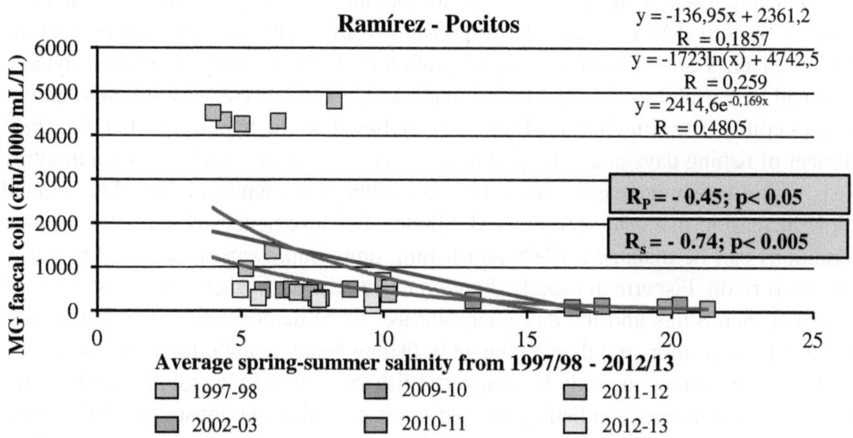

Fig. 6.9 Relationship (R_P and R_S) between average beach October–March salinity (1997/1998–2012/2013) and faecal coliforms (geometric mean) for two beaches (Ramirez and Pocitos) at Montevideo

criteria were selected which were ranked from 1: very low importance to 5: very high importance, and five actions were suggested (Table 6.2) which balance the need of information, monitoring, and management actions.

Here we focus on the fourth research line. A research agreement was achieved with the staff of the IdeM in charge of evaluating beach water quality with regard to the threat imposed by the cyclic occurrence of low salinities (<10) during beach season. The aim is to integrate information on the climatic variables that determine low salinity, into the monitoring strategy. Limit coliform thresholds (1,000 cfu/1,000 ml/L) associated with high river flow from the La Plata Basin and/or local rainfall, that occur often in coincidence with El Niño years, demand increasing surveillance and prevention of recreational use to access to bath (IdeM 2010). The relationship between average salinity and coliforms (Fig. 6.9) shows thresholds at salinities 5 and 10. Given that a seasonal average of 500–1,000 cfu/1,000 ml/L implies several days above the limit, any salinity within this range is a warning of threat.

Management Options

In order to research the simulation of management options we have revisited a previous work (Nagy et al. 2008b) about the fishing activity, catch and income in relation to the displacement of the EFS during ENSO years (1999–2000 and 2002–2003). The small fisheries settlements placed along the EFS (Fig. 6.10) depend on the accessibility (increased distance = increased effort) to the resource within the estuarine front (with typical salinities ca. 2–12), and both freshwater and marine waters are unfavourable conditions. Successive bad fishing years decreased catch/fishing effort ratios to unsustainable levels if the tendency was to continue. The implementation of an early warning System (EWS) is a public responsibility, thus, only costs to improve fishing units' efficiency are equated into the adaptation cost-benefit analysis. System dynamics could be a useful tool for exploring management options (www.system-dynamics-courses.com/). The simulations (Table 6.3) are based on the hypothesis that the actual number of fishing days and fishing units is about 80 % of the maximum on a monthly basis within a fishing season. Both the availability of resources and market demand allow increasing fish catch (Nagy et al. 2008b). Because neither climate nor weather conditions can be managed, EWS and fishing units improvement are the key variables to work on. Fishermen usually do not navigate the days when south-eastern wind is greater than 8 m/s and the day after. Norbis (1995) demonstrated that fishing was possible the day after, and then a change in fishing behaviour should allow increasing catch. This behaviour can only be changed if fishermen trust on a real time early warning system. The increase in fishing navigations is central to maximise catch/effort ratio which could be achieved with an increase in cost estimated to be about 20 %. Thus, adaptation options should improve benefits by ca. 60 % over the long-term income to compensate both the increase in costs and losses during bad fishing ENSO years.

This example under development will serve as an analogy to explore key variables and beach quality management options in a future step of this work in progress. Since there are no available tools to forecast water quality, staff is constrained to monitoring, preparedness and surveillance actions. The results of the monitoring program are published in the institutional web page every week (http://www.montevideo.gub.uy/ciudadania/desarrollo-ambiental/playas) and if fecal coliform values do not comply with the regulatory limits, a sanitary flag is posted in the affected beaches to communicate possible heath risks and prevent exposure of swimmers. Thus, any tool to forecast a high probability of observing fecal coliforms values above the limit, e.g., excess in local rainfall, would be extremely useful to facilitate awareness, and to define actions such as increase surveillance frequency for example.

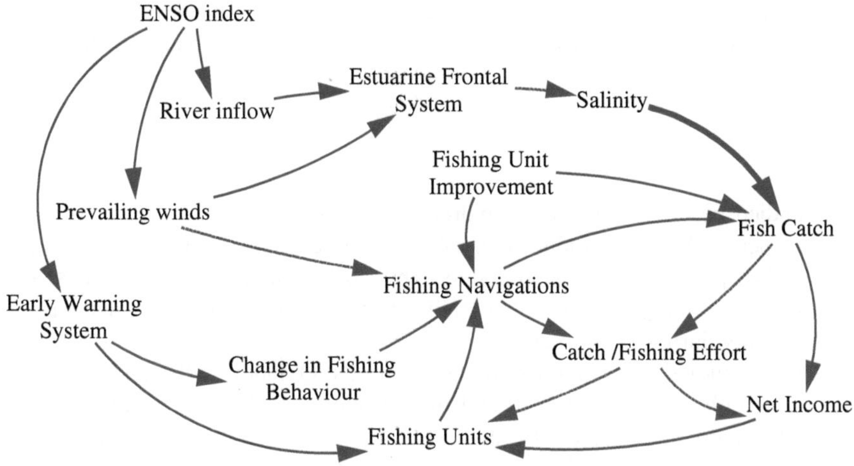

Fig. 6.10 Causal loop diagram for the fish catch simulation. (+) feed-back = *continuous line*; (−) feed-back = *dashed line*; (+ or −) = *thick line* (Color figure online)

Table 6.3 Fishing activity, catch and income: comparison between a good year (1998–1999), long-term average and model results for a low-typical year (1); a bad year (2); results with change in fishing behavior (3), and results with adaptive management alternatives suggested in this work (4). Modified from Nagy et al. (2008b)

	Observed 1998–1999	Long-term average	Model 1	Model 2	Model 3	Model 4
Number of boxes per boat per day	22	21	20	16	20	23–24
Fishing days	64	57	50	40	72	72–74
Fishing navigations	963	850	640	550	1,200	1,300–1,400
Boats per day	15	15	10	10	17	18–19
Increasing cost to adapt (%)	Without adaptation measures				Low cost easy actions	≥20
Income (% relative to 1998–1999)	100	85	70	45	120	140–150

We suggest implementing a colour-based early warning system (green: no-risk, yellow: moderate risk, orange: high risk) to provide a simple predictive tool to be used in combination with the monitoring program results in order to improve beach management practices. This should be supported by environmental education and awareness communiqués to make sure the population understands the relationship between hydro-climatic and meteorological conditions and water quality. Thus, a pro-active planning based on "advert before and explain why" could be incorporated to the current policy. To get this done, climatic and meteorological surveillance, modelling and simulation must be integrated and continuous. Such an adaptive flexible management approach could be supported by system dynamics simulations as a decision-making tool.

Conclusions

The studied period shows strong hydroclimatic interannual fluctuations, often in coincidence with ENSO induced variability. These fluctuations force river- and sea-ward displacement of the frontal system causing environmental impacts such as loss of fish catch during both El Niño and La Niña events, and decrease in beach quality during El Niño event.

Several studies show a good fit between ENSO index (MEI and SST 3.4) with river inflow and salinity, frontal displacements, and beach quality and fish catch. The correlation peak between ENSO index and impacts have usually a lag between three to five months, allowing the development of a monitoring-modelling-early warning system to support the implementation and timing of management options. Monitoring of ENSO and hydroclimatic variables (local rainfall and river flow) in this region allows forecasting the future behaviour (1–6 months) of the estuarine front displacement and salinity close to Montevideo.

The institutional agreements and consultation process followed by the Project have shown to be a successful way to select agreed adaptation actions. The involvement of partners since the early phases of the implementation of adaptation measures is a process of active participation which implies institutional stakeholders' ownership and facilitates the implementation.

A system dynamics approach under development will have a great chance to be used as a tool in order to simulate alternative management options to cope with increasing climate variability under climate change scenarios.

The next step will be to define an adaptive risk management approach based on agreed overall criteria to analyze climatic variables, their distribution, impact transfer functions, and thresholds of risk acceptability for each issue. Because of the accumulated uncertainty and varying lags of the causal relationships the use of risk classes (low, moderate, high) and a continuous update on monthly basis should be implemented. The time step of climatic monitoring and simulations should ideally change to weekly basis once the season begins.

References

Bidegain M, de los Santos M, Rodriguez C, Verocai JE (2011) Climate scenarios, extreme events, and dynamics of the frontal zone of the Rio de la Plata. Working paper, GEF-Project URU/07/G32 "Implementing pilot adaptation measures in coastal areas of Uruguay", UCC-DINAMA-UNDP, Facultad de Ciencias, Montevideo, Uruguay

IdeM (2010) Programa de Monitoreo de Agua de Playas y Costa de Montevideo. Informe 2009–10. Departamento de Desarrollo Ambiental. Servicio Laboratorio de Calidad Ambiental. Intendencia de Montevideo

Nagy GJ, Gomez-Erache M, Kay R (2013) A risk-based and participatory approach to assessing climate vulnerability and improving governance in coastal Uruguay, Chap. 16. In: Glavovic B, Kay R, Kelly M, Travers A (eds) Climate change and the coastal zone. Taylor & Francis, London
Nagy GJ, Severov DN, Pshennikov VA, de los Santos M, Lagomarsino JJ, Sans K, Morozov EG (2008a) Rio de la Plata Estuarine system: relationship between river flow and frontal variability. Adv Space Res 41:1876–1881 (Elsevier)
Nagy GJ, Bidegain M, Caffera RM, Norbis W, Ponce A, Pshennikov V, Severov DN (2008b) Fishing strategies for managing climate variability and change in the Estuarine Front of the Rio de la Plata. In: Leary N (ed) Climate change and adaptation. Earthscan, London, pp 363–370 (Chapter 20)
Nagy GJ, Pshennikov V, Robatto P (2003) Monthly variability of salinity at Montevideo, Rio de la Plata′s frontal zone, in response to ENSO fluctuations and to the River Uruguay Flow (1998–2000). Contribution Nº 1 in The Rio de la Plata: Research to Manage the Environment, Fish Resources and the Fishery in the saline front. D Vizziano et al. EcoPlata, Programme, Montevideo, Uruguay
Norbis W (1995) Influence of wind, behaviour and characteristics of the croaker (Micropogonias furnieri) artisanal fishery in the Río de la Plata (Uruguay). Fisheries Res 22(43–58):1995
Unidad de Cambio Climático (2013) Ficha Técnica Nº 7 "Medidas de adaptación en el sitio piloto Frente Salino", Proyecto GEF URU/07/G32 "Implementación de Medidas de Adaptación al Cambio Climático en Áreas Costeras del Uruguay", Unidad de Cambio Climático-UCC/DINAMA-MVOTMA/PNUD. Available at: http://www.adaptationlearning.net/

Part II
The Management of Climate Change Impacts

Chapter 7
Managing the Impacts of Climate Change in Latin America: The Need for Technology Transfer

Walter Leal Filho and Franziska Mannke

Abstract Due to its geo-political characteristics and social and economic features, the Latin American region is considered as being among the most vulnerable ones, as far as climate change is concerned. The combination of two further important elements, namely limited access to technologies and restricted adaptation capacity, may help to explain why the region is so vulnerable and is likely to remain so, unless fundamental changes in decision-making processes are implemented. From an objective point of view, decision-making processes may play a key role in facilitating the ways countries perceive and, as importantly, manage the impacts of climate change. Yet, there is a paucity of research which looks at the extent to which the sound management of the impacts of climate change may take place, across Latin America, in a systematic way. This paper addresses this need, by discussing the extent to which individual Latin American countries handle matters related to climate change, and by illustrating a number of the problems and deficiencies which have been limiting progresses over the past two decades. It also describes some of the recent and on-going initiatives from across the region, and introduces the project CELA, an initiative to promote technology transfer on climate change by means of cooperation between universities in the European Union and Latin American nations.

Keywords Climate change • Latin America • Vulnerability • Impacts • Technology transfer

---

W. Leal Filho (✉) · F. Mannke
Research and Transfer Centre "Applications of Life Sciences",
Hamburg University of Applied Sciences, Faculty of Life Sciences,
Lohbrügger Kirchstraße 65, 21033 Hamburg, Germany
e-mail: walter.leal@haw-hamburg.de

F. Mannke
e-mail: franziska.mannke@haw-hamburg.de

W. Leal Filho et al. (eds.), *International Perspectives on Climate Change*,
Climate Change Management, DOI: 10.1007/978-3-319-04489-7_7,

Introduction: Some Facts About Latin America

With an area of approximately 21,069,500 km^2, Latin America accounts for almost 3.9 % of the Earth's surface. This is more or less the equivalent of 14 % of the total land surface area. The population of the region exceeds 590 million, whereas its combined GDP of around US$ 5.16 trillion, attests its economic relevance. The Latin American economic growth rate has been of around 5 % in the last 3 years (OECD 2012). Throughout Latin America, a distinct influence of European culture is noticeable, especially Spanish and Portuguese and, to a lesser extent, English (Belize and Guyana are the only countries in the region where English is the official language) and French, spoken in Suriname and in some Caribbean islands.

As a whole, Latin America can be subdivided into several sub-regions based on geography, politics, demographics and culture. For the purposes of this paper, Latin America is interpreted as including the basic geographical sub-regions of North America (Mexico), Central America and South America. Figure 7.1 provides an overview of the population and size of the economies of the countries in the mainland in the region (i.e. excepting the islands).

It can be seen in Fig. 7.1 that Brazil, Argentina, Chile and Mexico have the largest economies, whereas countries such as Bolivia, El Salvador, Honduras, Guatemala and Nicaragua belong to the poorest nations in the region.

Climate Change and Its Impacts in Latin American Countries

Similar to what happens at other regions of the world, Latin America is facing the impacts of climate change in a variety of ways (IPCC 2007). What is different when compared to other world regions such as North America or Europe, is the fact that due to a combination of factors such as geographical position and topography, as well socio-economic aspects, this region is especially vulnerable.

In addition, there are many *problems and deficiencies* which have been contributing to the high degree of vulnerability of Latin American countries to climate change. Among others, the following ones may be listed:

- poor or non-existing climate change governance systems,
- limited awareness on the causes and consequences of climate change,
- endemic poverty,
- limited access to capital and global markets,
- continuous ecosystem degradation,
- complex disasters and conflicts,
- unplanned urbanization,
- limited capacity (personal and institutional) to address the problem and its many ramifications.

Country	Population (2013 July est.) Millions	GDP (PPP) (2012 est.) In US$
Argentina	41,2	746.9 bn
Bolivia	10,5	56.14 bn
Belize	0,3	1.56 bn
Brazil	201,0	2.39 tn
Chile	17,2	325,8 bn
Colombia	45,8	511,1 bn
Costa Rica	4,7	45,13 bn
Ecuador	15,4	155,8 bn
El Salvador	6,1	47,09 bn
Guatemala	14,4	79,97 bn
Guyana	0,7	6,26 bn
Honduras	8,5	38,42 bn
Mexico	116,2	1,79 tn
Nicaragua	5,8	27,1 bn
Panama	3,6	58,02 bn
Paraguay	6,6	41,55 bn
Peru	29,8	332 bn
Suriname	0,6	6,9 bn
Country	Population (2013 July est.) Millions	GDP (PPP) (2012 est.) In US$
Uruguay	3,3	54,67 bn
Venezuela	28,4	408,5 bn

Fig. 7.1 Overview of the population and economy of Latin American countries (Mainland). *Source* Index Mundi (http://www.indexmundi.com/argentina/)

Combined, these factors undermine the ability of many countries—and communities—to adapt to climate change. In addition, they prevent them from establishing a solid basis upon which a state of preparedness to climate change-related phenomena can be built.

According to a UNEP report, even though the contribution of Latin America and the Caribbean to greenhouse gas emissions represents only 8 % of global emissions—excluding those related to land use change-, the expected changes in climate during the present century are certain to have a significant impact on the region. As an example of this fact, it can be mentioned that the number of people in Latin America and the Caribbean affected by extreme temperatures, forest fires, droughts, storms and floods grew from 5 million in the 1970s to more than 40 million from 2000 to 2009. Overall, adverse weather conditions have cost the region more than US$40 billion in the last 10 years (UNEP 2010).

The Inter-American Development Bank, the Economic Commission of Latin America and the Caribbean and the World Wildlife Fund produced a compendium titled "The Climate and Development Challenge for Latin America and the Caribbean: Options for Climate Resilient Low Carbon Development", where the key questions of what are the key physical impacts and consequences of climate change that will most affect the region, what are the likely costs to the regional economies derived from these impacts, and what are some of the key adaptation measures that need to be deployed to minimize these adverse impacts. It also deals with the question of how and at what cost would the region be able to reduce its contribution to the global carbon footprint at a level consistent with climate stabilization goals (IADB, ECLA and WWF 2013). Even though there are uncertainties in current previsions and forecasts (Knutti et al. 2009) overall trends point towards one direction: immediate action is needed.

For a better understanding of the vulnerability situation in Latin American, countries may be classified in two main categories:

- **moderately vulnerable**: this refers to countries where operational climate change strategies with nationwide influence are available, where investments in at least basic climate adaptation projects are being made and resilience plans for vulnerable communities exist and are operational;
- **very vulnerable**: this refers to countries where national climate change strategies (including mitigation and adaptation components) do not yet exist or are not fully operational, where little investments in climate change adaptation are seen, and few resilience plans to support local communities are in place.

Table 7.1 presents the distribution of Latin American countries according to their vulnerability to climate change. It can be seen that even large economies such as Brazil, Chile or Mexico are not immune against the threats posed by climate change, and do not yet have a full ability to cope with its impacts. Table 7.1 shows, to the same measure, that all Central American countries find themselves in a difficult position, being very vulnerable and hence especially prone to a wide range of problems. This fact alone should act as a catalyst for immediate action from Government in that part of the world, motivating them one the one hand, but also international

Table 7.1 Classification of Latin American countries in respect of their climate change vulnerability

Status	Countries
Moderately vulnerable	Argentina, Brazil, Chile, Mexico, Paraguay, Uruguay
Very vulnerable	Belize, Bolivia, Colombia, Costa Rica, El Salvador, Guatemala, Guyana, Honduras, Nicaragua, Panama, Peru, Suriname, Venezuela

Source Authors

agencies and funding bodies on the other, to intensify their efforts so as to help them in the mitigation—but especially in their adaptation—efforts in the short term.

The fact that in many of the very vulnerable countries little systematic (and long-term) action in respect of climate change adaptation exists, characterizes what the authors regard as "**chronic climate change vulnerability**", i.e. a continuous state of lethargy in relation to the preparedness to handle future climate change. Often, those countries are not even prepared sufficiently to current impacts of climate variability. Chronic climate change vulnerability means that:

a. little investments which may lead to changes in land use, territorial planning or improvements in infra-structure (and which would, *inter alia*, reduce vulnerability) take place;
b. few or no components of institutional capacity-building at the relevant institutions at government level exist or are even considered, with a heavy dependency on international know-how;
c. many communities, especially in rural and remote areas, have limited or no access to assistance in cases where extreme events (droughts/floods) affect them;
d. little social capital, i.e. part -or sometimes even full- reliance on external (international) assistance and funds to support them, in the cases when extreme events take place.

The phenomena of chronic climate change vulnerability becomes very conspicuous by the devastating impacts of extreme events such as hurricanes and cyclones, which tend to draw much public attention and the attention from international agencies and donors (e.g. Hurricane Mitch in 1998 and diverse hurricanes since). It however tends to return to the same status soon thereafter–or immediately after the international media has identified other pressing issues to report–hence perpetuating itself.

As far as the effects of climate change and climate variability in Latin America are concerned, it is difficult to establish a regional pattern. This is so due to the many factors which influence the climate dynamics (e.g. intensive temperature increases or decreases) in the region as a whole and on its sub-regions in particular. As a result of this reality, one can see situations where dry spells tend to be more frequent—with substantial increases in the occurrence of forest fires—, but also occasions where decreases in temperature combined with suitable air conditions, may lead to snow storms in areas these may not have been expected, or never took place before. An example of this pattern is the severe snow storms in southern Peru in the summer of 2013. To the same measure, there are occasions where rain may become less frequent, whereas torrential downpours intensify their occurrence, leading to increased soil erosion, flash floods, as well to damages to agriculture through unseasonal runoffs.

A further impact of climate change is the aggravation of water stress, which many countries across the continent have already been experiencing. This is exemplified by the extreme droughts seen in 2012 in Brazil and in Argentina, where weeks of scant rainfall and unusually hot temperatures linked to the La Nina phenomena have severely damaged crops across both countries, which together account for almost half the world's soybean exports and about 24 % of corn shipments. More recently, severe droughts in Bolivia in mid-2013 ignited more than 47.000 isolated fires, damaging agriculture, destroying large portions of forests and creating health problems among the population, as well as affecting the country's air traffic. Finally, hurricanes regularly lead to wide damages in the region (Pielke Jr et al. 2013) but have led to little changes in policies or perspectives on how to handle such hazards.

From a critical perspective, a fact which illustrates the importance that should be afforded to the impacts of climate change in Latin America, is that the economies of all countries in the region are especially dependent on sectors that are most vulnerable to climate conditions, such as:

- agriculture,
- fisheries,
- forestry and
- tourism

Agriculture alone provides the livelihood for over 70 % of the population of the region, and at country level it accounts for half of the GDP in many nations. This sector is also vital to the regional economy if one bears in mind that vital revenues derive from the export of agricultural products such as coffee, soya beans, banana and many others. Disturbances in climate conditions leading to damages to agriculture production may not only lead to severe crop losses (Challinor and Wheeler 2008), but also to reductions in revenues. These, in turn, aggravate social problems such as unemployment and poverty. Bearing in mind that most Central American countries already suffer the negative impacts of poverty, they simply cannot afford to adapt to climate change. Climate change adaptation is a matter of utmost economic and social importance and, in this context, the management of resources such as water (Leal Filho 2012) and disasters risk management, can play a key role (Leal Filho 2013).

There are some concrete examples of efforts by many Latin American countries towards handling climate change, both in respect of mitigation and adaptation. As examples of work taking place at country level in respect of **mitigation**, the following may be mentioned:

- Argentina: the country has initiated a number of policies and measures that will lead to a reduction in emissions of greenhouse gases, measured against a business as usual scenario. The main focus of Argentina's climate-related legislation

is on energy, and the government has implemented a comprehensive and far-reaching strategy to increase energy efficiency—in recognition that reducing energy demand is cost-effective—and, at the same time, has set targets for renewable energy supply.
- Brazil: established a national greenhouse gas reduction target of roughly 36 % of projected emissions by 2020. Brazil's greenhouse-gas emissions fell nearly 39 %, with a 76 % drop in cumulative emissions from deforestation, between 2005 and 2010.
- Mexico: it was the first developing country to create a comprehensive climate change law in 2012 with targets to reduce GHG emissions by 30 % by 2020 and 50 % by 2050.
- Ecuador: the Yasuní-ITT Initiative seeks compensation for roughly half the estimated value of certain untapped oil deposits, in order to leave these resources untouched. The funds are earmarked to protect national parks and promote renewable energy.

As far as adaptation is concerned, some examples of current action at the country level include:

a. Bolivia: in the frame of the UNDP's Community-based adaptation programme six adapation projects focusing on water and soil, food security, agricultural adaptation and risk management;
b. Peru: in frame of an IFAD funding scheme, local communities in the Altiplano focus on livelihood diversification;
c. Nicaragua: GIZ and Café Direct support Nicaraguan coffee farmers in assessing climate risks for future coffee production and designing strategies and measures for improved adaptation;
d. Guatemala: UNDP adaptation projects focus on community-based natural disaster risk reduction activities in rural communities in globally important ecosystems and watersheds.

Apart from the fact that many parts of Latin America–as in other tropical areas-may become less suitable to agriculture because critical thresholds of adaptability are exceeded in most marginally suitable areas (Jarvis et al. 2012), a further factor that should be mentioned is the social implications of climate change and climatic variability on indigenous peoples and communities living in the highlands, lowlands, and coastal areas of Latin America. As stated by Kronik and Verner (2010), indigenous people across the region, already perceive and experience negative effects of climate change and variability. Since these groups have little access to finances and resources to pursue adaptive strategies, the find themselves in a high state of vulnerability, with little possibilities to cope with the pressures and threats posed to them.

Managing Climate Change: The Role of Technology Transfer

Much of the vulnerability of Latin American countries to climate change is a result of inadequate governance, limited access to technologies and limited access to capital for investments in mitigation and adaptation initiatives. In particular, current limitations seen in access to technologies and in the access to resources, hinder these countries to adapt successfully. Therefore, there is a need for more international cooperation to enable Latin America to cope with the many challenges climate change poses to them.

One of the means via which such cooperation may be reached, is by means of technology transfer. The **appropriate transfer of climate technologies**, i.e. technology transfer which takes into account the local ability to absorb them—which may include adjustments in materials, approaches and process to suit local conditions—can greatly support vulnerable countries to reduce their degree of exposure to climate phenomena. In this respect, the approaches pursued for building technological capacity are very similar to those which build adaptive capacity of a nation or society. It is very important here to reiterate the fact that technologies successfully used in one country may not necessarily yield the same success elsewhere. However, if due emphasis is given to local conditions and the particularities of a given region or site, they may prove very useful. Technology transfer in climate change can be advantageous in many ways, as outlined in Fig. 7.2.

Based on both the importance and relevance of technology transfer on climate change to Latin America, the project "Network of Climate Change Technology Transfer Centres in Europe and Latin America" (CELA) was started. The project entails a cooperation between European and a set of Latin American countries, being implemented over the period 2010–2013, as part of the ALFA Programme. ALFA is a programme of co-operation between

Higher Education Institutions of the European Union and Latin America. The CELA project partners are:

- Hamburg University of Applied Sciences, **Germany**
- Universidad Galileo, **Guatemala**
- Universidad Católica Boliviana, **Bolivia**
- Universidad de Ciencias Comerciales, **Nicaragua**
- Pontificia Universidad Católica del Perú, **Peru**
- Tallinn Technical University, **Estonia**

The purpose of the CELA project is not only to improve the academic quality of European Union (EU) and Latin American higher education institutions (HEIs), but also to strengthen their role so as to contribute to sustainable development and social cohesion. The rationale for the project CELA is the fact that HEIs can play a key role, for example in terms of research, consultancy and technology transfer as well as in the education and qualification of current and future employees.

A problem seen today is that, despite the importance of the topic climate change and the expertise available at Latin American HEI, the potential of these institutions to offer solutions for climate change adaptation and mitigation is not sufficiently developed. The CELA project works under the premise that international cooperation and the establishment of informal as well as formal structures in LA countries—which allow joint research collaboration, exchanges of experiences and lessons to be learnt among different institutions in the region—may help it

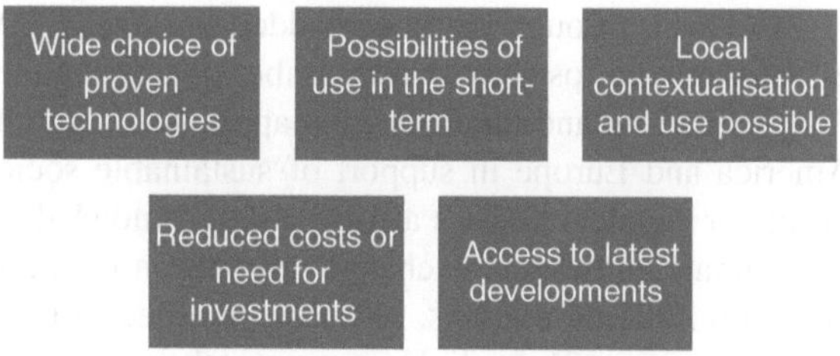

Fig. 7.2 Advantages of technology transfer in climate change

to better handle the interdisciplinary challenges posed by climate change. In this respect, technology transfer can occur on national as well as international level. Both transfer types are characterized by distinctive prerequisites and constraints which need to be considered in the design of technology transfer initiatives such as CELA (Mannke 2012).

The project CELA is meant to foster EU–Latin American applied research and technology transfer in the climate change sector. It does so by means of the set-up of an infra-structure called **Technology Transfer Centres,** which makes better use of the science and technology knowledge existing in the participating regions, and in setting-up **networks to intensify** joint EU–Latin American applied research in the field, in support of socio-economic development. Since lack of expertise and access to knowledge is a major impediment to tackle the challenges of climate change, the project also entails the development of **highly-qualified research staff** and research institutes, which will play a very important role in training people and **providing expert advice** (research and technology transfer) and thereby have a positive, long-term impact on the socioeconomic development in the participant Latin American countries and beyond.
The specific objectives of the CELA project are:

1. To increase the capacity and to improve the quality of research within the scientific and technology community in Latin America and the EU
2. To develop and establish a market-oriented research framework to better capitalise and disseminate research on climate change
3. To strengthen the link of EU and Latin American research communities with the regional market, business and legislation (policy) in the field of climate change
4. To develop and establish a market-oriented EU–Latin American network of Climate Change Technology Transfer Centres.

The set-up of a dedicated networking infrastructure of so-called Climate Change.

Technology Transfer Centres (CTTCs) resembles the backbone of the CELA project: these centres serve not only to make better use of the science and technology knowledge existing in the partnering countries Bolivia, Guatemala, Nicaragua, Peru, Estonia and Germany, but they have been growing local networks which link the initiating university with further universities, with industry and government actors, and also with the civil society.

Moreover, through being embedded in the EU-LAC network, the CTTCs have established and fostered a vast number of international academic community relations to start up and intensify joint applied research in the field and between Latin America and Europe in support of sustainable socioeconomic development. The virtual or physical CTTCs also resemble "knowledge hubs": they strengthen links and climate knowledge exchange not only in the academic sense but also beyond, i.e., with regional markets, business, and decision-makers (policy) in the field of climate change. These links are strengthened in the frame of capacity-and network-building actions on local as well as international level, targeting and actively involving the main beneficiaries of the project. CELA also comprises capacity-building actions for the research staff of the partnering universities. The ultimate goal of this comprehensive capacity-building programme serve to provide an enabling environment for technology transfer, i.e. creating technological as well as adaptive capacities by means of training staff and providing expert advice and latest climate knowledge to stakeholders from academia, industry, and politics (Mannke 2012).

Finally, the four Latin American project partners developed own pilot projects in collaboration with and to the benefit of civil society. Grounded in local needs, appropriate technology has been implemented in local communities or technological capacity has been built in cooperation with government:

- Guatemala: early warning system in a flood-prone area in Guatemala
- Peru: wireless monitoring network in the Peruvian Amazon Forest.
- Bolivia: sustainable forestry management centre for local communities
- Nicaragua: new water monitoring mapping and building technological expertise among government representatives.

This pilots, all of which included local capacity-building actions, combined with clever ownership of local stakeholders and involvement of university students can ensure proper maintenance of the new technologies and foster local commitment.

The project thus greatly assisted efforts in Latin America to provide a market-oriented research and technology transfer approach, complemented by establishing specific recommendations for the different regions of how to excel within their region and beyond. In doing so, the project addressed a set of issues seen in the Millennium Development Goals, such as fighting poverty and pursuing environmental sustainability. The final beneficiaries of the CELA project were:

- More than 180 research, teaching, administrative and management staff, which have benefitted either from new research, cooperation or networking opportunities kicked-off by CELA or from the provision of concepts and recommendations developed within CELA which supports them in continuously enhancing the teaching and research and technology transfer activities. Links with other EU programmes such as Horizon 2002 and Erasmus-Mundus will be pursued.

- Over 180 EU/Latin American enterprises and other private or public institutions in the Climate Change sector, which have received project information and/or benefitted from the local availability of expertise and from the research partners.

Finally, the general EU/Latin American population have benefitted economically and environmentally from the increased capability of key actors for tackling the challenges of climate change.

Conclusions

Climate change affects the Latin American region in special ways. The current degree of vulnerability seen in the region calls for immediate efforts and investments, in order to address the problem of chronic climate change vulnerability. It is important that mechanisms are put in place so as to allow the countries in the region in a better position to handle the many challenges climate change poses to them. The project CELA is an example of how Europe and Latin America can cooperate in the field of climate change and illustrates how the principles of applied research and technology transfer in the climate change sector may be implemented in practice. The infra-structure created as part of CELA, namely the "Technology Transfer Centres on Climate Change", will allow European and Latin American countries to continue to work together in support of climate change mitigation and adaptation, and ultimately, supporting the socio-economic development of both regions.

References

Challinor AJ, Wheeler TR (2008) Crop yield reduction in the tropics under climate change: processes and uncertainties. Agric For Meteorol 148(3):343–356

Inter-American Development Bank (IDB), Economic Commission of Latin America and the Caribbean (ECLAC) and the World Wildlife Fund (WWF) (2013) The climate and development challenge for Latin America and the Caribbean: options for climate resilient low carbon development. IDB, Washington DC

Intergovernmental Panel on Climate Change (IPCC) (2007) Fourth assessment report: climate change 2007 (AR4). IPCC, Geneva

Jarvis A, Ramírez-Villegas J, Tapasco J, Navarro C, Peterson CA, Zapata-Caldas E, Fisher MJ (2012) Are climate change adaptation and mitigation options eco-efficient? In: Hershey CH (ed) Eco-efficiency: from vision to reality. CIAT, Cali

Knutti R, Furrer R, Tebaldi C, Cermak J, Meehl GA (2009) Challenges in combining projections from multiple climate models. J Clim 23(10):2739–2758

Kronik J, Verner D (2010) Indigenous peoples and climate change in Latin America and the Caribbean. World Bank Publications, Washington, DC

Leal Filho W (ed) (2012) Climate change and the sustainable use of water resources. Springer, Berlin

Leal Filho W (ed) (2013) Climate change and disasters risk management. Springer, Berlin

Mannke F (2012) Fostering international technology transfer through climate technology networks between European and Latin American universities. Latin American Bus Rev 13(3):171–197. doi:10.1080/10978526.2012.730018

OECD (2012) Latin American economic outlook 2012—transforming the state for development. OECD, Paris

Pielke Jr RA, Rubiera J, Landsea C, Fernandez ML, Klein R (2013) Hurricane vulnerability in Latin America and The Caribbean: normalized damage and loss potentials. Nat Hazards Rev 4(3):pp 101–114

United Nations Environment Progamme (UNEP (2010) Vital climate change graphics for Latin America and the Caribbean. UNEP/GRID-Arendal, Arendal

Chapter 8
Strategic Contributions to Extreme Climate Change: The Innovation Helixes as a Link Among the Short, Medium and Long-Terms

Nelson Amaro, Cyrano Ruiz, Juan Luis Fuentes, Julio Miranda and Ericka Tuquer

Abstract Technology innovations have been introduced with a velocity unknown in the past during these last years. For that, the joint efforts of universities, industry and government have been vital. This approach has been called "Triple Helix". After this elaboration, a strategic approach of a "Quadruple Helix," has been proposed which also includes the civil society. This approach could be extended to include international organizations and the media in general, whose influences go beyond national borders, although they do have an important national incidence within any country. This document identifies the lessons learnt from the vulnerability of Guatemala and the activities needed to face the effects of extreme climate change. An adaptation strategy is outlined, based on the extended "Quadruple Helix". However, the actions towards mitigation measures are also to be well kept and efforts have to be made in the short and medium terms in the whole spectrum of climate change. The perspective elaborated here examines these strategic approaches and their application in Guatemala, the vulnerabilities of the country before extreme climate change, the attitudes of the different actors towards these threats, the coordination of strategies that take into account the four helixes

N. Amaro (✉) · C. Ruiz · J. L. Fuentes · J. Miranda · E. Tuquer
Network of Climate Change Technology Transfer Centres Project, CELA,
Galileo University, Edificio de Rectoría, 7ma. Avenida, Calle Dr. Eduardo Suger Cofiño,
Zona 10, Ciudad de Guatemala, Guatemala
e-mail: nelsonamaro@galileo.edu

C. Ruiz
e-mail: jcrc1962@hotmail.com

J. L. Fuentes
e-mail: juanfuentes@galileo.edu

J. Miranda
e-mail: jmiranda@galileo.edu

E. Tuquer
e-mail: erickatuquer@galileo.edu

W. Leal Filho et al. (eds.), *International Perspectives on Climate Change*,
Climate Change Management, DOI: 10.1007/978-3-319-04489-7_8,

approach, and the lessons that might be inferred from its implementation. Lastly, this reflection aims to design an appropriate agenda to face extreme climate changes in the short, medium and long-terms by showing their links throughout time.

Keywords Triple helix • Quadruple helix • Extreme climate change • Vulnera bility • Short-, medium- and long-term agenda • Vulnerability • Adaptation • Civil society • Strategy

Introduction

This reflection is the result of three years of work programmed by the CELA project (2011–2013). It responds to the need of networking with different people interested in the effects of climate change. An agenda for the future is defined, attributing the lack of preparation for these threats to the absence of short, medium, and long-term links in the perception of key different actors that should be called to make a difference through their decisions. For Guatemala, that means the link among extreme climate events that are suffered periodically, almost each four years, and solar warming forecasted throughout the 21st century in the whole planet. The action system proposed suggests narrowing this gap, based on an extended "Quadruple Helix," which includes the academia, the businessmen, the government, and the civil society, including international cooperation and the media. These forces are capable together of building a real front against these threats.

A Diagnosis of the Effects of Extreme Climate Change in Guatemala

Every effort in the climate change field has to be referred to the Sustainable Development Paradigm that should currently manage all progress efforts in the consensus reached around "The Millennium Objectives" (Amaro 2012). The Brundtland Commission of the United Nations defined the concept of Sustainable Development as follows:

> …a process of change in which resource exploitation, investment management, technological development orientation and institutional change are all harmoniously together and emphasize both present and future potential to satisfy the human aspirations and needs. All that means that human development must be done in ways that are compatible with biological processes that support de work in the biosphere (United Nations 1987).

The subject of climate change in Guatemala is intimately related to disasters in general, which have shaken up the country periodically. Table 8.1 shows the estimated death toll in many disasters that hit the country in 100 years. As can be seen in the numbers of the table, from higher to lower, earthquakes and floods are the

Table 8.1 The ten most catastrophic natural disasters in Guatemala

Disaster	Date	Death toll
Floods	1949	40,000
Earthquake	1976	23,000
Volcanic eruption	1902	6,000
Volcanic eruption	1929	5,000
Earthquake	1917	2,650
Earthquake	1902	2,000
Storm	2005	1,513
Volcanic eruption	1902	1,000
Flood	1982	620
Storm	1998	384

Source EM-DAT, The International Disaster Database, Centre for Research on the Epidemiology of Disasters-CRED, supported by the U.S. Agency for International Development, USAID. Found at: http://www.emdat.be/result-country-profile

disasters that have taken greater numbers of lives. Notice, however, that floods and storms tend to repeat more frequently and with more intensity in recent years. This is part of the predictions made by the organizations in charge of studying climate change in the world (IPCC 2007). Four(4) out of 10 disasters identified are floods and storms; 3 are earthquakes; and 2 are volcanic eruptions. Also, the 4 floods and storms with some important impact happened after 1949 and the last 3 happened from 1982 to 2005. This account shows that those phenomena more related to extreme climate changes have been more frequent in the recent past.

Table 8.2 focuses on the loss suffered by different sectors in the field of social and economic activities. The storms with the most impact in the past 15 years have been selected. The impact in the GDP fluctuates between 0.63 and 3.40 %. These numbers show the impact of extreme climate change events in the economic growth of Guatemala, especially when taking into account that, for example, for 2010 and 2011, the annual growth rates of the GDP have been 2.9 and 3.9 % respectively. Agatha practically eliminated any growth by itself (Bank of Guatemala 2013). To analyze Table 8.2 from the sectoral point of view, please notice the substantial loss in environment and transport infrastructure. In 2010 and 2011, the most recent dates, these categories are the most affected. It also reaches out to houses, education and agriculture, stockbreeding and fisheries.

Table 8.3 shows how Guatemala is ranked regarding the degree of risk to be affected by natural phenomena. This index not only determines the degree of exposure, but also the capacity to adapt to natural threats. As can be seen, researchers found that Guatemala is in the fourth place in the world among 173 countries analyzed. Vanuatu, which is a place where the United Nations recognized the first climate change refugees, takes up the first place. Central America, which is represented by Costa Rica, El Salvador and Nicaragua, also takes important places in the world ranking.

This index is made up by many variables that all added up together include the degree of exposure to natural disasters (earthquakes, cyclones, droughts and

Table 8.2 Quantity of damage by sector affected, according to extreme climate change in the form of storms (in millions of quetzals)

Affected sector	E12 2011	Agatha 2010	Stan 2001	Mitch 1998
Social	*523.14*	*1,556.40*	*1,135.30*	*316.12*
Houses	258.05	773.40	965.16	232.20
Human health	38.94	116.70	108.34	32.12
Education	218.64	655.30	61.80	51.80
Cultural heritage	7.51	1.00	–	–
Production	*344.79*	*1,033.40*	*2,050.86*	*3,769.45*
Agriculture, cattle, fisheries	215.80	646.80	593.59	3,244.00
Industry	105.97	317.60	432.60	406.65
Commerce	11.21	33.60	617.24	–
Tourism	11.81	35.40	407.43	118.80
Infrastructure	*946.47*	*2,896.70*	*3,410.47*	*765.11*
Transportation	875.77	2,624.80	3,280.77	592.58
Energy	31.60	94.70	38.91	66.08
Water and public health	39.10	177.20	90.79	106.45
Cross subjects	*806.77*	*2,418.00*	*309.00*	*33.60*
Environment	698.07	2,092.20	309.00	33.60
Impact in women	63.76	191.10	–	–
Risk management	44.94	134.70	–	–
Total in quetzals	Q2,621.17	Q7,904.50	Q6,905.63	Q4,884.28
Total in dollars	$338.22	$740.56	$983.00	$740.05
Impact on the GDP	0.63 %	1.54 %	3.40 %	1.54 %

Source Own elaboration made by CELA technical team, based on different charts exposed on the Economic Commission for Latin America and the Caribbean (CEPAL-ECLAC 2011). Regional summary of the impact of tropical depression 12-E in Central America. Quantification of damage and loss in the countries of the region, on October 2011. Mexico, D.F.: UN, Table 6; Guatemalan Government, ECLAC and UNDP (2010). Damage and loss assessment by sector and estimation of needs provoked by tropical storm Agatha and the eruption of the Pacaya volcano. Guatemala: Guatemalan government and ECLAC. Summary chart of loss and damage. ECLAC and the Planning General Office, Guatemalan government (2005). Effects of the torrential rain in Guatemala and the tropical storm Stan. Found at: http://www.segeplan.gob.gt/stan/docs/InformeGuatemala.pdf, Charts 34 and other. ECLAC (1998). Damage assessment of Hurricane Mitch. LC/MEX/L.370, Chart 8.3, page 23

Table 8.3 World risk index for the fifteen most affected countries in the world (%)

Country	Index	Country	Index
1. Vanuatu	36.31	8. Cambodia	17.17
2. Tonga	28.62	9. Timor-Leste	17.13
3. Filipinas	27.98	10. El Salvador	16.89
4. Guatemala	20.75	11. Brunei Darussalam	15.92
5. Bangladesh	20.22	12. Papua Nueva Guinea	15.81
6. Islas Salomón	18.15	13. Mauricio	15.39
7. Costa Rica	17.38	14. Nicaragua	15.36
		15. Fiji	13.60

Source United Nations University—Institute for Environment and Security, UNU-EHS and The Nature Conservancy (2013). World Risk Report 2012. Found at: http://www.ehs.unu.edu/file/get/10487.pdf

floods), regarding readiness to face them, mainly concerning infrastructure; the susceptibility that measures the probability of these non-controllable natural phenomena to happen; the ability to face these situations; and lastly, the strengths to face them (UNU-EHS and Nature Conservancy 2013).

All the data shown above highlight a quite critical condition in Guatemala in the short-term. It is not about the dangers that solar warming could present in the next century. For example, the IPCC predicts that the two degrees of temperature needed to avoid uncontrollable disasters could be overcome if the needed efforts are done. Regardless of those long-term phenomena, climate change is already present in the Guatemalan reality here and in the immediate future. This situation demands an urgent change of mentality to face natural disasters that are the consequence of extreme climate change.

Perception of the People Regarding Climate Change

This is according to surveys carried out in Guatemala by the CELA Project, whom first of all interviewed the people with most authority and the higher positions in related businesses, government, non-governmental organizations and universities. The beginning was a list of 531 agricultural businesses and 845 industrial plants, given by the most representative lobby organizations (Guatemala Exporters Association, AGEXPORT for its initials in Spanish; and the Guatemala Chamber of Industry, CIG for its initials in Spanish). With the help of three judges, those companies were sorted for their impact in the market. The people who gave the information as judges were asked to rank those companies from 1 to 5, being 5 the most important and 1 the least important.

The final score varied from 15 the highest and 3 the lowest. After that, the 40 most important companies were picked up from each list. Then, the person with the highest position in each business organization was contacted, but it was accepted that a specialized person with more experience in the activities of the company in climate change and social responsibility also could answer the applied questionnaire. At the end, 32 and 39 cases were obtained for agriculture and industrial companies respectively (Supervision and Technical Team 2009; 2011; 2012a, b).

When looking at the perception of the people surveyed, we noticed that it is not different from the perception of the rest of the world, which looks at climate change as something that is not happening in the immediate reality. That indifference has been called the "Giddens Paradox", which highlights the tendency human behaviour has towards long-term threats. Due to their non-immediate nature, there is an inclination that tends to minimize them, and postpone the needed decision-making process, making those threats negligible. Giddens says:

> According to this, since the dangers posed by global warming aren't tangible, immediate or visible in the course of day to day life, however, awesome they appear, many will sit on their hands and do nothing of a concrete nature about them. Yet waiting until they become visible and acute before taking serious action will, by definition, be too late (Giddens 2010: 12)

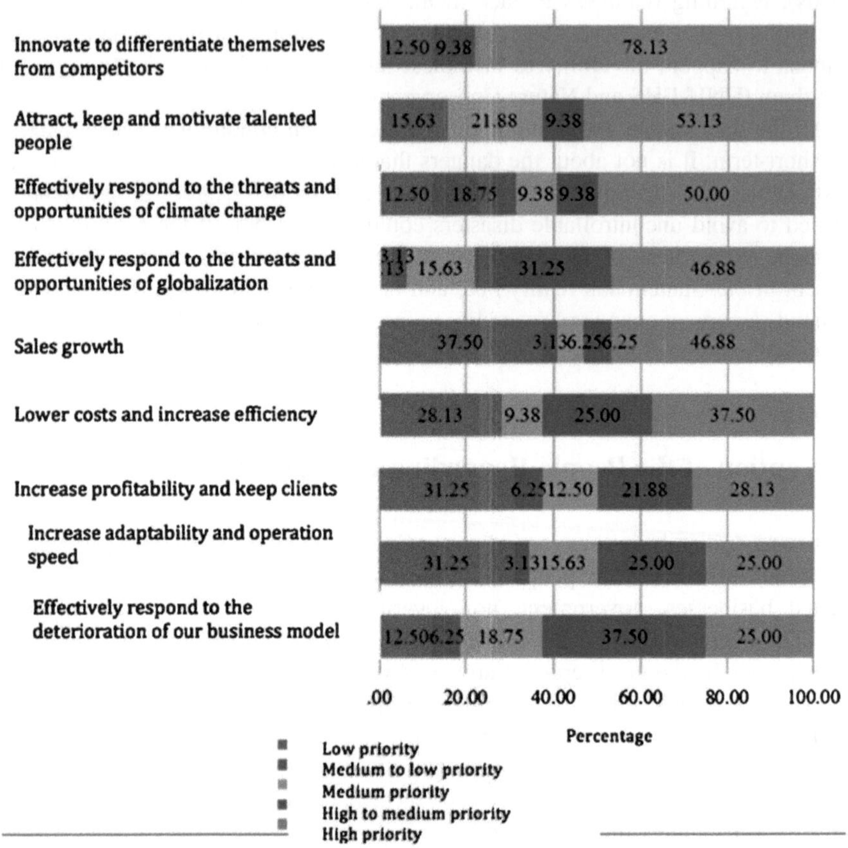

Chart 8.1 Perception of the most important challenges that agricultural companies will face in the short-term (2 years)

As can be seen on Chart 8.1, the impact of Climate Change is an important matter for agricultural businesses Directors, but it is only a priority to exactly half of them. This attitude might be contrasted with their industrial counterparts.

On the contrary, as can be seen on Chart 8.2, it is interesting to notice less concern about Climate Change in industrial than in agricultural companies in Guatemala. Curiously enough, when these data are compared to the world average, one finds out that this value is similar to the priority that North American industries give to the subject (McKinsey 2007). Comparing the industrial and agricultural sectors, according to the numbers in Charts 8.1 and 8.2, it is 50–31.77 % respectively. Although the study made does not allow to explain the reasons for such attitudes, a explanatory hypothesis might be the great connection around the same business ideas of both, the United States and Guatemala.

An example of this is the opposition to the Kyoto Treaty from some of the biggest and most influential companies of the United States, like Exxon and Chevron

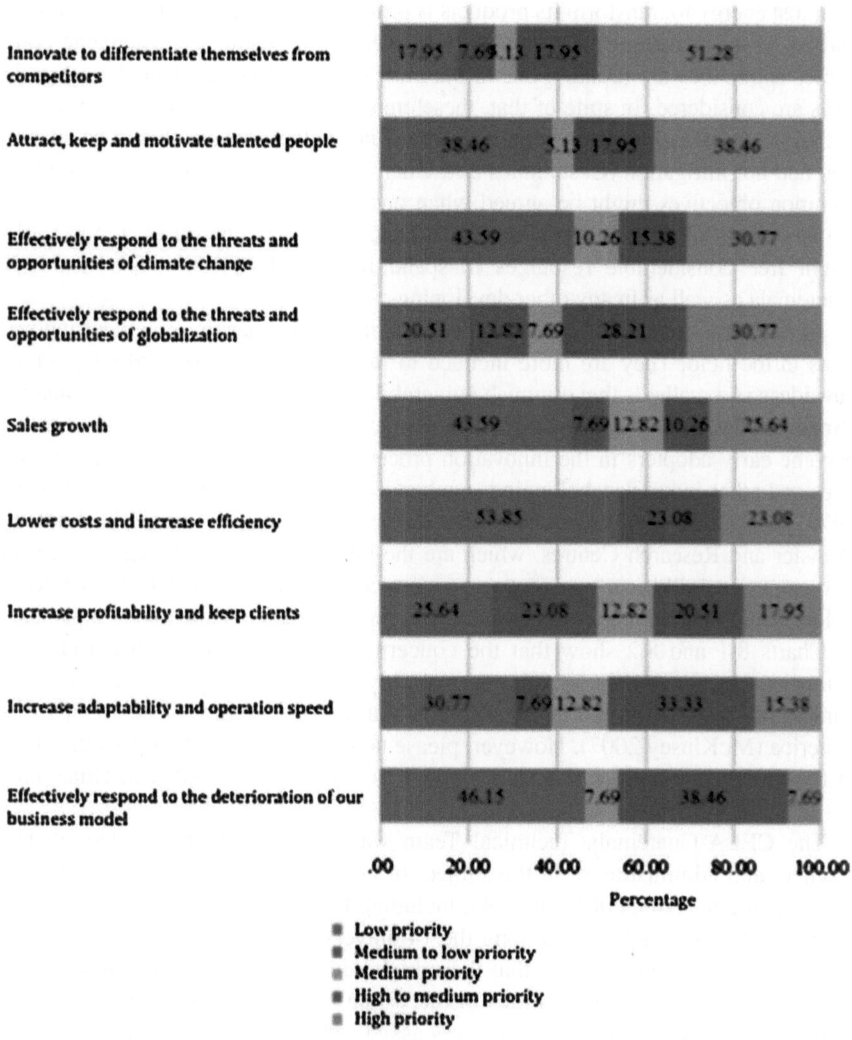

Chart 8.2 Perception of the most important challenges that industrial companies will face in the short-term (2 years)

(Levy and Kolk 2002), that have challenged and question climate science, due to the high cost of greenhouse effect gasses control, GHG, and have been lobbying against an emission control law. That has also influenced the no recognition of the CC effects.

Although in Guatemala some laws regarding GHG emissions have been promoted, the contribution of the country to that effect is minor because industry in our country is not such a strong sector and it is not quite representative worldwide. Also, there are no big iron and steel companies nor oil refineries. The company that uses the most energy to transform its products is probably the cement industry, which has carbon footprint measuring programs. However, the problem of Guatemalan companies is that they are found in one of the most vulnerable areas when climate problems are considered. In spite of that, these emissions barely reach 0.04 % of all the emissions in the world, which confirms that Guatemala's concern should be adaptation and not mitigation (Castellanos and Guerra 2009; UNDP 2007). Nevertheless, common objectives might be agreed when one observes the high figure that reach the so-called "oil bill" in the country's expenditures. Any control of this variable, might free considerable resources to spend in needed education and health in Guatemala as well as in any other developing country.

Agricultural companies appear to be more alert and eager to implement innovations in the field. They are more inclined to innovations diffusion and adoption of new ideas and methods that diminish vulnerability and increase the resilience against climate change problems. These data show that these agricultural companies might become early adopters in the innovation process of practicing climate change iniatives and other industries, belonging to other sectors, might act as followers (Rogers 1976). These data seems to indicate that the action of Climate Change Technology Transfer and Research Centres, which are the main objective of the CELA Project in 6 countries, will be more effective and less complex when transferring new technologies and methodologies to agricultural companies than to industrial companies.

Charts 8.1 and 8.2 show that the concern for climate matters in agricultural companies in Guatemala is superior, not only to the concern of industrial companies, but also to the world average concern and the average concern of Latin America (McKinsey 2007). However, please note that it does not surpass the 50 % of the surveyed people, which shows how it is not a priority neither in Guatemala nor in Latin America or in the world.

The CELA-Guatemala Technical Team was able to elaborate a list of 157 teachers and administrators in the subject of environment and climate change, in the 7 main universities of Guatemala, including the national public one. Although there is not comparable data among the 14 universities that exist in the country, a rough estimation can be made that these high education institutions cover almost the entire universe of these entities in the whole country. From this list, 64 cases were randomly selected. A question was asked to them that appears in Chart 8.3: "In your opinion, which is the situation of climate change in the teaching priorities and research and the high education institutions you work for?".

According to Chart 8.3, sixty-four percent of the interviewed people among teachers and administrators at the universities related to the environment, think that the subject is included in the agenda, but there are other more important subjects, 17.2 % think that it is a subject as critical as others, 10.9 % said it is always on the agenda, but it is only part of other subjects, 6.3 % think it is not on the agenda because they think it is just a temporary trend, and only 1.6 %, which is only one teacher, thinks it is a permanent item on the agenda and it is essential in the teaching and research plans.

Chart 8.4 works with three subsamples built based on thorough lists of Directors of programs related to environment and climate change in universities and

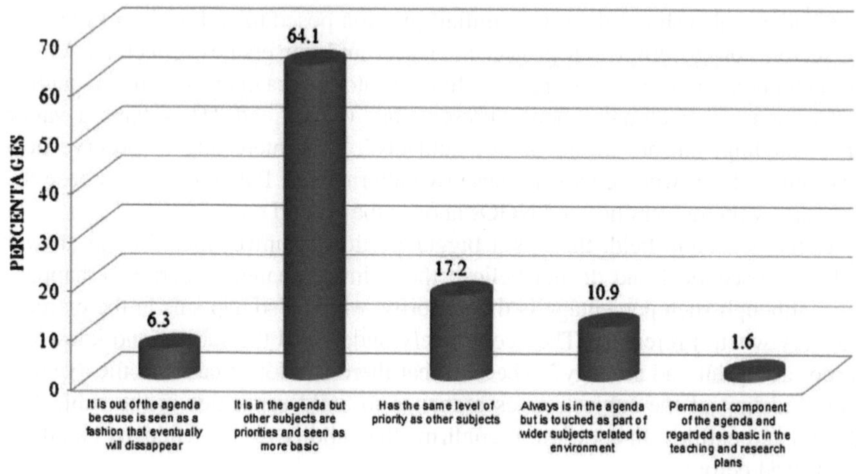

Chart 8.3 Perception of teachers and managers regarding teaching and research priorities in the higher educations institutions they work for

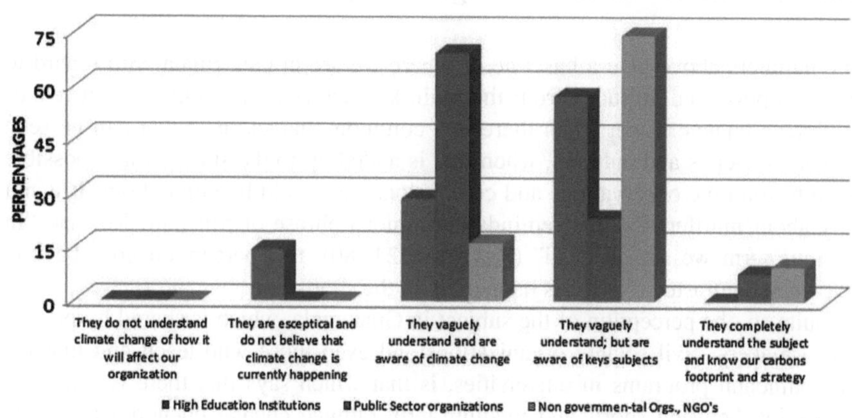

Chart 8.4 Perception of decision makers regarding the degree of awareness of their employees

departments in those universities, that were focused on climate change or intimately related to it; also with managers similarly related within the Guatemalan government; and finally, with people at the decision-making level in environmental NGOs with more influence on climate change. It was a sample of 193 cases (40 cases for each of the first two and 106 cases for the third). From this universe, 71 cases were randomly selected and 71 questionnaires were answered. While for the first two, there were made lists as complete as possible, based on the mere identification of their activities and informants, the third list had a selection with the help of three judges who established which were the ones with more influence, using scores similar to those of the companies mentioned before (Technical Team and Supervisor 2011).

Chart 8.4 also shows the same similar question posed to authorities of the public sector, non-governmental project managers and universities' directors. As can be seen here, the answers are mainly divided into two categories: "Understanding and awareness of climate change present, but vague" and "They have a vague understanding, but are aware of key subjects". It is interesting to observe how institutions differ when choosing these two alternatives. Public entities choose the first one, while universities and NGOs choose the second one.

In the academic field, there is a bigger portion in universities for the answer "They are sceptical and do not believe that climate change is currently happening," although such percentage is the minority, which also happens in the extreme category when it refers to "They completely understand the subject and know our carbon footprint and strategy". The fact that there is a lot of cases in the "vague" understanding of the subject, gives a chance to build capacities in the subject in all sampled entities. In addition, it confirms the validity of the "Giddens Paradox" mentioned before.

The Baseline of Climate Change Perception in Guatemala

The numbers above give a baseline of where we are in Guatemala with regard to the perception and attitudes from the main key actors in the country as it relates to climate change. Even when there is a common characteristic that can be seen in other societies and cultures, when that is added up to the surveys, it is possible to go beyond the observations and conclusions that could be pointed out. It is not only about indifference that reminds the famous phrase of Maynard Keynes: "In the long-term we are all dead" (Keynes 1924: 80); this perception goes beyond that when contrasted to what is happening in the short-term.

Quite so, the perception of the subject in Guatemala, which is shared by public-policy makers, civil society organizations and even those who teach and manage environmental programs in universities, is that which says that there is a lack of connection between short- and medium-term climate change phenomena on one hand and long-term phenomena on the other. Such disconnection allows for indifference from those institutions, which is transferred to ordinary people. Most people do not see this subject as connected, and decision makers are not aware of its urgency. That also influences politicians, with votes and decisions that establish a baseline of lack of preparation and connection in the short, medium and long-terms in relation to the adverse effects of climate change.

"Priority strategy to face extreme climate changes in Guatemala: The Three and Four Helix Action System"

To manage climate change it is imperative to create and consolidate a short-, medium- and long-term action system, able to face extreme climate change events and long standing problems. Such system should also be the manager of the adaptation process and its research and technological implications. A source

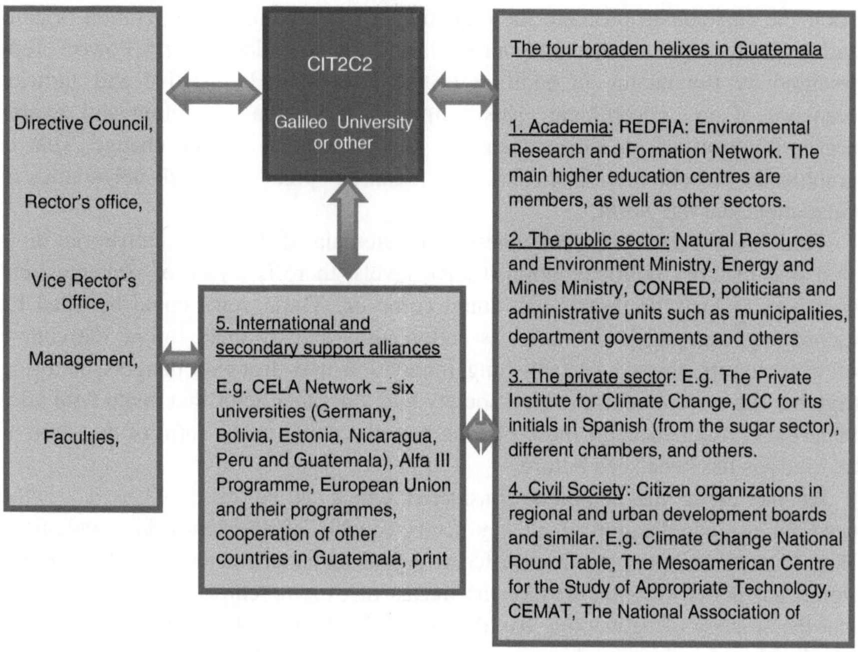

Chart 8.5 Alliances to make an action system based on the approach of the four helixes (Climate Change Technology Transfer and Research Centre, CIT2C2 for its initials in Spanish)

of inspiration for that objective is the Triple Helix Model (TH), which can be explained as follows:

> The TH model focuses on the analysis of mutual relationships and interactions between universities and scientific surroundings as the first blade of the helix, companies and industries as the second blade, and administrations or governments as the third blade. It deals with the interaction and communication between actors and institutions from the three blades of the helix, because it assumes that innovation comes from mutual interaction: the potential for innovating knowledge, economical resources and market possibilities, and innovation public policies regulations and incentives (Gonzalez de la Fe 2009, p. 740).

On the other hand, following the spirit of what is above, we have proposed a "Quadruple Helix" (Arnkil et al. 2010). The difference of this approach is that additionally to the relationship university-industry-government, civil society is also included, taking into account NGOs and other entities with a broader civil society belonging approach, and specifically with an orientation towards innovation attitudes and practices.

That would mean a fourth blade, with all its implications. It is aimed to distinguish the central government from regional and local governments, trying to cover all levels, making emphasis on those administrative divisions found nearest the citizen like local governments and municipalities. This is shown in Chart 8.5. Following this logic, we could suggest an extension of these helix blades, calling upon the capacity to promote the objectives of international cooperation organizations both in the country and out of it, and the so-called "Fourth Power", represented by the media, in addition to the Executive, Legislative and Judicial branches of any government. This is important for the perception and awareness promotion among citizens, on the implications of climate change. That is graphically presented on Chart 8.5, which can be spread out to all universities in Guatemala and the world.

This thought uncovers a weakness of Guatemala. If there is no territorial division that matches political/administrative levels, there is a lack of strategies and decisions to face the threats that could come up. These voids could be filled by identifying watersheds or basins as action units and by them follow the course of water or its absence, establishing inundation and droughts dangers, to bring together the government, the civil society and any other actor that could help alleviate the consequences of these phenomena that come in the form of disasters of all kinds as has been seen before.

Another aspect that should be considered, which can be seen in Chart 8.5, is the need for the coordination of all key actors capable of decision-making measures. How to get a commitment from different organizations and actors, together and for significant periods of time to face the threats of extreme climate change? It seems that the strategy of networking based in strategic alliances that could flexibly come together, based on certain objectives, and that could in a dynamic way broaden, innovate themselves and get a solid structure according to the needs of it, is an essential ingredient for a systematic approach to really adapt to the challenge of the events. The approach described on this picture has been started by the Climate Change Technology Transfer and Research Centre (CIT2C2) in Galileo University, and has been projected towards other universities and strategic sectors through the so-called Network of Environmental Capacity Building and Research, (REDFIA by its acronym in Spanish).

It is crucial to project this climate change management from universities by fighting the tendency to build "ivory towers" among academics that permanently work in universities. That leads towards changing the behaviour of teaching and administrative staff who often just isolate from society (Ruiz 2005). The same can be said of business corporations that have not received the call for social responsibility beyond their profit-oriented activities; and non governmental organizations, NGOs, that disperse their activities in different directions and objectives frequently duplicating activities and spending unnecessarily needed resources; and finally, joining government efforts whose officers often confuse **coordination** with a monopoly of the "truth", more akin to a hierarchy, to which all citizens should submit.

Conclusions

These conclusions would be framed in a way that may improve the decision-making process with regard to how should we face future extreme climate change events. The points raised in this paper are the following:

1. Constant extreme climate change that devastates Guatemala allow for a vision of more urgency that in other parts of the world but this does not imply to neglect medium and long-term alerts that address global warming. This means to establish research and decision-making connections in a way that effective links are made among short, medium and long-terms.
2. The most compelling argument to adopt a short-term approach is to put attention in extreme climate change events that are recurrent in certain sensitive territorial areas, leaving short-time periods between each other. The prediction derived from medium and long-term evidence is that the effects of global warming will be more intense and frequent throughout time. Therefore, they will occur in shorter periods of time. Research is needed to prioritize those territories, events and gather the needed data to address these issues.
3. The information gathered from these observatories would be the best argument against the mentioned "Giddens Paradox" when decision-makers are paralyzed based on the belief that climate changes are a problem for future generations and the next century.
4. This approach forces to invest resources and qualified people to monitor continually either prioritized basins or watersheds or "dry corridors" where significant environmental, social and economic losses occur on a continuous basis. The most compelling argument here is, as the Guatemalan case shows, that this kind of event may erase in days, significant amount of resources including, in some instances, the entire annual economic growth.
5. This evidence already proves that this is not an investment without returns and a careful defined adaptation process, that may start tomorrow, already is capable of preventing significant economic losses that delays in making these decisions could bring.
6. This approach also highlights that countries affected by these extreme events should change their traditional view of how the territory is administratively and politically divided. Extreme climate change events do not follow regions, state, province or local limits defined for other goals. They need new political and administrative divisions to make decisions work. Perhaps, they may involve several municipalities or provinces or regions… For making this approach sustainable, a realistic territorial division has to be defined with a different decision-making structure.
7. The people affected by significant extreme climate change events vary. There is no question that the poorest are the ones that suffer the most. Nevertheless, infrastructures, building and generalized damages are paramount. At the local level, especially, it does not distinguish among social strata and at the national level. Therefore, it calls for the cooperation of all citizens and often, the international community. These results force decision-makers to reach all sectors to cooperate

in laying the grounds to strengthen the needed adaptation processes. A "Four Helix" approach should be the correct vision to generate the innovations needed for consolidating this approach.

8. Particularly, universities that may contribute to research, technological transfers and training, should give steps to open itself to the external world and extend their collaboration to other sectors. "Ivory Towers" of which some segments of high education institutions are often fond of building should give their spaces to innovations and new actors in this field.
9. On the other hand, some subsectors of the private sector, frequently prefer not to pay attention to what they can do to make a difference. A more positive attitude is seen, however, in those corporations that emphasize social responsibility and engage in programs at all levels where a contribution is made.
10. Civil society organizations constitutes the fabric of social relations in any country. The essence of their activities consists of concerns around the public expression of private interests. They should also have their share of rights and obligations. They may have a role in these matters searching for the common good rather than specific special interests.
11. Governments should extend their coordinating hand to assemble any citizen and sector wishing to make a difference. In doing so, hierarchical attitudes should give way to dialogue and consensus in looking for the needed solutions to face the tremendous challenge that climate change represents for the entire nation in the present and future.
12. Innovating ideas will come forward from key actors in these coalitions. Universities, the private sector, national, local and intermediate governments, non-governmental organizations, the media and foreign cooperation should come together and reach all possible avenues open in the international world. Actually, the consolidation of such alliances, in itself, represents a formidable mitigation and adaptation agenda for the short, medium and long-term measures that are needed to effectively face the threats of extreme climate changes.

Acknowledgments The authors want to thank the ALFA III Programme of the European Union for its financial support throughout the CELA Project and Galileo University for the support.

References

Arnkil R et al (2010) Exploring quadruple helix outlining user-oriented innovation models. Final Report on quadruple helix. Research for the CLIQ project. University of Tampere, Institute for Social Research, Cofinanced by the European Regional Development Fund, INTERREG IVC Programme, Finland. Extracted from: https://www.google.com.gt/?gws_rd=cr&ei=ttM4UqrWGYiu2QWZmIGgAQ#q=University+of+Tampere

Amaro N (2012) Renewable energies in the light of development experiences in fifty years, 1960–2010. In: Leal Filho W, y Gottwald J (eds) Educational and technological approaches to renewable energy. Peter Lang GmBH, Frankfurt, pp 11–39

Banco de Guatemala (2013) Guatemala en cifras. Extracted from: http://www.banguat.gob.gt/inc/ver.asp?id=/publica/guatemalaencifras2011.htm&e=101267(in spanish)

Bank of Guatemala (2013) Gross domestic product (in millions of quetzals with prices from 2001) and gross domestic product (in millions of quetzals each year). Extracted from: http://www.banguat.gob.gt/inc/main.asp?id=51809&aud=1&lang=1

Castellanos E, y Guerra A (2009) El cambio climático y sus efectos sobre el desarrollo humano en Guatemala. Cuadernos de Desarrollo Humano (Climate change and its effect on human development in Guatemala), 2007/2008-1. UNDP, Guatemala

CEPAL-ECLAC (2011) Resumen regional del impacto de la Depresión Tropical 12-E en Centroamérica. Cuantificación de daños y pérdidas sufridos por los países de la región en el mes de octubre de 2011 (Regional summary of the impacto of tropical depression 12E in Central America. Quantification of the damage and loss in the countries of the region in October 2011). Mexico D.F, United Nations

ECLAC (1998) Damage assessment of Hurricane Mitch. Mexico City, LC/MEX/L.370

Giddens A (2010) La política del cambio climático (Climate change policy). Alianza Editorial, Madrid

González de la Fe T (2009) El modelo de triple hélice de relaciones universidad, industria y gobierno: un análisis crítico ARBOR Ciencia, Pensamiento y Cultura (The triple helix model, of a connection among university, industry and government: ARBOR critical análisis science, thinking and culture). CLXXXV 738 Jul–Aug, pp 739–755. Online early publication. doi:10.3989/arbor.2009.738n1049

Guatemala Government, Economic Commission for Latin America and United Nations Development Programme (2010) Evaluación de daños y pérdidas sectoriales y estimación de necesidades ocasionados por el paso de la Tormenta Tropical Agatha y la erupción del Volcán Pacaya (Assessment of the damage and sectorial loss and estimation of the needs derived from Tropical Storm Agatha and the eruption of Pacaya Volcanoe). Guatemala Government, Guatemala

IPCC (2007) Summary for policymakers. In: Solomon S, Qin D, Manning M, Chen Z, Marquis M, Averyt KB, Tignor M, y Miller HL (eds) Climate change 2007: the physical sciences basis. Working group I, contributions to the fourth assessment report of the Intergovernmental panel on climate change. Cambridge University Press, Cambridge and New York

Keynes J (1924) A tract on monetary reform. Macmillan, London

Levy DL, Kolk A (2002) Strategic responses to global climate change: conflicting pressures on multinationals in the oil industry. Bus Polit 4(3):275–300

McKinsey (2007) How companies think about climate change: A McKinsey global Survey. The McKinsey Quarterly 1–10

Rogers EM (1976) New product adoption and diffusion. J Consum Res 2(4):290–301

Ruiz C (2005) Títulos Universitarios: De la Nostalgia a la Competitividad (University titles: from Nostalgia to competitivity). Revista Futuro IDEA-Galileo, 137 edn, pp 6–7

Supervision and Technical Team (2009) Estudio sobre Energía Renovable y Mercado Laboral entre Universidades y Sectores Público y Privado de Guatemala, Guatemala (Study on renewable energy and labour market in universities and the public and private sectors of Guatemala). Galileo University/European Union, Alfa III Programme-Continental Impresos, Guatemala

Supervision and Technical Team (2011) Estudio sobre necesidades de tecnologías para cambio climático (Study on the needs for technologies for climate change). UGAL-Alfa. EU, Guatemala

Supervision and Technical Team (2012a) Plan Estratégico del Centro de Investigación y Transferencia Tecnológica en Cambio Climático, CIT2C2, con Apoyo del Proyecto CELA 2012 al 2014 (Strategic plan of the Climate Change Research and Technology Transfer Centre, CIT2C2, with the support of the CELA proyect 2012–2014). UGAL-Alfa, EU, Guatemala

Supervision and Technical Team (2012b) Proyecto Redes de Centros de Transferencia de Tecnología en Cambio Climático en Europa y América Latina. Estudio sobre necesidades de tecnologías para Cambio Climático (Climate change technology transfer centres in Europe and Latin America Project). CELA, European Union, Alfa III Programme-Continental Impresos, Guatemala

UNDP (2007) La lucha contra el cambio climático: solidaridad frente a un mundo dividido (The fight against climate change: solidarity before a divided world). Informe Mundial sobre Desarrollo Humano 2007/2008 (World report on Human development 2007/2008). Grupo Mundiprensa, Mexico

United Nations, Brundtland Commission (1987) Report of the World Commission on Environment and Development: our common future. Transmitted to the General Assembly as an Annex to document A/42/427—Development and International Cooperation: Environment

United Nations University—Institute for Environment and Security (UNU-EHS) and The Nature Conservancy (2013) World risk report 2012. Extracted from: http://www.nature.org/ourscience/ourscientists/world-risk-report-2012-pdf.pdf

Helix research for the CLIQ project co-financed by European regional development fund made possible by the INTERREG IVC (2010). Programme Tampereen yliopisto Yhteiskuntatutkimuksen instituutti Työelämän tutkimuskeskus. Extracted from: http://www.cliqproject.eu/en/products/research/quadruple_helix_research/?id=127

Chapter 9
What Contributes to Climate Change Adaptive Capacity? A Retrospective Case Study in a Caribbean Small Island Community

Jessica Jaja and Jackie Dawson

Abstract Climate change has been argued to be the single largest threat to human society and will be especially pronounced in Caribbean Small Island Developing States. Adaptive capacity—defined in the context of this chapter as the ability to design, develop and implement successful strategies that address climate change impacts—is dependent on a range of socio-political factors that dynamically interact and operate at different scales. While adaptive capacity is central to the theoretical foundations of climate change adaptation research, very few studies have attempted to understand how these different factors facilitate or hinder the ability of a community to successfully implement adaptation strategies. This chapter presents a preliminary analysis of how certain social, institutional, political and economic factors interact to facilitate and enhance adaptive capacity in a Caribbean small island community. The study also assesses the different perceptions surrounding the concept of "successful" climate change adaptation. The community of Paget Farm on the island of Bequia in Saint Vincent and the Grenadines has been chosen as a case study since it has both demonstrated high adaptive capacity and increased its adaptive capacity to future climate-related threats by having access to freshwater from the Caribbean region's first carbon–neutral desalination plant.

Keywords Climate change adaptation • Adaptive capacity • Socio-political dimensions of climate change • Successful adaptation • Caribbean Small Island Developing States

J. Jaja (✉) · J. Dawson
Department of Geography, University of Ottawa, Ottawa, Canada
e-mail: jjaja013@uottawa.ca

J. Dawson
e-mail: jackie.dawson@uottawa.ca

W. Leal Filho et al. (eds.), *International Perspectives on Climate Change*,
Climate Change Management, DOI: 10.1007/978-3-319-04489-7_9,

Introduction

The Caribbean region is already facing multiple and interacting impacts of climate change, which include sea level rise, ocean acidification, changes in extreme weather events and reduced freshwater availability (Mimura et al. 2007). The ability of different communities to successfully adapt to these impacts, however, is unequal and dependent on a range of socio-political factors that dynamically interact and operate at different scales (Adger 2003b; Brooks and Adger 2005). While adaptive capacity is central to the theoretical foundations of climate change adaptation research, very few studies have attempted to understand how these different factors facilitate or hinder the ability of a community to successfully implement adaptation strategies (Engle 2011). Also at issue are the questions: *What does it mean to successfully adapt? Are current strategies implemented successful and if so, successful according to who and at what spatial and temporal scales?* This research used a retrospective case study approach to pragmatically assess some of the socio-political factors that contribute to local-level adaptive capacity in Caribbean Small Island Developing States (SIDS). It also serves to analyse different perceptions of successful climate change adaptation strategies. This paper is the result of a preliminary assessment of data collected over a three-month period in the summer of 2013 and is based on a series of interviews with community members and key-informant stakeholders.

Background

Determinants of Adaptive Capacity

A limited number of studies have explicitly assessed the determinants of adaptive capacity, meaning the pre-existing characteristics of a system that facilitate and promote the adoption of successful climate change adaptation strategies (Adger 2003a; Engle 2011). This is partly because there are currently a limited number of fully implemented and operationalized structures or facilities that were purposefully established to mitigate the impacts of climate change and also because the focus of local-level climate change adaptation research has been largely on establishing an understanding of local-level vulnerability and in developing lists of site-specific adaptation strategies (Smithers and Smit 1997; Adger and Kelly 1999; Turner et al. 2003; Smit and Pilifosova 2003; Smit and Wandel 2006). The Third Assessment Report of the Intergovernmental Panel on Climate Change (IPCC) introduced some general determinants of adaptive capacity as (Smit et al. 2001):

- Economic wealth
- Technology
- Infrastructure
- Information and skills
- Equity
- Institutions

Many researchers and practitioners have highlighted the lack of attention to social determinants in the IPCC's Third Assessment Report and elsewhere (Adger 2003a, b; Eriksen and Kelly 2007; Engle 2011). Of the research that does exist, values, perceptions, customs, traditions and social capital appear to play a significant role in determining the ability of a system to adapt to climate change (Adger et al. 2007). Social capital refers to the network of relationships that exists within a society, relations of trust, reciprocity and exchange, and the evolution of common rules (Adger 2003a). It systematically describes not only informal relationships between individuals, but also the interactions between civil society and the network of formal institutions and governing bodies (Woolcock and Narayan 2000). It has been argued that some groups and communities may in fact be less vulnerable than modeling studies suggest since many determinants of adaptive capacity lie in social and institutional networks (Adger 2003a; Naess et al. 2005).

Successful Climate Change Adaptation

Considering that the premise of the proposed research is based on the distinction that the desalination plant implemented as a climate change adaptation strategy in the community of Paget Farm is indeed representative of a "successful" adaptation process, it was important to clarify the meaning of "success". Thus, in this study, successful adaptation is any adjustment that reduces the risks associated with climate change, or vulnerability to climate change impacts, to a predetermined level, without compromising economic, social, and environmental sustainability (Doria et al. 2009). In order to assess whether or not the desalination plant is truly successful, an evaluation of how well the project contributed to decreasing Paget Farm's vulnerability to freshwater scarcity was preformed. However, assessing only effectiveness is not sufficient since it may not take into consideration externalities at other spatial and temporal scales (Adger et al. 2005). In other words, benefits for Paget Farm may produce negative impacts for other social and ecological systems, and what may seem successful in the short term may not be in the long term. Following suggestions by Adger et al. (2005), an evaluation of success that also includes efficiency, legitimacy and equity was implemented. Failing to consider these dimensions of success could lead to maladaptation, whereby a strategy inadvertently increases vulnerability either to the target or non-target systems (McCarthy et al. 2001; Adger et al. 2005).

Case Study

Paget Farm is predominantly a fishing community of approximately 750 individuals on the island of Bequia in Saint Vincent and the Grenadines. Paget Farm was chosen as the case study community for this research because the first carbon–neutral desalination plant in the Caribbean region was implemented as a direct strategy to adapt to climate change in 2012 (CCCCC 2011). The island of Bequia shares many of the biophysical impacts of climate change as other islands in the region, but the primary threat has been identified as freshwater availability (Durrant et al. 2008). Due to its geology and size, Bequia has no surface water reserves and very limited, in terms of quantity and quality, groundwater resources (Greaves 2012).

Before the desalination plant was in operation, households in Paget Farm and the entire island relied almost exclusively on rainwater harvesting techniques to supply themselves with water for domestic use. However, climate variability and change in recent decades have severely challenged these techniques since there have been observed changes in rainfall patterns leading to prolonged periods of drought (Durrant et al. 2008). A particularly devastating example was the 2009–2010 drought, which lasted nearly 10 months in the Grenadines (Greaves 2012). The Vincentian government had to provide the residents of Paget Farm with weekly water supplies by boat from the island of Saint Vincent to Bequia. The water was provided at no cost, but was in limited supply and individuals had to either pay to transport the water by car or walk with water containers back to their homes (H. Belmar, personal communication, October 29, 2012).

In response to changing climate patterns, a pilot project was initiated to supply the community of Paget Farm with a safe and reliable source of freshwater for domestic use. The project was part of the 2007–2011 Special Program for Adaptation to Climate Change (SPACC) initiative, "*Implementation of pilot adaptation measures in coastal areas of Dominica, St. Lucia and St. Vincent & the Grenadines*", which was coordinated by the World Bank through funds made available by the Global Environment Facility. The regional implementing agency was the Caribbean Community Climate Change Centre while its national counterpart was the Vincentian Ministry of Health, Environment and Wellbeing (World Bank 2005). Multiple institutions were also involved in the planning, development and implementation phases of the project, including local community-based organizations, national government departments, industry, regional actors and international agencies. Several strategies were tabled, but ultimately a saltwater reverse osmosis desalination plant fully powered by renewable energy from a photovoltaic system was chosen as the adaptation strategy to be implemented (CCCCC 2011). The desalination plant has been in operation since late 2012, and the photovoltaic system on which it relies produces excess energy that is redirected to the island's central electricity grid. Water is collected in a 20,000-gallon holding tank located above the community of Paget Farm, which is then piped by means of a gravity-fed distribution system to three main water outlets within the community (CCCCC 2012).

Paget Farm is an ideal case study to increase our understanding of climate change adaptive capacity since prior to the desalination plant the community was highly exposed to droughts and was extremely sensitive to the associated impacts since there were little alternatives to accessing freshwater aside from rainwater. But climate change vulnerability, which is the degree to which a community is susceptible to effects of climate change, is not only shaped by exposures and sensitivities but also adaptive capacity (Smit and Pilifosova 2003; Füssel and Klein 2006). Through the implementation of the desalination plant, Paget Farm displayed not only high adaptive capacity since it responded to the observed changes, but has also increased its adaptive capacity to future climate-related impacts by decreasing its sensitivity to increasing drought exposure and thereby reducing its overall vulnerability. A premise to this statement, however, is the assumption that the adaptation strategy implemented is indeed successful.

Methodology

This research used the community of Paget Farm as a case study to understand how various social, institutional, political and economic factors can interact to facilitate and enhance local-level adaptive capacity in Caribbean SIDS. Although the focus was at the local level, a multi-scalar perspective was taken to situate international, regional and national factors and conditions within a local context. Data collection involved semi-structured and key informant interviews with both Paget Farm community members and actors directly involved in the planning, development and implementation of the desalination plant. A total of 40 interviews were audio-recorded in person or via Skype over a three-month period during the summer of 2013: 20 interviews with community members, and 20 interviews with project stakeholders at all scales (i.e., local, national, regional, international actors). The interview response rate for community members was 80 % (20 out of 25 individuals) and 77 % (20 out of 26 individuals) for actors directly involved in the project. Interview data has been collated and analyzed using a constant comparison methodology. The following section presents preliminary observations based on research that is still ongoing. Despite the early stage of analysis some significant findings have emerged, which are highlighted in the following section. Limitations include potential interviewees being out of town (either temporarily or permanently), project stakeholders no longer employed at the same institution, and easier access to local and national stakeholders than those considered at the regional and international scales.

Preliminary Results

Perceptions Regarding Level of Success

A preliminary assessment of perceptions surrounding the desalination plant's level of success is presented in this section prior to the factors contributing to adaptive capacity because "success" is the premise on which the following section's analysis can be made. The four previously identified criteria of successful adaptation by Adger et al. (2005) were used in our assessment: effectiveness, efficiency, legitimacy and equity.

Effectiveness

Based on interviews with both stakeholders and community members, the desalination plant appears to have been effective at addressing a decrease in the amount of rainfall due to climate variability and change. Most stakeholders perceived the plant as successful since it provided a safe, reliable and adequate supply of water to the community during the 2012–2013 drought. The following quote by a project stakeholder highlights his perception of the project's level of success:

> If I have to judge the project, I'd say it was 100 % successful. I am judging it on the fact that in 2012 we suffered a drought was that almost as severe as in 2010. In 2010, we had like 10–11 months of absolutely no rain and in 2012–2013, it went 7–8 months without rain. We pumped over 1 million gallons of water to the public, compared to 2010 when we had 32,000 gallons. So 1 million gallons compared to 32,000 gallons which were brought from Saint Vincent in 2010 says 100 % successful.

The stakeholder maintained the adaptation strategy was completely successful because Paget Farm suffered a drought in 2012–2013 of almost equal severity as the one experienced in 2009–2010, yet there was no need for the Vincentian government to bring water to the community by boat from Saint Vincent. The plant was able to provide enough water for the entire community for the full duration of the drought.

Efficiency

An evaluation of the efficiency of the desalination plant was based on four sub-criteria. First, there was a formal evaluation of alternate adaptation strategies to address a decrease in freshwater availability (Greaves 2012). The study examined a range of non-conventional water sources for the island of Bequia, including submarine pipelines, a dual distribution system, a government cistern project and a desalination system. The analysis was performed on the basis of financial viability, social acceptance, environmental integrity and technical abilities. It was concluded that a desalination system was the most suitable strategy to be implemented. Second, the implemented desalination plant represents one of the most advanced systems available on the market. Third, the photovoltaic system produces more energy than the desalination plant requirements to operate at full capacity. Fourth, although better measures could be put in place, there does not seem to be a significant overconsumption of the water produced at the plant.

Legitimacy

Almost unanimously, stakeholders and community members approved of the project and saw its value in terms of addressing water scarcity on the island. However, over half of the community members interviewed refused to drink the water produced from the desalination plant. In general, these community members would use the water to wash clothing and to bathe, but not for consumption or cooking purposes. The following two quotes highlight concerns from community members regarding the use of desalinated water for consumption.

> I use it mostly for washing purposes. People drink it and they say it drink good, but I never drink it. [...] Some people use it 'cuz when the dry weather, it real rough so some people drink it. But I never drink it.
>
> We still buy water to drink. That water we use to wash clothes and bathe our skin [...] I don't like the way it taste, I prefer the one straight from the sky. People drink that water, but I don't.

Despite the fact that the water is drinkable and of high quality, a local-level perception and skepticism does exist with respect to consumption. The legitimacy of the desalination plant, which is the extent to which a project is acceptable to stakeholders, is therefore affected since a primary aim of the project was to provide the people of Paget Farm with a safe and reliable source of drinking water. Although community members interviewed accepted the project as a source of water for general domestic use, over half did not view the water produced as a legitimate source for consumption. It is difficult to determine if this perspective will remain consistent over time or if perceptions will change.

Equity

Based on preliminary analysis of interviews conducted with community members, the adaptation strategy appears to be equitable. The water is currently being provided at no cost to individuals and seems to be accessible to all members of the community. The following is a quote from a community member highlighting certain aspects of adaptation equity:

> A couple months ago we had a shortage of water, we only have a black tank. We actually used the water when they were pumping up the water [...] our entire black tank was filled. And that helped us out a lot, you know, because we were left in a situation where we had to buy water, which is quite costly, or we had to go searching for water throughout the village. So believe me, it was quite helpful. It saved us a lot of money and time. [...] We didn't have to pay, the guys used a power pump to bring the water from the desalination plant to our house because our house is not far away from the project.

Multi-Scalar Factors Contributing to Community-Level Adaptive Capacity

Preliminary analysis of data collected in this study reveal several key factors that seem to have contributed to the enhancement of local-level adaptive capacity to climate change in Paget Farm. These factors occur at multiple scales including local, national, regional and international, and include the following:

Local

Community Champions and Local Knowledge: Community champions played a critical role in bringing the project to completion. Several challenges arose at the community level, most notably with regards to the construction of the 20,000-gallon holding tank in terms of land acquisition, potential for landslides and road access. Community leaders were instrumental in communicating with property owners, different government ministries and construction companies to facilitate the process.

Community Motivation and Timescales: The community was highly motivated to see the project to completion since they had been dealing for some time with worsening impacts of drought and inadequate water supplies. The 2009–2010 drought had clear implications on the community's perception of water needs—residents easily recalling the seemingly endless months of water shortages and the hardships that ensued. The importance of timescales is also evident. Individuals tend to be less motivated to deal with issues that are perceived to be in the distant future than the ones that are currently plaguing their community. Individuals were therefore very receptive to the project, often times facilitating the process by, for example, providing land and other resources at no cost.

Local Project Stakeholders: Many project stakeholders were either from the community of Paget Farm itself or the island of Bequia. They understood the severity of droughts in Paget Farm and the importance of water to the community because often they themselves had suffered from lack of water availability. For this reason, among others, local stakeholders often went above and beyond their mandated roles to bring the project to completion.

National

Institutional Collaboration: Multiple national government department and agencies were involved in the planning, development and implementation of the adaptation strategy. While it appears collaboration between various national institutions played a significant role in bringing the project to completion, additional analysis of this factor is required. In particular, a formal analysis of institutional networks using Social Network Analysis (SNA), a relatively new and powerful way of visualizing and assessing interactions between institutions within a defined network (Borgatti and Halgin 2011) would be extremely fruitful. SNA could provide a visual and quantitative representation of the institutions involved in the different phases of the project and would help further assess whether institutional collaboration contributed to adaptive capacity in Paget Farm.

In-Kind support: It is typical of Vincentian government departments to provide as many in-kind resources as possible to such projects when a financial contribution is not feasible. The time, expertise and services provided by these departments were critical to bring the project to completion.

Regional

Climate Change and Other Technical Expertise: The implementation of the adaptation strategy is the outcome of several large Caribbean-wide climate change adaptation initiatives, each building upon the previous to develop and implement concrete adaptation strategies. The desalination plant in Paget Farm is the result of clearly defined regional assessments of climate change impacts and vulnerability, which set the foundational work for this project.

Flexibility to Accommodate Local and National Needs: In the final stages of project development, the source of renewable energy was switched from wind to solar due to local concerns regarding aesthetics and lack of familiarity with the energy source at the national scale. Regional actors, determined to bring the project to completion, accommodated local and national actors and institutions by accepting the transition and helping seek additional funding to ensure the project remained powered by a renewable energy source, although solar was less cost effective than wind energy.

International

External Financial Resources: Although the project was not completed solely due to external financial resources, it was an important component and significantly contributed to Paget Farm's adaptive capacity. These resources were key not only in terms of monetary value, but also because of the stringent requirements that had to be met to access them. These requirements included, for example, an environmental impact assessment to ensure that the project did not cause significant harm to land and coastal ecosystems, and a geotechnical survey since the land on which the holding tank is located is prone to landslides. These additional steps helped ensure that the project, while seeking to increase community adaptive capacity, did not increase vulnerability to other systems.

Recognition of Disproportionate Climate Change Impacts: International recognition that Small Island Developing States are among the most vulnerable to and the least able to adapt to climate change lead to the accessibility of funds set in place by the United Nations Framework Convention on Climate Change (UNFCCC). These negotiations and established funds led to the creation of the Caribbean Community Climate Change Centre in Belize, which spearheaded three five-year climate change initiatives in the region and ultimately executed, with the help of national and local counterparts, the adaptation strategy in Paget Farm.

Concluding Remarks

Climate change is a substantial threat to human society and will be disproportionately experienced in Caribbean Small Island Developing States (Mimura et al. 2007). While there has been growing interest in climate change adaptation in the region, much of the research conducted thus far has focused on an assessment of the physical, social and economic vulnerabilities of Caribbean communities (Medeiros et al. 2011). These assessments have provided valuable insight to researchers and practitioners but fail to delve deeply into the underlying conditions that facilitate or hinder the ability of communities to implement successful adaptation strategies (Smit and Wandel 2006; Engle 2011). Using retrospective case studies, such as the one presented in this chapter, uniquely position researchers to more realistically assess and identify how certain social, institutional, political and economic factors interact to facilitate and enhance adaptive capacity. Preliminary analysis reveals the importance of key actors working at different scales, the value of functional institutions, and the role that motivation (and perhaps to some extent fear) among local community members play in enhancing local-level adaptive capacity. Additional analysis of data collected is required to more fully articulate and understand the factors that contribute to enhancing local-level adaptive capacity to climate change. Further retrospective research using different case study locations is also recommended in order to cross-reference and validate findings from this case study.

References

Adger WN (2003a) Social capital, collective action, and adaptation to climate change. Econ Geogr 79:387–404
Adger WN (2003b) Social aspects of adaptive capacity. In: Smith J, Klein J, Huq JS (eds) Climate change, adaptive capacity and development. Imperial College Press, London, pp 29–49
Adger WN, Kelly PM (1999) Social vulnerability to climate change and the architecture of entitlements. Mitig Adapt Glob Change 4:253–266
Adger WN, Arnell NW, Tompkins EL (2005) Successful adaptation to climate change across scales. Glob Environ Change 15:77–86
Adger WN, Agrawala S, Mirza MMQ et al (2007) Assessment of adaptation practices, options, constraints and capacity. In: Parry ML, Canziani OF, Palutikof JP et al (eds) Climate change 2007: impacts, adaptation and vulnerability. Contribution of working group II to the fourth assessment report of the intergovernmental panel on climate change. Cambridge University Press, Cambridge pp 717–743
Borgatti SP, Halgin DS (2011) On network theory. Org Sci:1–14 Articles in Advance
Brooks N, Adger WN (2005) Assessing and enhancing adaptive capacity. In: Lim B, Spanger-Siegfried E (eds) Adaptation policy frameworks for climate change: developing strategies, policies and measures. United Nations Development Program, Global environment facility. Cambridge University Press, Cambridge, pp 165–181
CCCCC (2011) Environmental impact assessment of the installation of a photovoltaic solar powered salt water reverse osmosis plant at Paget Farm Bequia. Technical report 5C/SPACC-11-08-02 prepared by the Caribbean Community Climate Change Centre, Belmopan, Belize
CCCCC (2012) Special programme for adaptation to climate change (SPACC) implementation of adaptation measures in coastal zones. Technical note 5C/SPACC-12-05-01 prepared by the Caribbean Community Climate Change Centre, Belmopan, Belize
Doria MDF, Boyd E, Adger WN (2009) Using expert elicitation to define successful adaptation to climate change. Environ Sci Policy 12:810–819
Durrant S, Nurse L, Stoddard A (2008) Monitoring the effects of below average precipitation on water resources in Paget Farm, Bequia, the Grenadines. Technical report no. 22 prepared for the Centre for Resource Management and Environmental Studies (CERMES), University of the West Indies, Barbados
Engle NL (2011) Adaptive capacity and its assessment. Glob Environ Change 21:647–656

Eriksen SH, Kelly PM (2007) Developing credible vulnerability indicators for climate adaptation policy assessment. Mitig Adapt Strat Glob Change 12:495–524

Füssel H-M, Klein RJT (2006) Climate change vulnerability assessments: an evolution of conceptual thinking. Clim Change 75:301–329

Greaves GE (2012) On water augmentation strategies for small island developing states: case study of Bequia, St. Vincent. Master's thesis submitted to the National Central University of China, China

McCarthy JJ, Canziani OF, Leary NA et al (eds) (2001) Climate change 2001: impacts, adaptation, and vulnerability. Contribution of working group II to the fourth assessment report of the intergovernmental panel on climate change. Cambridge University Press, Cambridge

Medeiros D, Hove H, Keller M et al (2011) Review of current and planned adaptation action: the Caribbean. Report for the International Institute for Sustainable Development

Mimura N, Nurse L, McLean RF et al (2007) Small islands: impacts, adaptation and vulnerability. In: Parry ML, Canziani OF, Palutikof JP et al (eds) Climate change 2007: impacts, adaptation and vulnerability. Contribution of working group II to the fourth assessment report of the intergovernmental panel on climate change. Cambridge University Press, Cambridge, pp 687–716

Naess LO, Bang G, Eriksen S, Vevatne J (2005) Institutional adaptation to climate change: flood responses at the municipal level in Norway. Glob Environ Change 15:125–138

Smit B, Pilifosova O (2003) From adaptation to adaptive capacity and vulnerability reduction. In: Smith J, Klein J, Huq S (eds) Climate change, adaptive capacity and development. Imperial College Press, London, pp 1–20

Smit B, Wandel J (2006) Adaptation, adaptive capacity and vulnerability. Glob Environ Change 16:282–292

Smit B, Pilifosova O, Burton I et al (2001) Adaptation to climate change in the context of sustainable development and equity. In: McCarthy JJ, Canziani OF, Leary NA et al (eds) Climate change 2001: impacts, adaptation and vulnerability. Contributions of working group II to the third assessment report of the intergovernmental panel on climate change. Cambridge University Press, Cambridge, pp 877–912

Smithers J, Smit B (1997) Human adaptation to climatic variability and change. Glob Environ Change 7:129–146

Turner B, Kasperson RE, Matson PA et al (2003) A framework for vulnerability analysis in sustainability science. Proc Natl Acad Sci 100:8074–8079

Woolcock M, Narayan D (2000) Social capital: implications for development theory, research and policy. World Bank Res Observer 15:225–249

World Bank (2005) Special program on adaptation to climate change project. Global Environment Fund (GEF) project brief to the Governments of Dominica, Saint Lucia, and St. Vincent and the Grenadines. Report no. 34152-LAC

Chapter 10
Taking a Micro-Perspective on the Global Challenge of Climate Change: The "Microenergy Systems" Research Focus at the Technische Universität Berlin

Jonas Van der Straeten, Kathrin Friederici and Sebastian Groh

Abstract In the past, fuel-based economies have led the way in economic growth, resulting in emissions that are now understood to cause climate change, having a major impact on countries of the Global South. It has become clear that effective climate change mitigation and adaptation not only require restrictions on energy consumption in the industrialized world, but also a decoupling of economic development from greenhouse gas emissions in the South. But can developing countries take a less carbon-intensive path, even though its outcome is highly uncertain? In many scenarios for a sustainable future energy supply, decentralized solutions play a crucial role. Such solutions are often based on renewable energies and can help to avoid lock-in effects based on fuel intensive energy systems. However, they are characterized by various aspects that conventional research paradigms have given little attention to, until now. These research challenges are reflected in the concept of Microenergy Systems, which serves as a theoretical basis for the Research Focus and Ph.D. program at the Technische Universität in Berlin. It brings together academics from various disciplines to take a micro-perspective on the idea of decentralized energy supply. This chapter introduces and discusses the concept of Microenergy Systems. It summarizes the development of the relatively young research group—the experiences of engineers, social and political scientists, economists and planners partaking in joint research and education in the field of decentralized sustainable energy solutions. Furthermore it highlights the attempt

J. Van der Straeten (✉) · K. Friederici · S. Groh
MES Research Group, Technische Universität Berlin, ZTG, Sekr. HBS 1, Hardenbergstraße 14-16, 10623 Berlin, Germany
e-mail: koordination-mes@mail.tu-berlin.de; Promotionskolleg-mes@mailbox.tu-berlin.de

K. Friederici
e-mail: friederici@ztg.tu-berlin.de

S. Groh
e-mail: groh@ztg.tu-berlin.de

W. Leal Filho et al. (eds.), *International Perspectives on Climate Change*,
Climate Change Management, DOI: 10.1007/978-3-319-04489-7_10,

to establish an international scientific community concerned with Microenergy Systems amongst academics who are scattered throughout different institutions and numerous universities worldwide.

Keywords Decentralized energy supply • User perspective • Transdisciplinary research • Knowledge transfer • Education

Introduction

Energy crises, climate change impacts, land degradation and high population growth are among the most critical challenges currently facing countries of the Global South. Steadily increasing energy demand and rising prices of fossil fuels have significantly increased the vulnerability of low-income populations, e.g. in Latin America (MIF 2012), and it has now been fully acknowledged that without major improvement in the quality and quantity of energy services to these countries, the Millennium Development Goals cannot be met (DFID 2002). The United Nations has therefore declared the decade 2014–2024 as the Decade of Sustainable Energy for All. Furthermore, it has become clear that climate change, energy supply and development cannot be tackled in isolation from one another, particularly if the lock in of carbon intensive energy infrastructure is to be avoided. Pathways for low carbon development which do not undermine the development goals of countries of the Global South are urgently needed in order to prevent dangerous, anthropogenic climate change in the future (Steckel et al. 2013).

Despite global clean energy investments totaling $244 billion in 2012, energy poverty continues to plague nearly half the world's population (UNEP-FS 2013). In bringing modern technology to the global poor, the energy sector has lagged behind other sectors such as the information and communications sector. For example, the number of mobile phone users in Africa has risen dramatically in the last decade, from 16 million in 2000 to 650 million users in 2011, most of which had never had a landline phone before (Yonazi et al. 2012). This technological leapfrogging remains a distant vision for the energy sector, despite indications for a similar potential. In Bangladesh, two million households are currently supplied with electricity from solar home systems, corresponding to 8.5 % of the rural population.[1]

Examples such as this suggest that decentralized energy technologies can provide a suitable means of leapfrogging to clean energy supply at the level of households and small enterprises. Still, energy povery continues to put a brake on people´s development path in the form of a energy poverty penalty (Groh 2014). Whilst encouraging results have been seen from energy inclusion activities of

[1] IDCOL (2013).

bottom-up initiatives and stakeholders outside the conventional energy sector, research on decentralized clean energy solutions for countries of the global South is in its infancy. One of the most pressing challenges is to find synergies among the development and the climate change research agenda.

Since current and past research on conventional centralized energy systems cannot be fully applied to sustainable decentralized energy solutions, the multidisciplinary Microenergy Research Group at the Technische Universität Berlin proposes the concept of Microenergy Systems. In the following sections the potentials and limitations of this concept will be discussed and a research agenda will be derived.

Decentralized Energy Systems and Their Relevance for Global Access to Energy, Development and Climate Change

In many scenarios for a sustainable future energy supply, decentralized solutions play a crucial role. In the *Energy for All Case*, a scenario for universal access to electricity by 2030 conceptualized by the International Energy Agency (IEA), only 30 % of rural areas are projected to be able to connect to centralized grids, mainly for reasons of cost-effectiveness, whereas 70 % of rural areas are projected to be connected either with mini-grids or with small, stand-alone off-grid solutions (OECD/IEA 2011). Small-scale decentralized systems represent a viable option for supplying modern and environmentally-friendly electricity to the 1.3 billion people worldwide without grid access, and a market for clean off-grid energy systems has already evolved in some countries of the global South—something which has never been actively advocated in the western world.

To what extent improved energy access can alleviate poverty and support economic development in countries of the global South is still a subject of debate in academia. So far the causality between energy supply and economic development remains inconclusive (Payne 2010). No systematic evidence has been found that shows increased energy use causes economic development in countries of the global South. However, it has been found that countries do not manage high levels of development, as measured by the Human Development Index, without crossing a border of 40 GJ/capita (Steckel et al. 2013). For decentralized energy systems, studies have found different impacts on poverty. While research findings for Kenya have demonstrated that the income threshold for modern energy use, including solar, tends to be very high (Lay et al. 2012), results from Bangladesh show strong advances in the pro-poor penetration of household based solar systems (IDCOL 2013). Groh (2014) argues that in the case of Peru energy poor people spend far more on energy sources even when income is kept constant which hinders or at least severly delays their development.

It remains to be seen whether decentralized energy systems have the potential to substantially contribute to climate change mitigation. As the World Bank stated in 2011, "recent evidence suggests that the climate-change debate needs to focus more on household energy" (World Bank 2011). For instance, the implications of cooking with biomass for climate change have only lately gained greater attention in the development community. Cooking with biomass was seen as climate-neutral for a long time, since biomass was considered primarily a renewable energy source; but this is only true if it was harvested in a sustainable way. If it is not renewable, biomass as a household fuel can be a net contributor to global warming. About 730 million tons of biomass are burned every year in developing countries,[2] amounting to more than 1 billion tons of carbon dioxide (CO_2) emitted into the atmosphere. With a new generation of advanced cooking stoves reducing CO_2 emissions by about 25–50 %, decentralized energy systems can make a considerable contribution to carbon mitigation (World Bank 2011).

This impact is less clear for off-grid electricity systems, as their total generation capacity is still marginal on the global scale. Several studies indicate that the carbon mitigation potential of household electrification is positive, but small (Kaufmann et al. 2000; Ybema et al. 2000). If climate change is taken as the primary concern, solar electrification is in fact less cost-effective than a number of alternative investments (Jacobson 2004). However, in a more long term perspective, decentralized energy systems can play a major role in avoiding the lock-in effect of carbon intensive energy systems which imply a high cost for reorientation once established (Jakob et al. forthcoming). Decentralized systems also have major potential for generating co-benefits, solving issues relating to energy access and development whilst simultaneously reducing CO_2 emissions. These co-benefits need more attention from the development and climate change research community.

The Concept of Microenergy Systems

In academia, the idea of small-scale, decentralized energy solutions was first advocated by the "Appropriate Technology"-movement, inspired by E. F. Schumacher's influential work, *Small is Beautiful* (Schumacher 1973; Hazeltine and Bull 1999). Since the mid-1980s decentralized energy technologies have become part of the development discourse thanks to Schumacher's neo-populist philosophy about small scale industries, local self-reliance and fulfilment of basic needs. For example, the focus of decentralized solar electrification programs has shifted from heavily subsidized, centrally planned projects in the 1970s and 1980s, towards distributed market approaches since the mid-1990s (Hankins 1993). Decentralized energy solutions have been increasingly regarded as part of an inclusive capitalism, addressing the four billion low-income people worldwide who constitute the "Base of the (economic) Pyramid" (Hammond et al. 2007). These shifts in thinking currently form the context for the implementation of decentralized energy technologies. They are characterized by various aspects which have received little attention from conventional energy research.

---

[2] WHO (2007).

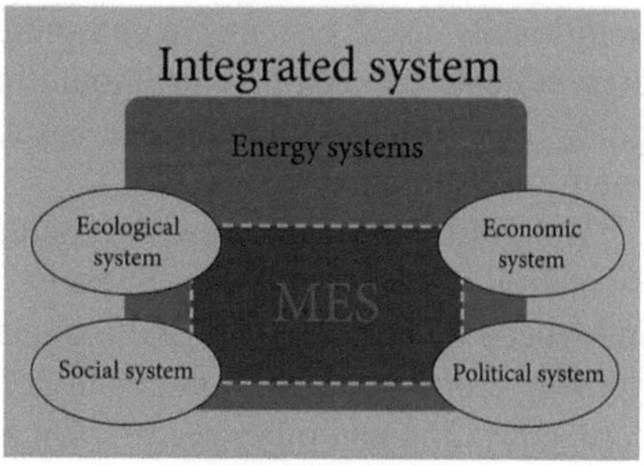

Fig. 10.1 Definition of microenergy systems (Phillip and Schäfer 2009)

These aspects are addressed in the Microenergy Systems research approach, which was first introduced by Phillip and Schäfer (2009). A MES is defined as a decentralized energy system based on small energy appliances, which provide households, public institutions, small businesses with energy and enables energy demand to be met by locally-based sources.

Its technical subsystem describes the transformation of energy, be it electrical or mechanical power, chemical energy, light or heat. This transformation can be further subdivided into energy generation (e.g. solar home systems, biogas plants or diesel gen-sets), energy storage (e.g. car batteries or water tanks) and energy consumption (e.g. LPG stoves, fridges or solar dryers).

The energy conversion unit is embedded in its operating environment and follows the requirements of its (potential) user or users. Producing energy close to where it is consumed results in context specific characteristics, dependent upon the local political framework and the interrelation of environmental, economic and social factors (see Fig. 10.1). Without considering these factors, it is highly unlikely that a MES will operate successfully.

The **ecological system** refers to the impact the MES has on the natural environment throughout its life-cycle, particularly during the production process, relating to the use of appropriate and locally available resources and equipment, as well as storage and disposal procedures. Implementation of MES based on renewable raw materials should take into account the risks and benefits, whilst ensuring sustainable usage wherever possible. When considering the entire system lifecycle it becomes clear that for some renewable energy technologies such as solar photovoltaics, environmental risk arises from energy-intensive production processes and transport (Kaufmann et al. 2000). Furthermore, consideration of the disposal phase is crucial, for instance of the proper disposal of batteries and recycling of PV modules.

Among the major aspects of the **economic system** is the availability of resources, which is determined by market prices and access to distributors, technology suppliers, and energy service providers. However, any analysis of energy supply should include economic considerations of costs paid directly or indirectly at the end-user level. This is especially important in regions of the Global South which are subject to far greater levels of energy poverty (DFID 2002; Nussbaumer et al. 2012). It is often important to make an economic comparison between traditionally used systems (e.g. candles, three-stone fire) and the new microenergy system. The comparison should take into account the capital investment, ongoing operation and maintenance costs, and the changing purposes and patterns of energy consumption (Mills 2005).

The **political system** sets the regulatory framework for the design and implementation of MES on multiple levels. The implementation of new technologies is influenced by both national and regional laws, including not only economic development, climate and environmental policies, but also policies deriving from health care, education and other sectors. On a global scale, the strategies of international organisations regarding development policies such as the UN MDGs and climate protection agreements such as The Kyoto Protocol heavily influence the availability of funding. This plays a significant role in determining which initiatives (e.g. renewable energies) are developed and to what extent implementation efforts follow market-oriented, state-run or community based approaches.

Defining MES as socio-technical systems refers to how the technology interacts with the surrounding **social system**. Crucial questions are: Who is promoting which type of energy supply? How can basic needs and interests of the respective population be met? How can (potential) end-users participate in decision making, design and operation processes? Is the acceptance and compatibility of new technologies predictable? How can the diverse range of impacts be evaluated?

A Research Agenda for MES

The interdependencies outlined above require a transdisciplinary MES research approach, combining different perspectives to achieve more effective product development, manufacturing and implementation practices. The following section introduces previous findings and on-going projects, showing how MES research addresses the multiple issues raised above:

- **Access to Energy and Potential Assessment**: The research group attempts to develop a broader and more inclusive understanding of energy access that goes beyond narrow conventional definitions (e.g. access to the national grid: yes/no) and addresses the complexity of energy use and demand. Meeting these multifaceted needs requires more adapted and demand-driven approaches to energy provision (Johnson 2013). To make these approaches feasible in the context of a market economy, it is crucial to understand the complex interdependencies between energy access and economic development, and to assess the market potential for different energy technologies. One of the main research goals of the group is to develop methods and criteria to model rural energy demand and assess the potential for decentralized solutions. These efforts are based on the assumption that the quality of energy supply is often negatively correlated with spatial remoteness—particularly in rural areas (Groh 2014).

- **Planning and governance of energy infrastructure**: With regard to the planning and governance of energy infrastructure, the concept of a decentralized energy supply at the household or community level holds new research challenges for planners and social scientists. This includes critical evaluation of the political and scientific conception of grid-based energy infrastructure, which predominantly follows the Western perspective of a centralized, supply-driven approach. This approach to electricity supply is based on the notion that state-owned monopolists or a few big utilities provide ubiquitous infrastructure services. These actors tend to assume that customers play a passive role and claim that the provision of infrastructure services is closely linked to and/or highly regulated by the state. Within this paradigm, rural electrification is usually only considered in terms of the extension of the centralized grid (Kirubi et al. 2009). In a dissertation project, the research group tracks the discourses in which this paradigm historically emerged in the industrialized West and how it was transferred and turned into infrastructure planning in the developing word. Based on the assumption that the provision of decentralized systems requires different infrastructure regimes and forms of organization, the research group explores research perspectives more attuned to integrated planning and governance of centralized and decentralized energy systems. In doing so, a focus is put on the users' specific needs and their active role in the provision of energy services.
- **Approaches for a bottom-up electrification—the concept of Swarm Electrification (Groh et al.** 2013): As an alternative to top-down electrification approaches through the extension of centralized transmission and distribution networks, the research group is currently conceptualizing a bottom-up approach that views electrification as a process with different phases. It addresses the major problems facing both mini-grid deployments and extensions of central grid connections to remote areas, above all the inability to recover associated capital costs. In analogy to the model of swarm intelligence this approach is referred to swarm electrification. It proposes a three-phase electrification process.

The first phase starts with the rapid and efficient deployment of SHS or other small-scale energy devices with integrated storage and intelligent control devices to households and businesses via end-user microfinance and the provision of long-term locally available technical services. In a second phase of micro-grid development, the individual nodes are connected together, thereby providing balancing of all systems, in that they are able to share the swarm of generation capacity from all individual systems. In a third phase this micro-grid can connect to other micro-grids and/or to central grid networks. Current research challenges include the conception and realization of individual intelligent control units, which can be implemented at the household and micro-enterprise level for a relatively low cost, and form the basis for an interconnected bottom-up micro-grid:

- **The User Perspective**: The evidence from a multitude of projects has shown that technical and economic feasibility are a necessary but by no means sufficient condition for the implementation of decentralized energy projects (Laufer and Schäfer 2011; Schweizer 1996; Stadler et al. 2002). The introduction and operation of new technologies, such as decentralized energy systems, does not solely involve the acquisition of specific knowledge, but also a myriad of interrelated values, attitudes and culturally determined behaviours. Among the preconditions of successful implementation is a broad pick-up of MES which requires a high degree of acceptance. However, at times habits and daily routines are not in favour of the adoption of systems based on modern energy resources (Ahmad 2001). According to Rose (2001), there is a clear need to develop a broader, more socially grounded understanding of energy users as well as their needs and to uncover the social meaning attributed to such technology. Research in the field of the acceptance and perception of renewable energies and climate change has been done predominantly in the global North (e.g. Devine-Wright and Devine-Wright 2009). For that reason, a current psychological research project analyzes the everyday understanding of energy and energy services that might influence the perception and acceptance of MES. This information will be critical to a successful assessment of needs as well as sustainable implementation strategies.

Where the value chains of decentralized energy systems or their components extend over cultural borders, the values, attitudes and perceptions of the various stakeholders involved will differ substantially. This becomes particularly apparent in projects involving technologies developed in the North but implemented in the South, as for example in the context of development cooperation. This is shown clearly in a previous research project which analyzed the differing perceptions of solar home systems among product designers, funders and users, and the inconsistencies, misconfigurations and misuse that result (Tillmans and Schweizer-Ries 2011).

- **Social Context**: Can these social, cultural and knowledge barriers be overcome? Looking at the introduction of biogas plants in Tanzania, one research project seeks to observe the design, adaptation and implementation process of a decentralized energy technology using ethnographic methods (Barthel 2013). The project brings context to models of user participation, ownership and empowerment. A postcolonial perspective enables a critical analysis of the power relations of the different actors that shape the broader socio-political context.
- **Methodologies for designing Energy Product-Service Systems**: The urgent demand for innovative design methods for decentralized energy products and services is shown in the various quality issues that the research group has unveiled in a number of technical evaluations. The results of these studies highlight the limited impact of classical quality assurance approaches which focus on product standardization and certification. Most quality problems originate from the design of the business and service model, rather than the technical hardware itself (Müller et al. 2009; Lindner 2011). This fact has led to a definition of quality that is neutral towards the technology and focuses instead on end user economic needs and value creation. It has been proposed that financing institutions are involved in the business model, having the ability to disburse small loans for energy products and services as well as having an economic incentive to assure sufficient quality for at least the credit period (Kebir

and Blessing 2009). To put these theoretical insights into product and service design, parts of the Product-Service-System (PSS) approach were adapted to decentralized energy systems in rural areas and an Energy-Product-Service (EPSS) Design Guideline was developed (Kebir and Müller 2010). It must now be investigated how these implementation models can be translated into economically viable business models, and how existing successful models can be replicated. One of the current research projects seeks to develop a tool for the design of products and services adapted to the requirements of decentralized energy systems in remote areas.

- **Economic Impacts**: In addition to the preconditions and means for a successful implementation of decentralized energy systems, economic and ecological impacts at different levels are one of the major research interests of the group. Through four case studies in Sri Lanka, the MES group has analysed different decentralized energy systems (solar home systems, small hydropower, gasification systems and energy efficient cooking stoves) in terms of implementation strategies and their contributions towards poverty alleviation in rural areas of the Global South (Laufer and Schäfer 2011). They found inconclusive effects on poverty when energy systems are used for consumption, but refer to positive results when the systems are implemented for productive use. In this case productive use was defined as directly generating income through the use of the energy system.
- **Ecological Impacts: MES and anthropo-natural cycles**: Regarding the ecological impact of energy systems, recent public debate has focussed heavily on the effect of energy consumption at the macro scale, particularly with respect to global warming. The micro-perspective, however, reveals that the direct impact of energy consumption on the local ecosystem often affects people's livelihoods far more severely. This holds particularly true for energy technologies that use the natural resources of the surrounding ecosystem for fuel. The intensive use of biomass (e.g. firewood) as cooking fuel can lead to deforestation, soil erosion and desertification. The need for sustainable soil-production systems is further underlined by climate change and other adverse anthropogenic activities which continue to negatively affect the soils. This is mainly attributable to man-made systems taking more substances from the ecosystem than giving back (Lal 2010). Thus, it is crucial that MES need to be considered embedded in the ecosystem. Therefore, material flow analysis should be undertaken (prior to implementation) in order to assess the ecological impact of MES.

For this reason, one research project looks into the possibility of balancing energy and material flows between small farmers and their natural environment. This can be done by introducing technological innovations which enable the users to close anthropo-natural cycles, e.g. by effective and efficient use of the by-products of energy production in agricultural practices, such as fermentation residues and charcoal, which are a valuable store of carbon and nutrients (Krause 2011). Another technical intervention currently explored by the research group is sanitation through the use of separation toilets in combination with MES, which allow the use of human faeces as fertilizer in agriculture, without running the risk of spreading diseases. Further work of the MES Research Group aims to develop scientific methodologies and practical tools for assessing the ecological impact of MES together with different stakeholders of the MES.

- **Ecological Impacts—Life Cycle Assessment**: These systemic approaches do not enable the quantification of the ecological impacts in a way that makes it possible to compare different decentralized energy systems. For this purpose, another research project analyses the complete life cycle of different technologies and quantifies the ecological impacts through all stages from the extraction of natural resources to the final disposal. Beyond this, the project seeks to identify the potential for ecological optimization as well as possible conflicts between ecological and economic targets. These conflicts are evident for example when recycling the lead acid batteries used in solar home systems. In the informal recycling sector of developing countries, up to 50 % of the lead of a battery can leak into the environment (Hoffmann and Wilson 2000). The final aim of the project is to provide advice on how unintended environmental impacts can be avoided, by integrating the disposal of MES into the business model for instance.

Since the introduced research concept is demand-driven and deduced from practical needs instead of pure theoretical considerations, it cannot be contained by a single research discipline. Gibbons and Nowotny (2000) call the type of research dealing with current societal problems 'transdisciplinary' or 'mode 2' research, differentiated from 'mode 1' research, which generates its questions solely within single disciplines in the scientific context. 'Mode 2' requires an iterative process, where societal problems are taken up in the field, reformulated as scientific problems and solved with contributions in an exchange between scientific knowledge and its practical applications (Schäfer et al. 2011).

The development of a comprehensive understanding of a MES embedded in its surrounding systems can be seen as an undertaking that requires a highly interdisciplinary approach. It includes in-depth knowledge in combination with customized solutions. The MES research group at the TU Berlin aims to concentrate expertise and provide a framework for communicating findings across disciplinary borders.

This means consolidation of the knowledge accumulated in the various fields of research through regular exchange as well as joint exploration and education. It should also enable the transfer of strategies, programs and tools to different local contexts, in close cooperation with partners acting in the field.

The MES Research Group

The Research Group, funded by the Hans-Böckler-Foundation, combines the microenergy systems research activities from multiple departments of Technische Universität Berlin and other universities:

- Center for Technology and Society
- Institute of Energy Technology

- Institute for Landscape and Environmental Planning
- Department of Environmental Technology
- Institute for Machine Tools and Factory Management
- Chair of Economics of Climate Change
- Department of History of Technology, Technische Universität Darmstadt
- Institute of Psychology, Otto-von-Guericke-Universität Magdeburg
- Research Unit in Engineering Science at the University of Luxemburg
- Reiner-Lemoine Institute, Berlin

The initiation of the MES Research Group in 2007 was based on the observation that projects for decentralized energy supply, though becoming increasingly popular in international development organizations and national development programs during the last three decades, have produced few success stories. Systematic evaluations of the experiences and lessons learned from projects for decentralized energy projects are rare, and knowledge transfer between scientific disciplines, the Global South and North and between universities and implementing organizations remains difficult (Schäfer et al. 2011). Since international and national development organisations are under constant pressure to prove the effectiveness of their projects, the incentive to investigate and share knowledge about failures, open questions and identified bottlenecks is low.

Academia offers a more neutral forum that allows an open and critical exchange, away from the constant need of securing funding for implementation projects. However, scientific research on decentralized energy systems is scattered across various institutes in different universities worldwide. The MES Research Group has committed itself to bringing together and consolidating this research community by providing forums for exchange, the biggest of which were two international scientific conferences on "Micro Perspectives for Decentralized Energy Supply" (MES) in April 2011 and February/March 2013 in Berlin (cf. the conference proceedings: Schäfer et al. 2013). With more than 400 participants at the MES 2013 conference and its side-events, the format has proven to be attractive for academics from various disciplinary backgrounds as well as practitioners working in the field.

Within these efforts, the research group has prioritized establishing partnerships and exchange programs with Asian, African and South American universities. Because of their geographical and cultural proximity to the implementation context for most projects, it is crucial to strengthen their links to academic networks and discourses on decentralized energy supply in Europe and North America. This includes discussions on a common language and use of terminology, as well as defining standards, e.g. for the evaluation of case studies in different cultural and geographic contexts.

Analogous to scientific research, systematic training and education in the field of decentralized energy supply is still in its infancy, with a strong theoretical and technological focus that sometimes has limited relevance for practical implementation. The research group has established its own teaching formats, such as the course "Energy Entrepreneurship and Rural Electrification", which provides comprehensive training to international master's students on the development of business models for decentralized energy technologies in countries of the Global South.

Founded on the basis of experiences derived from project implementation, the Research Group has put a considerable effort into creating an innovative set-up which enables a beneficial interplay between academic research and practical implementation. It has maintained its strong links with partners in the field of development cooperation (e.g. Engineers without Borders and LAREF in Granada and Nicaragua), local enterprises (e.g. Mobisol in Tanzania and Kenya), manufacturers of renewable energy technologies—with which several workshops on Energy-Product-Service Systems are held, and platforms for knowledge transfer like energypedia.info. These links also extend to consultancies, such as MicroEnergy International, a Berlin based energy and finance consulting company that was a spin-off from the Technische Universität Berlin. Such cooperation facilitates direct field-testing of the latest research results, as has been the case with the Energy Inclusion Initiative in Peru.

The Energy Inclusion Initiative in Peru: An Exemplary Case for Cooperation with the Private Sector

The Research Group took part in the Energy Inclusion Initiative, initiated in 2010 by Appui au Développement Autonome, a Luxembourg-based non-governmental organisation, and MicroEnergy International. The primary objective of this initiative is to examine ways to successfully incorporate energy products into the portfolios of Microfinance Institutions. In this manner, the Energy Inclusion Initiative aims to broaden access to clean energy with the goal of increasing the economic, social and environmental well-being of small- and micro-sized businesses in rural areas of developing countries. It was founded on the assumption, that the utilisation of Microfinance Institutions human and capital infrastructure, flexible financing options, and detailed quality-focus is a uniquely successful platform for the distribution of energy systems to communities with limited access to energy (Groh 2013). The EII further aims to reduce the vulnerability to the effects of climate change through effective adaptation measures. In Peru, for example EII works closely with different universities and private laboratories, research centres, and independent institutions in order to test and validate energy technologies. Therefore, a close collaboration was established between the Research Group MES with la Universidad Nacional de Ingeniería (UNI) and la Universidad Nacional San Agustín (UNSA). Continuous customer feedback and technical product testing by these universities has been essential for product optimisation regarding design and functionality. Together with the Research Group MES, a methodology for the life cycle assessment of different decentralized energy products has been developed. Based on these successful experiences and given the plan to replicate the EII in Central America, there are conversations underway with a range of universities in order to set-up joint test laboratory facilities for product validation and certification with a regional focus.

Conclusions and Outlook

As countries of the global South strive to reach the economic status of more developed countries, they will add to the challenge facing the world's climate system. This chapter has demonstrated that decentralized energy solutions based on renewable resources have a high potential to confront poverty in a climate friendly way, since the provision of energy access is considered to be fundamental for economic development. However, there is still a lack of empirical evidence regarding the impacts of MES implementation, both on the economical and the ecological level—not to mention the lack of knowledge about the long term social effects. Following the MES approach, further research and sustainable distribution strategies require a deep understanding of the complex interdependencies between technologies for decentralized energy supply and their environment. To meet these challenges, insights should be gathered through the cooperation of researchers from different disciplines and cultures, including both academics and practitioners. There is a largely untapped potential for new perspectives and innovative approaches for decentralized energy supply through knowledge transfer from the Global South to the North.

The first steps towards bringing the MES research community together have been made, however this consolidation is still in its infancy. Further efforts and resources have to be invested to find a common language and to compile findings and experiences systematically to be able to draw transferable conclusions. This chapter has introduced some of the completed and ongoing projects of the MES research group and outlined potential areas for further research. It should be read as an invitation for joint research projects to tackle the multifold challenges decentralized energy solutions are still facing. These projects should go beyond the attempt to analyse and replicate the many piecemeal approaches that already exist. They should have the courage to offer visions of large scale energy supply based on decentralized systems and provide concepts for scaling up existing solutions to this point.

Acknowledgments The authors thank Prof. Dr. Dr. Martina Schäfer, Noara Kebir, Daniel Philipp, Ariane Krause, Bettina Barthel, Klara Lindner and Alexander Batteiger from the postgraduate program Microenergy Systems for their contributions to this chapter, Richard Kunert from Max Planck Institute for Psycholinguistics and Donders Institute for Brain, Cognition and Behaviour and Daniel Hinchliffe from Microenergy International GmbH in Berlin for their comments and corrections.

References

Ahmad B (2001) Users and disusers of box solar cookers in urban India: implications for solar cooking projects. Sol Energy 69:209–215

Barthel B (2013) Developing a small-scale biogas digester in rural Tanzania. An ethnographic approach for the analysis of an untypical innovation journey. In: Schäfer M, Kebir N, Philipp D (eds) Micro perspectives in decentralized energy supply. Conference Proceedings 2013, pp 133–136

Department for International Development (DFID) (2002) Energy for the poor. Underpinning the Millennium Development Goals, Glasgow (UK), Future Energy Solutions. Available (March 2013) www.dfid.gov.uk

Devine-Wright H, Devine-Wright P (2009) Social representations of electricity network technologies: exploring processes of anchoring and objectification through the use of visual research methods. Br J Soc Psychol 48:357–373

Gibbons M, Nowotny H (2000) The potential of transdisciplinarity. In: Klein HTh et al (eds) Transdisciplinarity: joint problem solving among science, technology and society. Birkhäuser, Basel, pp 67–80

Groh S (2013) The synergy potential of financial and energy inclusion. Accion-center for financial inclusion blog, Washington

Groh S, Philipp D, Edlefsen B, Kirchhoff H (2013) Swarm electrification-suggesting a paradigm change through building up microgrids bottom-up. Conference "Energy Systems in Transition: Inter-and transdisciplinary contributions". 9th–11th of October 2013 in Karlsruhe, Germany

Groh S (2014) The role of energy in development processes-The Energy Poverty Penalty: Case study of Arequipa (Peru). Energy for Sustainable Development. Vol 18, February 2014, Pages 83–99

Hammond AL, Kramer WJ, Katz RS, Tran JT, Walker C (2007) The next 4 billion: market size and business strategy at the base of the pyramid. World Resources Institute and International Finance Corporation, Washington DC

Hankins M (1993) Solar rural electrification in the developing world. Four case studies: Dominican Republic, Kenya, Sri Lanka and Zimbabwe, Solar Electric Light Fund (SELF), Washington, DC

Hazeltine B, Bull C (1999) Appropriate technology: tools, choices and implications. Academic Press, New York

Hoffmann U, Wilson B (2000) Requirements for, and benefits of, environmentally sound and economically viable management of battery recycling in the Philippines in the wake of Basel Convention trade restrictions. J Power Sources 88:115–123

IDCOL (2013) Infrastructure Development Company Limited. Available: (June 2013) http://www.idcol.org/news_detail.php?id=147

Jacobson A (2004) Connective power: solar electrification and social change in Kenya. University of California, Berkeley

Johnson O (2013) Universal energy access: moving from technological fix to poverty reduction. Briefing Paper 6/2013, German Development Institute. Available (June 2013) http://www.die-gdi.de/CMS-homepage/openwebcms3.nsf/%28ynDK_contentByKey%29/%20ANES-96F95S/$FILE/BP%206.2013.pdf

Kaufmann S, Duke R, Hansen R, Rogers J, Schwartz R, Treler M (2000) Rural electrification with solar energy as a climate protection strategy. Renewable Energy Policy project, No. 9. Available (January 2013) www.repp.org

Kebir N, Blessing L (2009) Development of a certification process for microenergy systems. In: Proceedings, international conference on engineering design

Kebir N, Müller P (2010) Energy product-service systems. E-PSS design guideline. GTZ and MicroEnergy International, TU Berlin

Kirubi C, Jacobson N, Kammen D, Mills A (2009) Community-based electric micro-grids can contribute to rural development: evidence from Kenya. World Dev 37(7):1208–1221

Krause A (2011) Hydrothermal carbonization as innovative technology in sustainable sanitation in Tanzania. In: Schäfer M, Kebir N, Philipp D (eds) Micro perspectives in decentralized energy supply. Conference Proceedings 2013, pp 238–240

Lal R (2010) Managing soils and ecosystems for mitigating anthropogenic carbon emissions and advancing global food security. Bioscience 60(9):708–721

Laufer D, Schäfer M (2011) The contribution of microenergy systems towards poverty reduction: case study of an implementation strategy for solar home systems in Sri Lanka. Energy Sustain Dev 15(3):330–336

Lay J, Ondraczek J, Stoever J (2012) Determinants of renewable energy use: solar home systems and lighting fuel choice in Kenya. Energy Econ

Lindner K (2011) Quality Issues in the market based dissemination of solar home systems. In: Schäfer M, Kebir N, Philipp D (eds) Micro perspectives in decentralized energy supply. Conference Proceedings 2013, pp 220–228

Multilateral Investment Fund (MIF) (2012) Climatescope 2012: assessing the climate for climate investing in Latin America and the Caribbean. MIF in collaboration with Bloomberg New Energy Finance, Washington

Mills E (2005) The specter of fuel-based lighting. Science 308(5726):163–164

Müller P, Kebir N, Stark R, Blessing L (2009) PSS layer method—Application to microenergy systems. In: Sakao T, Lindahl M (eds) Introduction to product/service-system design, pp 3–30

Nussbaumer P, Bazilian M, Modi V (2012) Measuring energy poverty: focussing on what matters. Renew Sustain Energy Rev 16(1):231–243

OECD/IEA (2011) Energy for all. Financing access for the poor, special early excerpt of the World Energy Outlook 2011, Paris

Payne J (2010) Survey of the international evidence on the causal relationship between energy consumption and growth. J Econ Stud 37(1):53–95

Philipp D, Schäfer M (2009) Interdisziplinärer Forschungsansatz zur Analyse von Mikroenergie Systemen, ZTG/TU Berlin. Available (June 2013) http://www.umweltpruefung.tu-berlin.de/fileadmin/fg123/Mikroenergiesysteme/ Downloads/Diskussionspapier_zu_Mikroenergie-Systemen.pdf

Rose DA (2001) Reconceptualizing the user(s) of—and in—technological innovation: the case of vaccines in the United States. In: Coombs R, Green K, Richards A, Walsh V (eds) Technology and the market, demand, users and innovation. Edward Elgar Publishing Inc, Cheltenham/Northampton, pp 68–88

Schäfer M, Kebir N, Neumann K (2011) Research needs for meeting the challenge of decentralized energy supply in developing countries. Energy Sustain Dev 15(3):324–329

Schäfer M, Kebir N, Philipp D (eds) (2013) Micro perspectives in decentralized energy supply. Conference Proceedings 2013

Schumacher EF (1973) Small is beautiful: economics as if people mattered. Blond & Briggs, London

Schweizer P (1996) Psychologische Faktoren bei der Nutzung regenerativer Energien: Eine Studie zum Einsatz von Solartechnik im zentralen Himalaya. Aktuelle Psychologische Forschung (Bd.19), Pabst, Lengerich

Stadler I et al (2002) Use of appliances in stand alone PV power supply systems. International Energy Agency, Photovoltaic Power Programme

Steckel J, Brecha RJ, Jakob M, Strefler J, Luderer G (2013) Development without energy? Assessing future scenarios of energy consumption in developing countries. Ecol Econ 90:53–67

Tillmanns A, Schweizer-Ries P (2011) Knowledge communication regarding solar home systems in Uganda: the consumers' perspective. Energy Sustain Dev 15(3):337–346

UNEP Collaborating Centre Frankfurt School of Finance and Management (2013) Global trends in renewable energy investment 2013, Frankfurt. Available (September 2013) http://www.fs-unep-centre.org

WHO (2007) WHO household energy database. Available: (June 2013) http://www.who.int/indoorair/health_impacts/he_database/en/

World Bank (2011) Household Cookstoves, environment, health and climate change: a new look at an old problem, Washington DC. Available (August 2013) http://climatechange.worldbank.org/climatechange/content/cookstoves-report

Yonazi E, Kelly T, Halewood N, Blackman C (eds) (2012) The transformational use of information and communication technologies in Africa. Available (September 2013) http://www.afdb.org/fileadmin/uploads/afdb/Documents/Publications/The%20Transformational%20Use%20of%20Information%20and%20Communication%20Technologies%20in%20Africa.pdf

Ybema JR, Battjes JJ, Jansen JC, Ormel FT (2000) Burden differentiation: GHG emissions, undercurrents and mitigation costs. Netherlands Energy Research Foundation ECN

Chapter 11
The Impacts of Climate Change on the Livelihoods of Coastal People in Bangladesh: A Sociological Study

Joydeb Garai

Abstract This paper is an attempt to assess the impacts of climate change on the livelihoods of coastal people in the south-western part of Bangladesh. This study has been conducted by using quantitative method with semi-structured interview questionnaire for data collection in purposive manner. The paper examines the impacts of climate change on agricultural productivities, food securities and institutional challenges that the coastal people face. This paper finds that the frequency and severity of natural disasters have increased in recent years, that threat on food securities by inundating low agriculture land, restricting economic activities, decreasing employment opportunities, expanding different health diseases, destructing houses, crops and other infrastructures in Bangladesh. The findings indicate that the most climate change induced vulnerable and risky people are women, children, elderly and disabled people as they cannot easily cope with the unfavorable environment during disasters. There has not been conducted much empirical research about the impacts of climate change, so policy makers can get comprehensive view about this concern by this study and implement policy for the survival of the climate change induced affected coastal people.

Keywords Climate change • Climate change impacts • Natural disasters • Livelihoods • Coping strategies • Coastal people

Introduction

Global warming and climate change are unequivocal which are agreed by all scientists around the world. Observing the global average air and ocean temperatures, scientists agree with that the world climate is gradually changing (IPCC 2007). Experts

J. Garai (✉)
Department of Sociology, University of Chittagong, Chittagong, 4331 Bangladesh
e-mail: garai.joydeb@gmail.com

W. Leal Filho et al. (eds.), *International Perspectives on Climate Change*, Climate Change Management, DOI: 10.1007/978-3-319-04489-7_11,

and Environmentalists from across the world collectively recognized the latest findings of the Inter-Governmental Panel on Climate Change (IPCC) that the average temperature might increase up to about 5.5 °C with respect to current mean values by 2100 in Asia (Cruz et al. 2007). The rising temperature is responsible for climate change and different kinds of natural calamities in the world. The UNDP has identified Bangladesh to be the most vulnerable country in terms of cyclone and the 6th most vulnerable country in terms of floods. The intensity of vulnerability is further extending by climate change. The country faces six key climate change-induced challenges i.e. cyclones, floods, droughts, riverbank erosions, erratic rain fall and salinity intrusion into the coastal land. Scientists have indicated that climate change is not causing anything new in Bangladesh but the frequency and severity of events are increasing day by day (Siddiqui et al. 2010). Currently, climate change possesses a new threat to the lives and livelihoods of the people in Bangladesh because over the last 10 years the frequency and intensity of natural hazards such as cyclones, floods, tidal surges etc. increased that brought unprecedented miserable situations to the people, especially the coastal people of Bangladesh.

The coast of Bangladesh is particularly vulnerable to sea level rise as 12 out of 19 coastal districts are directly exposed to the sea. The inland coast of Bangladesh has a population density of 1,200 persons/km^2, while the exposed coast has a density of 570 persons/km^2. The exposed coast is a hazardous zone in terms of frequent floods, cyclones and tidal surges (Akter 2009). As a result, the livelihoods of coastal people become vulnerable as they struggle with nature through the whole years. This work tends to shed light on the livelihoods of coastal people and their coping strategies during and after the natural hazards in Bangladesh.

Coastal Zone in Bangladesh

The coast of Bangladesh is known as a zone of vulnerabilities. It is prone to natural hazards like cyclone, tidal surge and flood. The combination of natural and manmade hazards, such as erosion, high arsenic contamination in ground water, water logging, earthquake, salinity in water and soil, various forms of pollution, risks from climate change etc. have adversely affected to the livelihoods of coastal people and slowed down the pace of social and economic developments in this territory (GOB 2005).

This study has been conducted in 8 villages of two unions named Burigoalini and Gabura under Shymnagar upazila of Satkhira district in the costal part of Bangladesh. Shymnagar upazila is bounded by Kaliganj and Assasuni upazila in the North, Sunderbans (Mangrove forest) and Bay of Bengal in the South, Koyra and Assasuni upazila in the East and West Bengal of India in the West (see Fig. 11.1). In this upazila about 65 % households depend on cropping livestock, forestry and fishery and 26.82 % are agricultural laborers. The overall literacy rate is about 28 % whereas the national average is 32 % (BBS 2006).

The logic behind choosing this area as study site is that it is one of the most vulnerable zones of natural disasters in Bangladesh. The type, intensity and frequency of natural disasters such as cyclone, tidal surge, soil salinity, and flood are common phenomena in this area. Moreover, being located adjacent to the Bay of Bengal and the Sundarbans people's vulnerabilities have increased to a great extent.

The Methodology

The study is quantitative in nature. For conducting this study a semi-structured interview questionnaire has been used to obtain quantitative data from eight villages in Shymnagar upazila of Satkhira districts in Bangladesh (Fig. 11.1 shows the location of the study area). With an aim to arrange this work in a representative manner, 400 samples were selected purposively from the study area. Each male and female (age 15–85) who lives in the selected area is the unit of the study. Statistical analysis used SPSS Windows program (version 17.5) to process the data in this work. Relevant information was also collected from secondary literatures including books, journals, annual reports, newspaper and magazines etc.

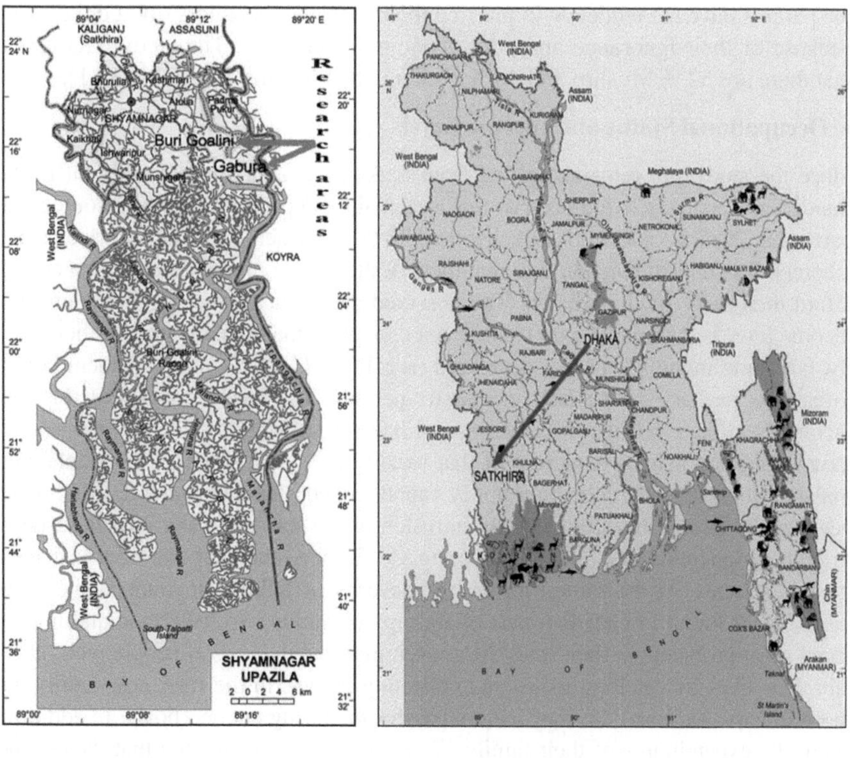

Fig. 11.1 Maps show the location of the study sites at Shymnagar upazila of Satkhira district in Bangladesh

Climate Change and its Impacts on the Livelihoods of Coastal People in the Context of Bangladesh

The major focus of this study is to explore the impacts of climate change on the livelihoods of coastal people. The study also delves into the climate change-induced risk and vulnerability of coastal people especially on women, children, elderly and disabled people.

- **Educational Condition**

There are 12 districts out of 19 coastal districts in Bangladesh are directly exposed to cyclone, storm surge and other natural hazards. Being one of the districts of the 19th coastal districts, the people of Shymnagar upazila in Satkhira always struggle with natural hazards; as a result, they can get limited scope to study much. Among the samples, only 40 % has completed primary education and 1 % accomplished post-graduate. The total illiteracy rate of that locality is 28 % (see Table 11.1).

- **Marital Status and Religion**

The observed data indicate that among the 400 samples, 79 % are male and total married persons are 86 %. In the study area it is found that a large proportion of the parents have the tendency to married their sons and daughters off at childhoods because of their ignorance and poor economic conditions. The study data reveal that there are 52 % Muslim and 48 % Hindu in that locality (see Table 11.1).

- **Occupational Status and Income Levels**

Since the study site remains remote from Satkhira upazila and are adjacent to the Sundarbans people's livelihoods of this area mainly depend on natural resources. In particular, 26 % people work as fishermen, 20 % as farmers and 21 % work as agro-laborers. It is unfortunate but true that most of the people of the selected area cannot afford their daily foods properly. The observed data indicate that 79 % respondents income level is maximum 3,000 taka is not enough to maintain the cost of their family, but some months it is also decreased (see Table 11.1). So many of them bound to eat less or starve, especially in disaster periods. The inherent cause of this low income is climate change, because natural hazards such as floods, tidal surges etc. have destructed natural resources of that locality to a great extent. The people who maintain their lives by collecting honey cannot collect that due to the destruction of beehives by natural hazards. Salinity intrusion in the soil and water affects severely to crop productions and livestock rearing, so farmers and agro-laborers lost their income sources. At the same time, due to severity of salinity in water, white spots diseases are found in GHER (place of shrimp cultivation), so fishermen and traders faced great problems in their lives. Moreover, many of the coastal people who maintain their lives by catching fishes and collecting prawns in the river now hardly get those in expected level. So they cannot pay for the hiring net and boats. In addition, about the expenditures of their family 67 % respondents mentioned that they spend maximum 3,000 taka for their livelihoods and 32 % spend maximum 5,000 taka per

Table 11.1 Socio-economic conditions of the respondents of coastal people

Socio-economic variables	N = 400	%
Education of the respondents		
Illiterate	110	27.5
Primary	159	39.8
Secondary	91	22.8
Higher Secondary	16	4.0
Graduate	19	4.6
Post Graduate	5	1.3
Sex of the respondents		
Male	314	78.5
Female	86	21.5
Marital status		
Married	343	85.7
Unmarried	47	11.8
Separated	2	0.5
Deserted	8	2
Religion of the respondents		
Hindu	191	47.8
Muslim	209	52.2
Occupations of the respondents		
Farmer	80	20.0
Teacher	6	1.5
Agro-laborer	82	20.5
Fisherman	104	26
Service holder	39	9.7
Student	16	4
Unemployment	6	1.5
Tailor	1	0.3
Businessman	10	2.5
Housewife	46	11.5
Elderly people	10	2.5
Income of the respondents (in BDT)		
0–3,000	317	79.3
3,001–5,000	58	14.5
5,001–7,000	7	1.8
7,001-above	5	1.3
Nothing	13	3.1
Expenditure of the respondents (in BDT)		
0–3,000	266	66.5
3,001–5,000	126	31.5
5,001-above	8	2.0

Source Field works

month (see Table 11.1). Among them a large proportion of the respondents asserted that they cannot earn enough money to support their family. So they borrow money from different sources and run their family deficiently through the year. However, most of the respondents said that before natural disasters their incomes were reasonably well but the devastating disasters have destroyed their working sources and decrease their incomes.

- **Restriction of Economic Activity**

Being an agro-based country the women of Bangladesh traditionally play a vital role to collect foods, fodders and fibers for their family that have an importance in the economy. Due to climate change, women face severe problems to do their traditional activities. Different natural hazards destruct natural resources i.e. forest creepers, roofs, tubes from the forest, honey, fuel woods, mineral water and so on, that are the collecting materials for women in the coastal area. In addition, a large proportion of women in the coastal area who collect fishes and prawns in the river cannot find that any more as natural hazard contributes to decline their productivities. Salinity intrusion around the home ground restricts women to do horticultures, livestock rearing and other economic activities. Women also cannot do agro-based processing activities such as harvesting crops, boiling paddy, drying cereals etc. during disaster periods. They also face problems in moving to the flooding kitchen garden to prepare foods and other household activities. All of these situations contribute to make women vulnerable economically and also psychologically. The study findings reveal that climate change limits women economic activities in disaster periods and 87 % respondents agreed with it (see Table 11.2). Moreover, 53 % respondents said that they cannot collect food, fodder and fiber for their livelihoods. Approximately 33 % respondents said that they cannot do horticulture and 12 % said that they cannot collect forest creepers, roofs, tubes from the forest. In addition, 21 % respondents mentioned that they cannot collect larvae and marshes from the rivers due to climate change that limit their economic activities in the family (see Table 11.3). At the same time, man cannot find job during and post disaster periods, as natural hazards destruct the sources of income in the coastal area.

- **Number of Natural Disasters Occurred During the Last Five Years**

From the prediction of environmentalists, it is found that the frequency and severity of natural disasters have increased to a great extent in recent year. Global warming has melted the ice of Himalaya and risen water in the sea which results in frequent coastal floods, salinity intrusions and other natural hazards in the country. The present research finds the same scenarios in the coastal areas of Shymnagar in Satkhira, as the environment being disrupted due to climate change causes to bring natural hazards in the locality. The survey data show that in the last five years 53 % respondents said that maximum 2 times, 39 % said that maximum 5 times and 8 % respondents mentioned that more than 5 times natural hazards occurred in the locality (see Table 11.2).

- **Nature of Disasters in the Coastal Area**

Bangladesh is a disaster prone country and her coastal people always struggle with natural hazards. The major causes of these natural hazards according to the respondents are deforestations, carbon emissions and other environmental

Table 11.2 Climate change and its impacts on the livelihoods of coastal people

Climate change affects coastal people's livelihoods	N = 400	%
Natural disasters have negative effect on family income		
Yes	340	85.0
No	60	15.0
Climate change limits woman's economic activity		
Yes	346	86.5
No	54	13.5
Number of natural disasters occurred during the last five years		
1–2 times	211	52.8
3–5 times	156	39.0
5-above	33	8.2
Nature of disasters occurred in the locality		
Flood	58	14.5
Tidal surge	45	11.2
Cyclone	214	53.5
Soil salinity	56	14
Other natural hazards	27	6.8
Degree of suffering after natural disasters		
Extreme	232	58.0
High	78	19.5
Moderate	88	22.0
Low	2	0.5
Degree of suffering before natural disasters		
Extreme	18	4.5
High	24	6.0
Moderate	172	43.0
Low	174	43.5
Nothing	12	3.0

Source Field works

pollutions in the country. The respondents of the study area claimed that they face different kinds of natural disasters in the locality as 54 % said that cyclone occurred in the study area. Significant number of respondents 15 % said that flood and 11 % claimed that tidal surge occurred in the area that contribute to create miserable situation on their living (see Table 11.2).

- **Natural Hazards and Migrations**

The findings of the present study indicate that the frequency and severity of natural disasters enforced people to migrate elsewhere and 37 % of the respondents agreed with it. The main causes of this migration of the affected people are ruined houses or factories, loss of trade, damage of agro-products, death or illness of family members, fear etc. due to the occurrence of natural hazards in that particular locality. This work also finds that the affected people losing their properties migrate to other districts in search of work. Some of them come back to their

Table 11.3 Natural disasters and its impacts on coastal people

	Multiple variables	%
Suffering from different kinds of diseases due to climate change		
Measeales	50	12.5
Malnutrition	56	14.0
Pneumonia	16	4.0
Diarrhea	305	76.3
Fever/Cold	177	44.3
Bacterial infection	28	7.0
Contagenious diseases	18	4.5
Jaundice	28	7.0
Dysentery	104	26.0
Accidental disability	64	16.0
Disaster and its impacts on coastal people's livelihoods		
Out migration	148	37.0
Ruined houses/factories	288	72.1
Loss of trade	45	11.3
Damage of agro products	234	58.5
Death/illness of family members	42	10.5
Fear	23	5.8
Damage of standing crops	155	38.8
Most vulnerable people during disaster periods		
Women	269	67.3
Children	185	46.3
Elderly people	173	43.3
Disabled people	54	13.5
Ways of limiting women's economic activities and facing vulnerability		
Cannot collect food, fodder and fiber	211	52.8
Cannot do horticulture around their houses	133	33.3
Cannot collect larvae and marshes from the river	84	21
Cannot collect fishes and prawns	60	15
Challenging in moving to flooding kitchen garden	149	37.3
Cannot collect forest creepers, roofs, tubes, and other things	49	12.3
No scope in agro-based processing activities	80	20

Source Field works

locality after some months back, in fact many of them do not return to their locality for the fear of frequent natural hazards (see Table 11.3).

- **Climate Change Induced Risk and Vulnerability for Different Group of People**

In the study area, it is found that women, children, elderly and physically challenged people are the most vulnerable during natural disasters. Approximately 67 % respondents claimed that women are the most vulnerable during natural disaster because of their household activities they cannot get information easily about the upcoming natural hazards and face problems. In addition, they have to save their homestead resources i.e. livestock and other properties during disaster periods, as a result they become victim of natural hazards. Moreover, they face problems to save their children and also face problems in the current of water by getting struck their cloths (sari) with sticks or other materials in the water. A significant number of respondents (46 % and 43 %) mentioned that children and elderly people are vulnerable during natural hazards respectively because they cannot easily cope with the hazard situations for their physical weakness (see Table 11.3).

- **Health Problems and Access of Medical Facilities to the Coastal People**

Due to climate change the frequency and intensity of natural hazards increased in the coastal area, as a result people easily affected from various kinds of diseases in the locality. Furthermore, because of the intrusion of saline water to the ground water, people face severe scarcity of pure drinking water and so they bound to drink impure water that also liable for spreading various kinds of diseases in the locality. In addition, due to low incomes many of the people cannot afford nutritious foods for their family, so they suffer from malnutrition for long time. The observed data indicate with 76 % responses that they suffer from diarrhea and a significant number of respondents (44 %) said that they suffer from fever as well as cold. On the other hand, 16 % mentioned that they suffer from accidental disability (see Table 11.3).

The study findings reveal that the people of the study area faced institutional challenges in terms of hospital, clinic and other medical facilities because there is no hospital or clinic in their localities especially in the area of Gabura union. They have to cross the Chunu river and come to Shymnagar sadar for medical treatment. To cross the river for medical purposes and others they have to be the victims of great problems especially in case of emergency patients and pregnant women. They also claimed that they could not get boats all the time to cross the river. About the degree of suffering of diseases due to climate change before natural disasters, only 5 % respondents said that they suffer extremely and 6 % said that they suffer highly. To share about the degree of suffering after natural disasters 58 % said that they suffer extremely and 20 % said that they suffer highly. So the findings of this study reveal that the degree of suffering of various diseases increase after natural disasters of that locality (see Table 11.2).

- **Problems of Crop Productions and Fish Cultivations**

Agricultural productivity has declined to a great extent in the coastal area of Bangladesh due to climate change. The inherent cause of this problem is saline water intrudes into farm land and hampers usual growth of crops. Moreover, in saline-intruded farmland livestock cannot grow properly as it suffers from different diseases. At the same time, flooding water in the coastal area inundates farmlands that refrain farmers from crop cultivation causing a threat for food securities of coastal people as 43 % respondents mentioned (see Table 11.4). Homestead forests, horticultures, livestock also cannot find suitable environment for its growth in the flooding situation. Agriculture and fish cultivations are the main occupations in the coastal area. Besides the agricultural problems they also face problems in fish

Table 11.4 Climate change and food insecurity of the coastal people

Climate change affects agricultural productivities	N = 400	%
Problems of crop productivities		
Yes	383	95.7
No	17	4.3
Factors affecting agricultural productivities		
Soil erosion	22	5.5
Drought	13	3.3
Flood	171	42.6
Soil salinity	116	29.0
Cyclone	43	10.8
Tidal surge	23	5.8
Sand cost	12	3.0
Perception of food consumption and food insecurity		
Deficit of food through whole year	152	38.0
Seasonal deficit especially dry lean periods	126	31.5
During disaster periods	83	20.8
Neither deficit nor surplus	33	8.2
Surplus	6	1.5
Access of food immediately after natural disasters		
Yes	124	31.0
No	276	69.0
Access of pure drinking water after natural disasters		
Deep tube well water	49	12.2
Pond water	312	78
Rain water	39	9.8

Source Field works

cultivation because of the salinity in water and soil which always create various kinds of diseases such as white spots disease leads to hamper fish cultivation especially shrimp cultivation.

- **Access to Pure Drinking Water and Sanitation System**

The study area located adjacent to the Sundarbans. The water condition of this area is salty and people are facing problem to drink salty water. The deep tube well as well as the pond water is also affected by salinity. So the people of this locality use pond water by using **Pond Sand Filter (PSF)** method as 78 % respondents claimed this. On the other hand, 10 % respondents use rain water by preserving it in rainy season. Before natural hazards such as tidal surge, cyclone etc. sanitation system was in satisfactory state but after the hazards, it was destroyed and people face problem to use their sanitation system (see Table 11.4).

- **Access to Foods, Livestock and other Natural Resources**

The coastal people of the study area frequently face acute problem of food scarcity during and immediately after the disaster. Due to poor communication and management system they cannot get access to food or other commodities in this period. Moreover, the prices of daily foods hike swiftly during this period, so poor people are not possible to afford their foods from the market and bound to starve frequently. Many of the people suffer from food deficits through whole year because; they cannot overcome or cope with natural hazards easily.

In the coastal area, natural disaster frequently swept away the livestock such as hens, ducks, cattle and at the same time, homestead crops and trees are destroyed which are found to contribute economical vulnerability to the people during disaster. The study indicates with only 31 % responses, they have access to food after natural disaster and 69 % said that they do not have access to food immediately after natural disaster. However, 38 % respondents have food deficit through whole year and 32 % respondents have seasonal food deficit in their family, especially in dry lean periods (see Table 11.4).

- **Coping Mechanisms and Sustainable Livelihoods in Coastal Area of Bangladesh**

Tyndall Center for Climate Change Research (2004) has provided two approaches namely Institutional Approach to Livelihood Adaptation and Adaptation to Livelihoods Context which emphasize on institutional and organizational support, generic knowledge and collective action of the people for adaptation. In the approach taking collective action, understanding the use of endowment, applying generic knowledge sustainable livelihood can be brought in the study area. In the study area this approach is rarely found as many of the respondents cannot get sufficient supports from the institutions and organizations during the disaster periods. The findings indicate that only 16 % respondents receive relief as support for coping with the situation during disasters (see Table 11.5). They receive foods, house making materials, medicines, cash money etc. as relief for coping with the situations.

As a coping strategy of the coastal people approximately 28 % respondents borrow money from different NGOs to overcome the losses incurred by natural disasters. In addition, approximately 17 % respondents sell livestock and 12 % respondents borrow money from local money lenders with high interest to overcome the losses after disasters. Moreover, the adaptation system of the locality is not good enough as 71 % respondents said that they do not have any training on adaptation strategies. Only 29 % of them receive training for adaptation of that particular locality.

Sustainable livelihood is important in coastal area but it is not prevailed in the study area as 66 % respondents claimed it. However, to make sustainable livelihoods in the coastal area people should emphasize on saving environment from destructions as 40 % respondents said this. Significant number of respondents (20 %) said that by receiving training about disaster management and 12 % said that by utilizing the existing resources, sustainable livelihoods can be brought in the study area (see Table 11.5).

In order to adopting sustainable livelihoods by minimizing various kinds of shocks and stress and reducing social, economic and natural losses, there is need

Table 11.5 Coping mechanisms and sustainable livelihoods of coastal people in Bangladesh

Coping mechanisms of coastal people	N = 400	%
Coping strategies of coastal people		
Borrowing money from local moneylender with high interest	47	11.8
Borrowing money from NGOs	110	27.5
Borrowing money from relatives	70	17.5
Selling livestock	66	16.5
Selling properties	42	10.4
Relying on relief	65	16.3
Types of relief get coastal people from different organizations		
Foods	251	62.8
House making materials	72	18
Medicines	44	11
Cash money	16	4
Others	17	4.2
Having sustainability to the livelihoods of coastal people		
Yes	136	34.0
No	264	66.0
Having training on adaptation		
Yes	116	29
No	284	71
Ways of bringing sustainability to the lives of coastal people		
Being aware about responsibilities	54	13.5
Receiving training about disaster management	81	20.2
Developing infrastructures in the locality	58	14.5
Saving environment from destructions	159	39.8
Utilizing the existing resources	48	12

Source Field works

to incorporate some assets such as natural capital i.e. (land, water, wildlife etc.), physical capital i.e. (water, sanitation, energy, transportation, communication etc.), human capital i.e. (knowledge, skill, information, ability to labor etc.), social capital i.e. (relationship of trust, membership of groups, network etc.,) and financial resources i.e. (regular remittances or pensions saving etc.), (Quoted by Tyndall 2004 developed by DFID). By incorporating these assets along with being aware, receiving training and changing structure sustainable livelihoods can be brought in the disaster prone coastal area of Bangladesh.

Conclusions

Bangladesh is considered one of the most vulnerable countries of the world. The geographic location i.e. Bay of Bengal to the south and Himalayas to the north, this two different environments along with gradual climate change make Bangladesh vulnerable. This work aims at to find out the impacts of climate change on the livelihoods of coastal people. The important finding of the work is that climate change affects severely to the lives and livelihoods of coastal people in terms of agricultural productivities, food securities and people's vulnerabilities. Due to climate change, floods, tidal surges and cyclones have become common phenomena for the life of coastal people and to minimize the loss and cope with the situations coastal people use their own strategies. However, the work indicates that to cope with the hazard situations coastal people should give emphasize on taking disaster management training, being aware about their responsibilities and utilizing existing resources, so that there can be brought sustainable livelihoods in the coastal area of Bangladesh.

Acknowledgments I would like to acknowledge my cordial gratitude and deep indebtedness to my esteemed teacher Professor Dr. Khondokar Mokaddem Hossain who supervised me throughout this path breaking research by his expertise, scholasticism and valuable suggestions. I also take this opportunity to express my indebtedness to my respected teacher Professor Dr. Mahbuba Nasreen for her valuable suggestions in this work. My special thank goes to the Department of Sociology, University of Chittagong for providing me logistic supports to writing the paper. My special thanks also go out to my friends, colleagues and respondents whose worthwhile supports help me to conduct the study smoothly. Finally, I would like to express my heartiest profound appreciation and gratefulness to my beloved parents who are my continuous source of inspiration in the successful accomplishment of this work.

References

Akter T (2009) Climate change and flow of environmental displacement in Bangladesh. Unnayan Onneshan, Dhaka, Bangladesh

BBS (2006) Population census wing. Bangladesh Archived from the original on 27 March 2005. http://web.archive.org/web/20050327072826/, http://www.bangladeshgov.org/mop/ndb/arpc91_v1/tables04.htm

Cruz RV, Harasawa H, Lal M, Wu S, Anokhin Y, Punsalmaa B, Honda Y, Jafari M, Li C, Ninh NH (2007) Climate change 2007: impacts, adaptation and vulnerability. In: Parry ML, Canziani OF, Palutikof JP, van der Linden PJ, Hanson CE (eds) Contribution of working group ii to the fourth assessment report of the intergovernmental panel on climate change. Cambridge University Press, Cambridge

Government of Bangladesh (2005) Coastal zone policy. Dhaka, Bangladesh

IPCC (2007) Climate change, 2007: the physical science basic. Contribution of working group I to the fourth assessment report of the intergovernmental panel on climate change, Switzerland

Siddiqui T, Abrar CR, Alamgir M, Ali MS (2010) Climate change and community adaptation: case study from Bangladesh. Monash Sustainable Institute, Monash University, Australia

Tyndall (2004) Adaptive, adaptation to climate change in vulnerable environment. A Tyndall Center for Climate Change Research Project at the University of Sheffield, UK

Chapter 12
From Rainforests to Drylands: Comparing Family Farmers' Perceptions of Climate Change in Three Brazilian Biomes

Gabriela Litre, Stéphanie Nasuti, Catherine A. Gucciardi Garcez, Diego Pereira Lindoso, Flávio Eiró, Jane Simoni, Carolina Joana da Silva and Cristiane Lima Façanha

Abstract Risk perceptions influence individual and collective adaptive responses to climate hazards. Up to now, the majority of literature addressing climate change perception and adaptation has been location-specific. Such an approach is limited with respect to the construction of a generalized theory around why and how people perceive and act towards climate change risks. This chapter seeks to contribute to overcome this limitation by offering a cross-sectional study of

G. Litre (✉) · S. Nasuti · C. A. Gucciardi Garcez · D. P. Lindoso · J. Simoni
Centre for Sustainable Development, University of Brasilia, Campus Darcy Ribeiro, Brasília, Brazil
e-mail: gabrielalitre@yahoo.com

S. Nasuti
e-mail: steph.nasuti@gmail.com

C. A. Gucciardi Garcez
e-mail: catigucciardi@hotmail.com

D. P. Lindoso
e-mail: diegoplindoso@gmail.com

J. Simoni
e-mail: jane.simoni@gmail.com

F. Eiró
Centre Maurice Halbwachs, École des Hautes Études en Sciences Sociales, Paris, France
e-mail: flavioeiro@gmail.com

C. J. da Silva
Centro de Pesquisa de Limnologia, Biodiversidade e Etnobiologia do Pantanal, Universidade do Estado de Mato Grosso, Cáceres, Brazil
e-mail: ecopanta@terra.com.br

C. L. Façanha
Rede Bionorte, Universidade Federal do Amazonas (UFAM), Manaus, Brazil
e-mail: crisfacanha@gmail.com

W. Leal Filho et al. (eds.), *International Perspectives on Climate Change*,
Climate Change Management, DOI: 10.1007/978-3-319-04489-7_12,

climate change risks perceptions among smallholder farmers settled in three contrasting biomes in Brazil: the Amazon (rainforest); the Caatinga (semi-arid); and the Cerrado (savanna). By articulating regional, local and micro scales of comparison, common traits in the perception of climate variability are identified. It is not intended, at this stage, to validate particular theories of climate change, but rather to contribute to a better understanding of climate change as a trans-regional and socially embedded environmental phenomenon. This study shows that, in spite of existing perceptive barriers, smallholders settled in dramatically different contexts share perceptions about risks linked to the following phenomena: (i) changes in the timing of seasons, (ii) decrease in rainfall levels; (iii) temperature rises. Moreover, there are specific adaptation strategies to climate change, like the timing of seeding, which appear to be addressed independently by smallholders of the three biomes. Public policies intended to support adaptive measures and the increase of food security must take subjective risk perception into account within the cultural and environmental contexts of the actors involved.

Keywords Smallholders • Climate change perceptions • Adaptation • Food security • Amazon • Caatinga • Cerrado

Introduction

Within the context of the growing challenges and opportunities presented by global change, there is an urgent need to reduce the vulnerability and enhance the adaptive capacity of communities that will potentially be most affected, thereby protecting their right to develop in a sustainable way (Adger 2003; Bursztyn 2008).

In agriculture, there is a need for enhancing adaptive capacity (Howden et al. 2007) in order to improve the prospects of food security (FAO 2012a). This can be made possible through the integration of climate change with other risk factors, such as market shocks, climate variability and overcoming social barriers to adaptation (Moser and Ekstrom 2010; Adger et al. 2009a).

Latin America and the Caribbean are expected to experience an increase in the frequency of droughts, as well as desertification and soil salinization (IPCC 2007). An increase of pests and diseases in crops and a greater competition for irrigation water is also expected, which will lead to an increase in the cost of agricultural production is estimated. In this vulnerable scenario, it is expected that poverty will increase, especially in rural areas (Kaztman 2005).

Brazil's territory contains approximately 30 % of the world's tropical forests (FAO 2012b). Over half of the country is covered by native forests. The Amazon, with approximately 3.5 million km^2, is the most extensive of the predominantly forest biomes in Brazil, while the Cerrado (savanna), in central Brazil, occupies about 2 million km^2 (Verburg et al. 2014, IBGE 2012a, b). The Caatinga is a semi-arid region, and in turn occupies about 850,000 km^2 (approximately 10 % of the country).

It is the only, exclusively Brazilian biome and is considered to be the most fragile in country.

For centuries, smallholders have been adapting to the seasonal climatic variability characteristic of the Amazon, Cerrado and Caatinga ecosystems (Pinedo-Vasquez and Padoch 2009; Lindoso et al. 2011; Simoni 2010). However, in recent years, traditional peoples have to face evidence of changes in the climate, bringing unpredictability and changing the seasonal, routine activities. The most obvious manifestation of this phenomenon is the proliferation of extreme weather events: the winter of 2012 was marked by severe drought in the Caatinga and record flooding in the Amazon, events that came in the wake of a memorable flood in 2009 and two major droughts in 2005 and 2010 (Marengo et al. 2011).

For biophysical systems, the impacts of climate change are reflected directly in changes in productivity, quality, or population figures. However, in social systems, the impact can be measured in a gain or loss of income, illness, mortality or other measures of social welfare (Parry and Carter 1998). It is necessary to understand the complexity of the interactions between smallholders[1] and their production methods, ecosystems, and also with regards to external influences, notably linked to public policies. Against a backdrop of vulnerabilities and threats posed by climate change, there is the need for new alternatives for regional development, translated into opportunities for income generation and improved quality of life through the resilience and adaptive capacity of socio-productive systems. Among those alternatives for sustainable development, understanding how risk perceptions influence individual and collective adaptive responses to climate hazards and support to mitigation public policies is vital (Lorenzoni et al. 2007). So far, the significant volume of literature examining people's understanding of climate change demonstrates de importance of individuals' perceptions about climate change-related risks (Wolf and Moser 2011). However, scholars suggest that current levels of awareness of and knowledge about climate change are insufficient in leading to effective behavioural change, mainly due to adaptation barriers (Wolf and Moser 2011). An analysis of adaption barriers point out the fragilities and entry points for adaptive adaptation (Ford et al. 2010). Adaptive barriers can vary in its nature: environmental, technological, informational, economical, physical and social (Adger et al. 2009b).

Among those alternatives for sustainable development, an understanding how risk perceptions influence individual and collective adaptive responses to climate hazards as well as support mitigation policies is vital (Lorenzoni et al. 2007). So far, the significant volume of literature examining people's understanding of climate change demonstrates the importance of individuals' perceptions about climate change-related risks (Wolf and Moser 2011). However, scholars suggest that current levels of awareness of and knowledge about climate change are insufficient in leading to effective behavioural change, mainly due to adaptation barriers (Wolf and Moser 2011). An analysis of adaption barriers points out both the fragilities

[1] In this article the term smallholders will be used as a synonym of small-scale producers and family farmers. The definition that has been adopted will be presented in the Methodology.

and entry points for adaptive adaptation (Ford et al. 2010). Adaptive barriers can vary in their nature, such as; environmental, technological, informational, economical, physical and social (Adger et al. 2009b).

These barriers constitute an important set of limitations that make adaption difficult, or, in other words, they represent the reverse side of adaptive capacity. Both are equally important in vulnerability assessment, depending on the analysis' objectives. Ultimately, the decision to adapt is individual and based on a set of subjective factors, such as risk perception, but also on other factors such as age, gender, personal preferences, social status (Di Falco et al. 2011; Cavatassi et al. 2011).

For this study, smallholders' perceptions of climate change have been identified through the application of standardized surveys combining closed and open questions in selected municipalities and communities in each of the three biomes. Smallholder farmers in Brazil are an example of the urgent need to integrate an analysis of climate and socioeconomic risks and overcome barriers to adapting to global changes. They suffer in a particular way from a double exposure to climate change and the dramatic swings in markets and institutions (Leichenko and O'Brien 2008; Leichenko et al. 2010). This vulnerability principally affects the food security of these communities, with increased risks related to agricultural production.

This study responded to two main objectives:

1. Discovering if Brazilian smallholder farmers share perceptions of climate change, thereby extending beyond their diverse environmental and cultural contexts.
2. Identifying specific adaptation strategies to climate change that appear to be addressed by Brazilian smallholder farmers in the three contrasting biomes.

The present research was conducted by the sub-network on Climatic Change and Regional Development of the Brazilian Network for Global Climate Change Research (Rede CLIMA), a multi-year, multi-sited project examining rural vulnerability to climate change and adaptation strategies among smallholder farmers. Since 2009, the sub-network has conducted research in four biomes: Amazonia, Cerrado, Caatinga and Pantanal.

Literature Review

Adaptation to climate change has been addressed by various scientific approaches, with the goal of understanding how human systems are affected and respond to environmental disturbances. However, within the academic research agenda, the vulnerability approach stands out. Its roots go back to the research of the 1930s' in Geography on natural hazards, in particular to the risk-hazard school of the U.S.A. (Marandola and Hogan 2004). Risk-hazard approach addresses risk as the probability of impact about a particular natural hazard (Cardona 2004). In early literature, risk and impact were considered as a result of external factors (i.e. the magnitude of the natural event). Human systems were relatively passive exposure units, with limited adaptive capacity.

Later, in the 1960s and 1970s, geographers realized that internal, human factors are as important as the environmental determinants in building risk (Alcántara-Ayala 2002). In the following decades, economists, sociologists, anthropologists, political scientists and ecologists interacted with risk-hazard geographers to give birth to the vulnerability approach (Eakin and Luers 2006; Gallopín 2006). Socioeconomic, cultural and political-institutional aspects became key factors of vulnerability analysis. At the same time (1980s), global environmental changes were emerging in international political agenda as one of the greatest challenges in 21st century, such as climate change (Nordhaus 1975). The vulnerability agenda incorporated climate change as pertinent issue of study (Timmerman 1981; Bogard 1988; Dow 1992; Kates 1985; Susman et al. 1984), which was enhanced with the establishment of the IPCC and UNFCCC.

In early climate change vulnerability and adaption research, the focuses were two: impact-driven scenarios and the establishment of a conceptual-theoretical framework. Recently, the implementation of adaptation policy agenda increased the need for scientific support on local-based vulnerability and adaptation process (Berrang-Ford et al. 2011). Some of the questions that are posed: How do people respond to climate impact? What are the key determinants of vulnerability? What triggers the adaption response? In addition, what delays or inhibits it? Risk perception underlies the answers to all this questions.

Vulnerability can be defined as the degree of susceptibility to harm or the incapacity to avoid or cope with impacts (Smit et al. 2000). Adaptation is the process or feature that reduces the degree of susceptibility or improves the adaptive capacity; it is also the ability to take advantage of opportunities that come with climate change (Brooks 2003; Barnett 2010) and is frequently associated with risk management (Howden et al. 2007).

Risk perception is at the crux of the matter, once it reflects how a hazard is perceived. If it is not perceived, it is not an issue. If it is not an issue, no response is undertaken. Thus, risk perception is a subjective dimension of vulnerability. It is the aspect that links a hazard risk and adaptive action. Understanding how perception is built and how it determines institutional and individual adaptation is a highly relevant issue in climate change research. Leichenko and O'Brien (2008) contend that although social and natural scientists have developed frameworks to measure vulnerability, resilience, and sustainability, these analyses do not allow for integrated approaches to global environmental and economic change.

This approach is shared by most risk sociologists, to whom the environmental and technological risks of modernity are responsible for a global wide risk perception that shapes individual behaviours (Beck 1995; Giddens 1994). Beck admits that he is more interested in cultural perceptions and definitions of risk, and not in their reality, from a perspective that fits in environmental constructivism (Lidskog 2001; Smith 2001). Following Beck's approach, this chapter takes into account individual perceptions about climate change related risks that do not need to match, necessarily, real climatological data. According to Beck, risks are real if they are experienced as such (Beck 2010, p. 95). In other words, adaptation strategies depend on how risks are perceived (or not), which leads us to the following issue: adaptive barriers.

Climate Change Communication

Effective adaptation and mitigation strategies have become crucial to securing livelihoods and community development, especially in developing countries. A critical element in promoting effective and successful adaptation and mitigation strategies among family farmers is communication (Harvey et al. 2012). Effective communication among stakeholders can help to identify problems, raise awareness, encourage dialogue, and influence behavioral change (Moser 2010; Nerlich et al. 2010). However, in order to communicate effectively on climate change and climate change adaptation and mitigation, an understanding how differently situated individuals and communities perceive, interpret, and discuss climate change drivers and impacts is needed (Harvey et al. 2012; Africa Talks Climate 2010).

We define climate change communication as mediated information (through the media, institutions, etc.). The challenges for effectively communicating climate change related risk in ways that are appropriate to specific audiences are enhanced by the complex and uncertain nature of climate change, as well as its timescale, which is often well beyond that which shapes decision making at local levels. In this sense, is vital to climate change risk communication to be aware of its audience, especially through the understanding of social differentiation (including age, gender, educational level, contexts, livelihoods, social status, spoken language, etc.) and of the importance of time and space scales, since localized contexts are key to successful climate change communication (Harvey et al. 2012). Scholars have recently adopted approaches aimed at better understanding how to engage people at an effective, emotional level (Harvey et al. 2012). The local knowledge of some audiences provides a framework to explain the relationships between particular climatic events and livelihood activities, such as farming (Kalanda-Joshua et al. 2011).

Communication is vital to enhancing adaptive capacity. Adaptive capacity is the set of human, financial, institutional and environmental resources that allows for adaptation. Such a perspective allows one to diagnosis how "healthy" the adaptive context is. Perceptions of climate and socioeconomic risks integrate the first of three phases of the adaptation process: (i) understanding of risk through communication, (ii) adaptation planning, and (iii) management of adaptation (Moser and Ekstrom 2010). Perceptions contribute to the identification of the problem and the most useful adaptation options (Moser and Ekstrom 2010; Maddison 2006; Gbetibouo 2009). However, adaptation barriers often impose difficulties for successful climate change communication strategies.

Methodology

So far, most studies on climate change perceptions have relied heavily on summarizing and comparing findings of previous surveys (Bord et al. 1998; Lorenzoni and Pidgeon 2006; Brechin and Bhandari 2011) Just a few have conducted standardized data collection with locally-derived and/or locally-relevant indicators across diverse contexts (Crona et al. 2013). This study aims at contributing to overcoming those limitations by comparing the perceptions of climate change by smallholder farmers settled in three dramatically different Brazilian biomes: the Amazon, Caatinga and Cerrado, as shown in Fig. 12.1. Surveys were performed at the level of the rural household.

Smallholder farming is defined by the Brazilian Law[2] 11.326 (2006), which takes into account factors such as the size of the rural property, the importance of the production in the household income and the proportion of family labour within the property activities and management. Using the official definition is appropriate not only for instrumental purposes, but also for its theoretical relevance. This concept has been debated and changed over the past decades, progressively incorporating traditional ways of life and land use, satisfactorily responding to the sociological need of delimitation of this segment of society. Indeed, this definition of smallholders is particularly suited for inter-biomes comparative studies, as it includes not only farmers, but also traditional forest extractivists or fishermen.

In this specific case, respondents offered their personal recollection of past and present climatic events that marked their livelihoods. Seeking a greater external validity, data collection at the household level was performed through the application of surveys in the three targeted biomes (Fig. 12.1). Survey questionnaires were divided into seven sections: (i) personal information about respondents, such as gender, age, level of study, residential history, life time in the biome (ii) productive activity, (iii) economical characteristics, (iv) social participation, (v) political and institutional characteristics, (vi) environmental characteristics, and (vii) climatic characteristics.

Fieldwork was conducted in collaboration with partner universities from the targeted regions. In order to generate consistency and to reduce the bias in the interpretation of the questions, team leaders from the Centre for Sustainable Development from the University of Brasilia trained the interview team about how to formulate questions.

In each rural community, members from 10 to 20 households (one per household, mostly men) were interviewed. Interviewees were chosen randomly by visiting one every each *n* households, n going from 1 to 3 according to the size of the community. For data processing and statistical analysis, statistical software SPHINX IQ was used. Due to the abovementioned reasons, the research sample cannot be considered as fully probabilistic.

The first fieldwork was conducted in the state of Acre in July 2010, a year of exceptional drought for the Amazon. The second case study, in Bahia, was performed during the months of June and July in 2011. 2011 was a year of relatively generous

[2] According to the criteria established in Brazil by The Smallholder Farming Act (No. 11,326, of July 24, 2006), the (a) smallholder (family farmer) or family entrepreneur must meet both requirements: (a) "does not hold in any capacity, area greater than four (4) fiscal units" (b) "use mostly labour of his or her own family in the economic activities of the establishment or undertaking" (c) "household income has predominantly originated from economic activities linked to the establishment or enterprise itself" (d) "manage his or her own establishment or undertaking with his or her family."

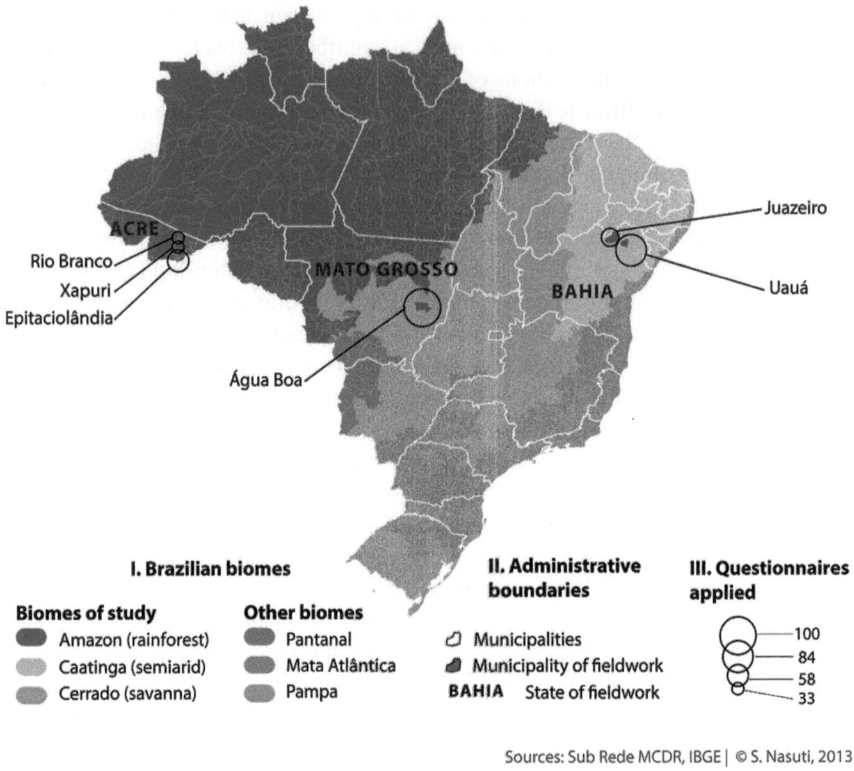

Fig. 12.1 Locations and sample sizes of the case studies

rainfall levels in the region, followed by 2 years of great drought (2012 and 2013). The third case was performed in the Cerrado biome of Mato Grosso in April 2012.

Results

Survey results allowed to identify at least three common perceptions of climate change associated risks by smallholders across the Amazon, Caatinga and Cerrado biomes. Those perceptions can be summarized as follows:

1. Observed changes in the timing of seasons
2. A decrease in rainfall levels
3. Temperature rises

Each region has as specific insights but in the three biomes, the global perception of environmental changes followed similar trends (Fig. 12.2):

- 87.5 % perceived changes in the rainy season;
- 52.4 % noticed changes in the phenology of local plants;

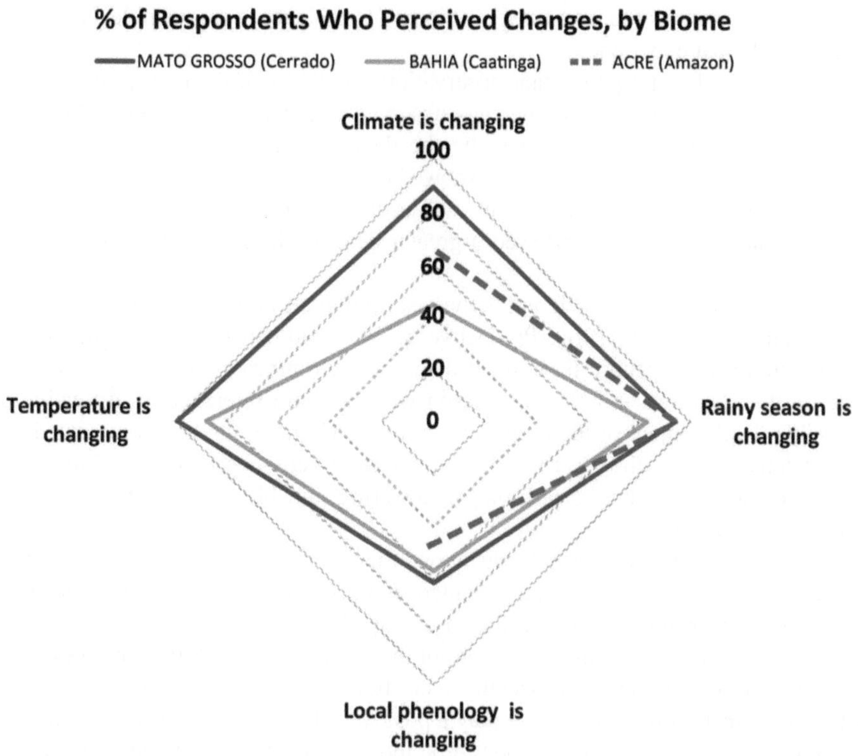

Fig. 12.2 Percentage of respondents who perceived changes, by Biome

- 71 % reported changes in the behaviour and quantity of wild animals;
- 58.4 % are reckoning that the climate is not the same than it used to be. The majority (66.1 %) feels it is worsening. A correspondence analysis showed that this perception is shared independently of the age group or the level of education.

Before detailing the reported perceptions, a profile of the interviewed smallholders will be presented.

Smallholders: Sample Characterization

More than half of the respondents (65.5 %) in the three biomes were men. Wives frequently participated in the interview, mostly as observers. The authors of this study did not seek to analyse, at this stage, the link between climate change perception and gender, though it is admitted that gender can be a key factor when trying to understand people's ideas about climate change (Crona et al. 2013). However, other factors have been relevant when analysing climate change risk perceptions, such as the interviewee's biome of residence, age group, level of education and local experience.

Survey results and participant observation revealed that men spend more time working on the fields, as Brumer has observed in her studies (Brumer 2008). It also showed that men continue to be the decision-makers in most of the rural establishments in Brazil. Surveys showed that men's involvement with agro-extractivist activities provided them with a closer contact with the natural environment and, accordingly, allowed a greater deal of climate change observations among them.

Respondents perform different activities, despite all falling under the definition of smallholders (Table 12.1). Productive activities influenced the perceptions, as each activity is a specific basis for evaluating environmental changes. For example, in the Caatinga, the planting season was the main reference of evaluating a possible displacement of the rainy season. In the Cerrado, farmers commented on other activities not directly related to agricultural activities to speak of the time displacement rain, such as phenology, flowering and blooming of fruit such as mango and cashew.

Despite the lack of formal education, the majority of the respondents had already been sensitised to climate change. Expressions such as "climate change" and "global warming" were familiar to at least 50 % (in the Caatinga) and 70 % (in the Cerrado and the Amazon region) of the respondents in all the 3 biomes.

Age and social capital (especially education and participation in groups and other local institutions) seemed to play a positive role in climate change communication, since these concepts seemed to be more familiar to people under 62 years of age having higher levels of formal education and links to local institutions.

The state of Bahia (Caatinga biome) has the lowest education levels among family farmers; one in three (33.1 %) respondents informed that they had received no formal education at all and 51.9 % informed that they attended school for less than 4 years. The term climate change was still unknown to almost half of family farmers, with 48.4 % admitting never to have heard about it. Interestingly enough, another concept: global warming, seemed to be much more familiar, with 6 out of 10 respondents from the Caatinga indicating that they knew the term.

In the Cerrado biome, with a slightly higher educational level (though 19 % is illiterate, 25 % finished primary school and 14 % finished secondary school), 75 % of respondents said to be aware of climate change and global warming. The same high level of awareness was identified in the Amazonia (where 70 % of the respondents said they were familiar with the terms climate change and global warming, even though educational levels remained low, with 23.4 % of respondents declaring themselves illiterate and 44.4 % as not having finished primary school).

Regarding the sources of information, respondents from all the three biomes mentioned multimedia (TV, radio and Internet) as their main source of information. This is due to the recent access to electricity brought by a state programme for rural electrification ("Luz para todos"). Multimedia seemed to be far more important to climate services provision than civil society representatives (like

Table 12.1 Sample distribution and institutional interviews by case study

Biome	State	Climate description	Smallholder systems	Period of survey
Amazon	Acre	Humid tropics: a maximum average temperature of 33 °C, an annual average of 24 °C, and a lower average of 17 °C. Total annual rainfall varies between 1,500 and 2,800/3,000 mm, (INMET 2012); rainy season (October–April)	Rural activities are a combination of agro-extractivism and small ranching • 66 % practice subsistence agriculture (cassava, maize and beans) • 68,5 % raise animals, mainly chickens and pigs, small cattle r • 32 % practice extractivism (rubber/latex and Brazil nut) • 8 % are fishermen • 15 % regularly hunt for subsistence • 62 % don't receive technical assistance • 75 % receive some kind of external income (mainly retiree pensions and social benefits)	July 2010
Caatinga	Bahia	Semiarid region: average annual precipitation: 400 mm. Average temperature: 24.2 °C. High rain variability	• 70 % practice rain-fed agriculture. Subsistence crop production (beans, corn, chicken, cassava) are mainly for animal feed and human consumption	June and July 2011
		Dry season (April–October) and raining season (November–March)	• Sheep/goats herds are the main economic activity: 72 % raise sheep/goats, 24 % raise cattle • 55,2 % don't receive technical assistance • 87 % receive some kind of external income (mainly retiree pensions and social benefits). Of them, 40 % considers that external income is higher than income from the internal	
Cerrado	Mato Grosso	Savanna: annual, average temperature 24 °C, maximum temperature 27.9 °C and minimum 14 °C; annual rainfall is approximately 1,600 mm	• 98 % declared cattle ranching as its principal activity; especially beef cattle (median of 45 animal per rural establishment) • 71 % practice rain fed agriculture. Main products are sugar cane, corn, cassava and pasture • 86 % don't receive technical assistance • 83 % receive some kind of external income (mainly retiree pensions and public job)	April 2012

Source Field surveys performed by the sub-network on Climate Change and Regional Development of Rede CLIMA

NGOs). In the case of the Caatinga, multimedia was mentioned by 7 out of 10 farmers (72.8 %) while with civil society, it was less relevant;1 out 10 farmers (9.2 %) public institutions (such as schools and state agencies) by less than 5 % of the sample (4.4 %) and the Church by 3 out of 10 respondents (2.8 %).

In the Cerrado biome, where 50 % of respondents from Mato Grosso said they had heard about climate change through multimedia, it was interesting to see the poor role played by civil society in communicating climate change. A striking 75 % of respondents said they had participated of a civil society group at some moment of their lives, although most of them declared to be unsatisfied with the role of their civil society representatives, who did "nothing", in their words, for the development of family farmers in the region.

In this regard, Climate services should ensure that the best available climate science is effectively communicated to agriculture, water, health, and other sectors, to develop and evaluate mitigation and adaptation strategies that are, easily accessible and timely. However, with the exception of radio or TV weather information, such communication is seldom available to the family farmers targeted in this study. Climate services, involving the production, translation, transfer, and use of climate knowledge and information in climate-informed decision-making and climate-smart policy and planning, are scarce in the areas of our fieldwork.

Changes in Rainfall

According to the smallholders' perspective, how exactly did rainfall evolve? In general, 87.5 % of respondents noted changes in the rainy season. This perception is widespread among the three biomes analysis, with a slight emphasis in Bahia and Acre (Fig. 12.2). But observations did vary: in Acre, the respondents overwhelmingly pointed to a shift of the rainy season. The rainy season is more concentrated (more rains from December to May), or, in other words, the dry season is longer (increased from 2 to 3 months). Respondents also pointed out: (i) a later onset of rains ("before we planted in September, now only in late October", or "before the rainy season began in February, today in March") and (ii) the rains end earlier, or, in other words the dry season starts earlier. It was also mentioned, to a lesser extent, that precipitation is weaker.

In Bahia, the most frequently observation is a decrease in the intensity of rainfall, and to a lesser extent, the unpredictability of the rain. As in the case of Acre, a displacement of the rainy season is observed.

In the Cerrado of Mato Grosso, the respondents pointed out that "the rains came late" and that the rain is becoming less intense.

Thus, the displacement of the rainy season and to a lesser extent, a decrease in intensity of rainfall phenomena are common perceptions observed in the three biomes, albeit emphasized differently. The Caatinga is characterized by precipitation of less intensity and more unpredictable, and the Amazon region by a contraction of the rainy season.

Changes in Temperature Levels

Due to formal differences in the questionnaire applied in Acre, the Amazon biome was excluded from the analysis of temperature perception. Accordingly, results from Bahia and Mato Grosso will be considered.

As far as temperature is concerned, the perceptions by smallholders from the Caatinga and Cerrado were converging (Fig. 12.2). A suggestive proportion of respondents from those two biomes (91.2 %[3]) perceived similar temperature changes. They noted hotter temperatures and an increase in the number of hot days during the year.

In Acre, results cannot be compared by using the same criteria; however an increase in the temperatures was spontaneously mentioned by 27.4 % of the respondents, as they were talking about general changes in the seasons. Some of the interviewees declared that they had to change their daily routine (i.e. by waking up earlier to avoid suffering the abnormally intense heat from 10 a.m. onwards).

Perception of Extreme Events

Respondents from each region identified the occurrence of severe climate events. In the Caatinga, 1960 and 1985 were years remembered as being very rainy, while 2002 and 1997 were also mentioned to a less extent. In the Cerrado, 2004 and 2012 were remembered as rainy years. In the case of drought, 1993 is a principle reference year in the Caatinga, while in the Cerrado, 2010 and 2011 were years considered to have the most striking episodes.

It is interesting to note that although respondents remembered similar trends in temperatures and precipitations, the climatic events that generated those perceptions differed, both temporarily and physically, from one region to another. In Acre, for example, the climate perceptions of riverside communities were very much related to the levels of the river. In Mato Grosso, the majority of respondents lived in the region for no more than 20 years, so that their climate references were more short-term or were associated with their place of origin.

Perceptions were also convergent as far as the behaviour and quantity of wild animals was concerned: 71 % of the total sample reported changes. More specifically, the principal trend is a reduction in the quantity of wild animals. This perception seems more accentuated in Bahia than in Mato Grosso. In Acre, in parallel with a decrease of the quantity of animal, respondent also reported that some animals[4] are being more audacious, coming closer to the houses to eat corn or chicks.

It is important to notice that in all regions, the causes mentioned are not specifically linked to changes in the local climate, but more frequently refer to deforestation, fires and excessive hunting, that reduce the habitat of the animals and provokes a lack of food.

[3] N = 227.

[4] Monkeys, agouti, paca, boars, capybaras, hawk.

Changes in the phenology of local plants are a phenomenon that received less emphasis, as only 52.4 % reported changes. In the Cerrado, 61 % of respondents made reference to fruit trees such as cashew and mango.

Do Convergent Perceptions Lead to Similar Behaviours?

Respondents from the Amazon, Caatinga and Cerrado shared convergent perceptions of rainfall levels and temperature trend. However, the events leading to those perceptions were different, as it was to be expected given the contrasting socio-environmental characteristics of each region. As a consequence, climatic events perceived as risks to agricultural production differ in each region. Those events were related not only to the climatic context, but also to the socio-political circumstances experienced in each region. In that regard, Bahia is particularly challenged by climate and access to water, while in Mato Grosso, smallholders declared to suffer more from difficulties to access public funding and lines of credit. In Acre, the production is more constrained by the distance and the quality of the roads, especially problematic for accessing markets. The respondents also complained of the lack of public institutions and help, and paradoxically of an excess of legislation/control, that restricts deforestation. The poor quality of lands and a reduction of productivity seem to equally affect all the focused areas.

As far as only climatic factors are involved, all regions combined, 70.7 % of the sample has already experienced some type of loss linked to climate stresses (Fig. 12.3), mainly losses of animals (40.2 %) and losses of crops (37.6 %). Smallholders in Bahia seem to have been slightly more affected than those from the two other regions (80.3 % vs. 65 % in Acre and Mato Grosso). Respondents from the Caatinga and the Cerrado identified droughts as the major cause for the loss of cattle. Loss of crops were more frequent in the Amazonian and Caatinga case studies.

How Did the Respondents Answer?

Except for the Acre sample, the proportion of those who did not make changes in the way they produce is slightly more important than those who did (Fig. 12.3). In all regions, the more common argument for those who didn't make any changes is: "it is not necessary" (23.4 %) or that they lack money to do so (7 %).

In all three biomes, the main adaptive strategy was a change in the timing of the seeding: as the rainy season started later, smallholders waited for the arrival of the first rains to seed. We can also notice that in all regions, but more importantly in the Cerrado, some crops were abandoned due to their inadequacy to the changing climatic conditions.

Some other adaptive strategies were determined in each biome. In the Amazon, respondents adopted a more proactive behaviour: 8.9 % tried to protect the

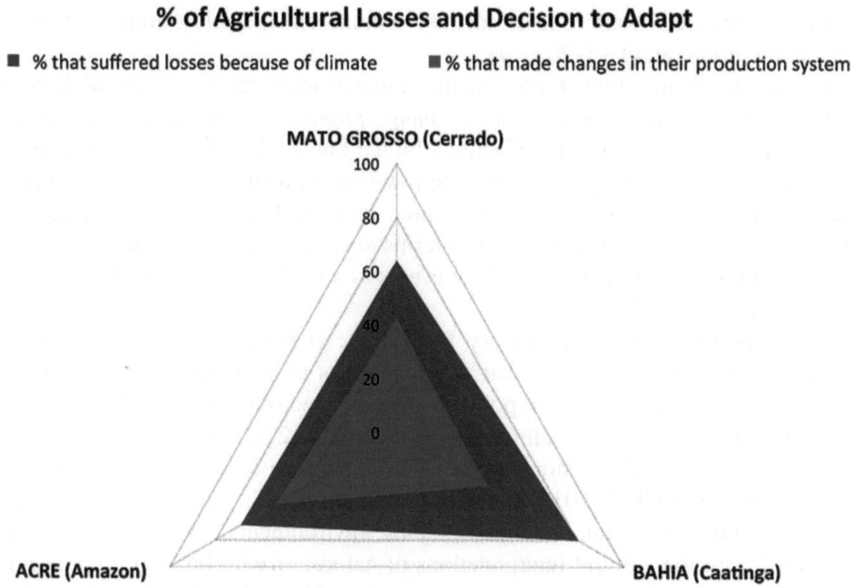

Fig. 12.3 Percentage of agricultural losses and decision to adapt

weakest cultures (vegetables) or animal (chickens) from the heat or the sun by providing some kind of shade or even planting trees. In the Caatinga, a small proportion of farmers (5.3 %) invested in the storage of fodder, an adaptation strategy specific to a region where natural hazards can seriously challenge cattle feeding.

Discussion

This chapter aimed at identifying common traits in smallholders' perception of climate variability. The intention was to gain a better understanding on a trans-regional and socially embedded environmental phenomenon such as climate change. It is believed, however, that these results may contribute to an understanding of how similar perceptions can be manifested in dramatically different Brazilian biomes.

The focus of this chapter has been the perceptions of rainfall and temperature levels. Regarding phenology and animal behaviour, comparison has been limited due to some discrepancies in the applied collection instruments.

The time of the year (rainy vs. dry season, for example) had a strong impact on perceptions, which are strongly influenced by the temporal proximity of a climatic event. The authors of this chapter recognize that immediate weather conditions influenced the memory and perceptions of past events. As each biome was facing specific weather conditions when fieldwork was conducted, we took special care to underline the major weather events that contextualized each case study, as shown in the Methodology. We conclude that, more than climate change long-term trends; interviewees were able to identify climate change micro-trends within a perception time span of 15–20 years.

One of the factors that impressed the research team most was the high convergence of the perception of climate change related risks in all the three studied biomes (even though in the Caatinga biome knowledge about climate change was low). Another striking factor was the low level of adaptation strategies implemented among smallholders from the three biomes. Even though the majority of the interviewees were familiar with terms such as climate change and global warming, most of them perceived those events as a fatality towards which very little (or nothing) could be done.

The importance of Beck's theory for the social sciences and for an understanding of natural phenomena that present threats to societies lies in regarding these phenomena as arenas for political debate, precisely because the risks are open to social processes of definition (Beck 2010, p. 27). For Beck, the absence of social sciences in discussions about environmental risks often leaves a gap of utmost importance (Beck 2010). As scientific knowledge is relevant to public recognition and dissemination of some problems, environmental sociology should be concerned with the cultural interpretations of nature, or with how different social groups perceive and define an environmental problem (Lenzi 2006). Recently, Beck presented a definition of risk very useful for this work: "Risk is not the same the catastrophe, but the anticipation of the future catastrophe in presence. As a result, risk leads to dubious, insidious, would-be fictitious, allusive existence: it is existent and non-existent, present and absent, doubtful and real" (Beck 2009, p. 3)

Climate change is a real, biological, chemical and geological phenomenon, not purely a social construct. Nevertheless, this is also a real dimension of the phenomenon, therefore merits the same degree of attention. The social construction of climate change is induced by a social organization, while at the same time is directly influenced by it (Grundmann and Stehr 2010).

The attention given by this study on local risk perceptions involves understanding better how communities (and social groups within communities) understand and respond differently to risk (Harvey et al. 2012). Risk is defined distinctly by different people and depends on the outcomes at stake, and such emotions and social processes play a key role in response to risk (Harvey et al. 2012). Studying the social reality of the perception of risk does not mean embracing the entire complexity of the experience of insecurity or fear, especially in the case of modern risks as pointed out by Beck (2010). The experience of risk perception is not something uniform, and may take confusing and contradictory forms even within the same group. However, exactly because it is a social theory, Beck assumes that personal and subjective experiences are marked by widely shared cultural characteristics. Moreover, to assume that risk perception is a social fact means accepting that its distribution may be, to greater or lesser extent, directed by fundamental social characteristics. Consequently, it is important to identify some of those fundamental social characteristics and to extract some general information from local experiences, so as to get "the bigger picture" of the main challenges of climate change risk communication and adaptation strategies. In this regard, this study has sought to overcome the specificity of local contexts and the heterogeneity of social groups to extract some lessons about climate change perceptions and adaptation strategies among family farmers situated in biomes ranging from rainforests to drylands. Some limitations, however, call for future research, such as including new elements in future surveys that may be applicable to other countries or regions. This would make our results more useful to an international leadership. As Wolf and Moser rightly point out, surveys are not enough. Future research should also seek to discern how individuals explain the causes and impacts of climate change. It should strive to understand how they process information, form their views and come to change their climate-relevant behaviour (rather than just expressing willingness to do so), and what the deeper motivations for, and barriers to, actual behavioural changes and other forms of engagement are (Wolf and Moser 2011, no page/online document).

In short, the theories discussed here allow for a partial understanding of the cultural reality of risk perception. They do not, however cover the entire complexity of how people perceive and react to everyday risks. Still, understanding how a society perceives climate change as a risk is a strategy to understand their relationship with the environment and related institutions.

Conclusion

This study confirms that, despite of the intrinsic temporality and subjectivity of climate change perceptions, there are still common trends in how people see and understand climate variability. This is also true regarding the way people seek to adapt (or not) to climate change risks. Those perceptions go beyond specific contextual characteristic such as the natural and socioeconomic environments.

The results presented in this chapter confirm that barriers to adaptation can be somehow overcome by a better understanding of climate change perceptions through interdisciplinary, bottom-up approaches.

In the context of agriculture, most of the recommendations made by policy decision makers about which techniques to use originated in the natural sciences. In order for such knowledge to translate into action by stakeholders, it is necessary to generate instrumental knowledge to help change attitudes, not only that of individuals but also of institutions. The social sciences can provide this link as they seek to understand how people perceive the changes in their environment, how they act about them.

Barriers to adaptation could be overcome through adequate dialogue and communication between policy makers and smallholders on the issues of global changes, their consequences and possible solutions. Effective communication about climate and socioeconomic change becomes critical to understand, educate, provide continuity and constructively engage decision makers in the most vulnerable sectors. Nevertheless, to be effective, communication about climate and socioeconomic risks must depart from a more refined understanding (and acceptance) of social barriers to adaptation, values and perceptual limitations of the target audience.

References

Adger WN (2003) Social capital, collective action, and adaptation to climate change. Econ Geogr 79(4):387–404
Adger WN et al. (2009a) Are there social limits to adaptation to climate change? Climatic Change 93(3):335–354
Adger WN, Lorenzoni I, O'Brien K (eds) (2009b) Adapting to climate change: thresholds, values, governance. Cambridge University Press, Cambridge
Africa Talks Climate (2010) Africa talks climate: executive summary research report. Africa talks climate: the public understanding of climate change in ten countries. BBC World Service Trust
Alcántara-Ayala I (2002) Geomorphology, natural hazards, vulnerability and prevention of natural disasters in developing countries. Geomorphology 47:107–124
Barnett J (2010) Adapting to climate change: three key challenges for research and policy—an editorial essay. Wiley Interdisc Rev: Clim Change 1(3):314–317
Beck U (1995) Ecological politics in an age of risk. Polity Press, Cambridge
Beck U (2009) World at risk. Polity Press, Cambridge
Beck U (2010) Sociedade de risco: rumo a uma outra modernidade, 34th edn. São Paulo
Berrang-Ford L, Ford J, Paterson J (2011) Are we adapting to climate change? Glob Environ Change 21(1):25–33
Bogard W (1988) Bringing social theory to hazard research: conditions and consequences of the mitigation of environmental hazards. Sociol Perspect 31(2):147–168
Bord RJ, Fisher A, O'Connor RE (1998) Public perceptions of global warming: United States and international perspectives. Clim Res 11:75–84
Brechin SR, Bhandari M (2011) Perceptions of climate change worldwide. Wiley Interdisc Rev: Clim Change 2(6):871–885
Brooks N (2003) Vulnerability, risk and adaptation: a conceptual framework. Tyndall centre working paper 38, pp 1–16. Available online http://www.tyndall.ac.uk/sites/default/files/wp38.pdf
Brumer A (2008) Gender relations in family farm agriculture and rural-urban migration in Brazil. Lat Am Perspect 35(6):11–28
Bursztyn M (2008) Sustainability science and the university: towards interdisciplinarity. CID graduate student and research fellow working paper no. 24. Harvard University, Boston
Cardona OD (2004) The need for rethinking the concepts of vulnerability and risk from a holistic perspective: a necessary review and criticism for effective risk management. In: Bankoff G, Frerks G, Hilhorst D (eds) Mapping vulnerability disasters development and people. Earthscan, London, pp 37–51
Cavatassi R, Lipper L, Narloch U (2011) Modern variety adoption and risk management in drought prone areas: insights from the sorghum farmers of eastern Ethiopia. Agric Econ 42(3):279–292
Crona BI, Wutich A, Brewis A, Gartin M (2013) Perceptions of climate change: Linking local and global perceptions through a cultural knowledge approach. Clim Change 119:519–531
Di Falco S, Veronesi M, Yesuf M (2011) Does adaptation to climate change provide food security? A micro-perspective from Ethiopia. Am J Agric Econ 93(3):825–842
Dow K (1992) Exploring differences in our common future(s): the meaning of vulnerability to global environmental change. Geoforum 23(3):417–436
Eakin H, Luers AL (2006) Assessing the vulnerability of social-environmental systems. Annu Rev Environ Resour 31:365–394
Food and Agriculture Organization (FAO) of the United Nations (2012a) The state of food insecurity in the world. Available online 22 June 2013: http://www.fao.org/docrep/016/i3027e/i3027e.pdf
Food and Agriculture Organization (FAO) of the United Nations (2012b) Forestry. Available online 6 June 2012: www.fao.org/forestry
Ford JD, Pearce T, Duerden F, Furgal C, Smit B (2010) Climate change policy responses for Canada's Inuit population: the importance of and opportunities for adaptation. Glob Environ Change 20(1):177–191

Gallopín GC (2006) Linkages between vulnerability, resilience, and adaptive capacity. Global Environ Change: Hum Policy Dimensions 16(3):293–303

Gbetibouo GA (2009) Understanding farmers' perceptions and adaptations to climate change and variability. The case of the Limpopo Basin, South Africa. IFPRI discussion paper no. 849. International Food Policy Research Institute, Washington, DC

Giddens A (1994) Lliving in a post traditional society. In: Beck U, Giddens A, Lash S (eds) Reflexive modernization: politics, tradition and aesthetics in the modern social order. Stanford University Press, Stanford

Grundmann R, Stehr N (2010) Climate change: what role for sociology? A response to constance lever-tracy. Curr Sociol 58(6):897–910

Harvey B, Ensor J, Carlile L, Garside B, Patterson Z, Naess LO (2012) Climate change communication and social learning—review and strategy development for CCAFS. CCAFS working paper no. 22. Available online 3 Oct 2013: http://hdl.handle.net/10568/24456

Howden SM, Soussana J-F, Tubiello FN, Chhetri N, Dunlop M, Meinke H (2007) Adapting agriculture to climate change. Proc Natl Acad Sci 104:19691–19696

Hulme M, Adger WN, Dessai S, Goulden M, Lorenzoni I, Nelson D, Naess LO, Wolf J, Wreford A (2007) Limits and barriers to adaptation: four propositions. Tyndall centre for climate change research, Tyndall briefing note no. 20. Available online: http://www.tyndall.ac.uk/sites/default/files/bn20.pdf

IBGE (2006) Instituto Brasileiro de Estatisticas e Geografia 2006, Censo Agropecuário 2006. Available online: http://www.ibge.gov.br/home/estatistica/economia/agropecuaria/censoagro/default.shtm

IBGE (2010) Censo Demográfico. IBGE. Rio de Janeiro. Censo demográfico 2010. Primeiros resultados

IBGE (2012a) Instituto Brasileiro de Estatisticas e Geografia 2012, Indicadores de desenvolvimento sustentável—Brasil 2012, Estudos and Pesquisas, n 9. IBGE, Rio de Janeiro. Available online: ftp://geoftp.ibge.gov.br/documentos/recursos_naturais/indicadores_desenvolvimento_sustentavel/2012/ids2012.pdf

IBGE (2012b) Instituto Brasileiro de Geografia e Estatística. Available online 18 June 2012: www.ibge.gov.br

INMET (2012) Brazilian National Institute of Meteorology. Available online 3 June 2013: www.inmet.gov.br

IPCC (2007) Contribution of working group II to the fourth assessment report of the intergovernmental panel on climate change. In: Parry ML, Canziani OF, Palutikof JP, van der Linden PJ, Hanson CE (eds) Cambridge University Press, Cambridge, UK, pp 976

Jones L, Boyd E (2011) Exploring social barriers to adaptation: insights from Western Nepal. Glob Environ Change 21(4):1262–1274

Kalanda-Joshua M, Ngongondo C, Chipeta L, Mpembeka F (2011) Integrating indigenous knowledge with conventional science: enhancing localised climate and weather forecasts in Nessa, Mulanje, Malawi. Phys Chem Earth 36:996–1003

Kates RW (1985) The interaction of climate and society. In: Kates R, Ausubel J, Berberian M (eds) Climate impact assessment: studies of the interaction of climate and society. Wiley, pp 3–36

Kaztman R (2005) Vulnerabilidad y Exclusión social: una propuesta metodológica para el estudio de las condiciones de vida de los hogares. CEPAL, Chile

Lane A, Jarvis A (2007) Changes in climate will modify the geography of crop suitability: agricultural biodiversity can help with adaptation. J Semi-Arid Trop Agric 4:1–12

Lash S, Szerszynski B, Wynne B (1996) Risk, environment and modernity: towards a new ecology. Sage Publications, London

Leichenko R, O'Brien K (2008) Environmental change and globalization: double exposures. Oxford University Press, Oxford

Leichenko RM, O'Brien KL, Solecki WD (2010) Climate change and the global financial crisis: a case of double exposure. Ann Assoc Am Geogr 100:963–972

Lenzi C (2006) Sociologia ambiental: risco e sustentabilidade na modernidade. EDUSC, Bauru, SP
Lidskog R (2001) The re-naturalization of society? Environmental challenges for sociology. Curr Sociol 49(1):113–136
Lindoso DP, Rocha JD, Debortoli N, Parente IIC, Eiró F, Bursztyn M, Rodrigues-Filho S (2011) Climate change and vulnerability to drought in the Semiarid Region: the case of smallholder farmers in the Brazilian Northeast. In: Motta RS, Hargrave J, Luedemann G, Guiterrez MBS (eds) Climate change in Brazil: economic, social and regulatory aspects. IPEA, Brasília, pp 235–256
Lorenzoni I, Pidgeon N (2006) Public views on climate change: European and USA perspectives. Clim Change 77:73–95
Lorenzoni I, Nicholson-Cole S, Whitmarsh L (2007) Barriers perceived to engaging with climate change among the UK public and their policy implications. Glob Environ Change 17(3):445–459
Marandola E, Hogan DJ (2004) Natural hazards: o estudo geográfico dos riscos e perigos. Ambiente Soc 2:95–109
Marengo JA, Tomasella J, Alves LM, Soares WR, Rodriguez DA (2011) The drought of 2010 in the context of historical droughts in the Amazon region. Geophys Res Lett 38(12)
Ministerio de Ganaderia, Agricultura y Pesca de Uruguay/FAO (2013) Estudio de Percepción de los productores agropecuarios y sus organizaciones sobre el Cambio Climático y las opciones de adaptación. Available online 10 Oct 2013: http://www.mgap.gub.uy/gxpfiles/mgap/content/audio/source0000000083/AUD0000070000001966.pdf
Moser S, Dilling L (2012) Communicating climate change: closing the science action gap. Oxford Handbooks Online. Retrieved 19 Feb 2014: http://www.oxfordhandbooks.com/view/10.1093/oxfordhb/9780199566600.001.0001/oxfordhb-9780199566600-e-11
Moser SC, Dilling L (2004) Making climate hot: communicating the urgency and challenge of global climate change. Environment: an official publication of Scientists' Institute for Public Information, vol 46, pp 32–46
Moser SC, Ekstrom J (2010) A framework to diagnose barriers to climate change adaptation. Proc Natl Acad Sci (PNAS) 107(51):22026–22031
Naab JB, Koranteng H (2012) Gender and climate change research results: Jirapa, Ghana working paper no. 17. CGIAR research program on climate change, agriculture and food security (CCAFS). Nairobi, Kenya. Available online 10 Oct 2013: http://ccafs.cgiar.org/sites/default/files/assets/docs/ccafs-wp-17-gender_adaptation_ghana.pdf
Nerlich B, Koteyko N, Brown B (2010) Theory and language of climate change communication. WIREs Clim Change 1:97–110
Nordhaus W (1975) Can we control carbon dioxide? IIASA, Laxenburg
Ostrom E (2005) Understanding institutional diversity. Princeton University Press, Princeton
Parry ML, Carter TR (1998) Climate impact and adaptation assessment: a guide to the IPCC approach. Earthscan, London, p 166
Pinedo-Vasquez M, Padoch C (2009) Urban, rural and in-between: muiti-sited households, mobility and resource management in the Amazon floodplain. In: Alexiades MN (ed) Mobility and migration in indigenous Amazônia; contemporary ethnoecological perspectives. Berghahn, Oxford
Pinto HS, Assad ED (orgs) (2008) Aquecimento Global e Cenários Futuros da Agricultura Brasileira. Embrapa/Unicamp, Campinas, Contrato FCO-GOF: PGL GCC 0214, British Embassy, Brasil, 80p
Rao KPC, Ndegws WG, Kizito K, Oyoo A (2011) Climate variability and change: farmer perceptions and understanding of intra-seasonal variability in rainfall and associated risk in semi-arid Kenya. Expl Agric 47:267–291
Simoni J, Lindoso D, Debortoli N, Ibiapina I, Eidt G (2010) Mudanças climáticas, vulnerabilidade e capacidade adaptativa de territórios da Amazônia: Pará. Estudo Técnico, PNUD/SAE
Smit B, Burton B, Klein RJT, Wandel J (2000) An anatomy of adaptation to climate change and variability. Clim Change 45(1):223–251
Smith M (2001) The face of nature: environmental ethics and the boundaries of contemporary social theory. Curr Sociol 49(1):49–65

Susman P, O'Keefe P, Wisner B (1984) Global disasters: a radical interpretation. In: Hewitt K (ed) Interpretations of calamity from the viewpoint of human ecology. Allen and Unwin, Boston, pp 264–283

Timmerman P (1981) Vulnerability, resilience and the collapse of society: a review of models and possible climatic applications. Institute for Environmental Studies, University of Toronto, Toronto

Ungar S (2000) Knowledge, ignorance and the popular culture: climate change versus the ozone hole. Public Understand Sci 9:297–312

VanderMolen K (2011) Percepciones de cambio climático y estrategias de adaptación en las comunidades agrícolas de Cotacachi (Debate Agrario-Rural). En Ecuador Debate. Problemas y perspectivas del extravismo, vol 82. Centro Andino de Acción Popular CAAP, Quito, pp 145–157

Verburg R, Rodrigues Filho S, Lindoso D, Debortoli N, Litre G, Bursztyn M (2014) The impact of commodity price and conservation policy scenarios on deforestation and agricultural land use in a frontier area within the Amazon. Land Use Policy 37, 14–26

Vignola R (2010) Estudio de la percepción y actitudes de la población costarricense sobre Cambio Climático. Informe para Programa de Naciones Unidas para el Desarrollo de Costa Rica. Available online 10 Oct 2013: http://www.cambioclimaticocr.com/biblioteca-virtual/cat_view/2-publicaciones-sobre-cambio-climatico?start=15

Wolf J, Moser SC (2011) Individual understandings, perceptions, and engagement with climate change: Insights from in-depth studies across the world. Wiley Interdisc Rev Clim Change 2(44):547–569

Part III
The Role of Policy-making and the Contribution of Information, Communication and Stakeholders' Involvement

Chapter 13
Precipitation Variability and Adaptation Strategies in the Southern Yucatán Peninsula, Mexico: Integrating Local Knowledge with Quantitative Analysis

Sofía Márdero, Birgit Schmook, Zachary Christman, Elsa Nickl, Laura Schneider, John Rogan and Deborah Lawrence

Abstract Climatic variability, including droughts, has long affected the Mayan Lowlands. Therefore, farmers have developed coping strategies to mitigate these impacts. In the past, however, records of these effects and responses were largely anecdotal. In modern times, the perceptions of farmers, especially those practicing rain-fed agriculture, combined with the increased availability of accurate historical climatic records and forecasts, can provide useful information regarding periods of decreased precipitation and strategies employed to resist and respond to drought effects. As part of the multidisciplinary and inter-institutional project, *New Knowledge about Ecosystem Level Response to Increased Frequency of*

S. Márdero (✉) · B. Schmook
El Colegio de La Frontera Sur, Chetumal 77014, QRoo, Mexico
e-mail: ssmardero@ecosur.edu.mx

B. Schmook
e-mail: bschmook@ecosur.mx

Z. Christman
Rowan University, Glassboro 08028, NJ, USA
e-mail: christmanz@rowan.edu

E. Nickl
University of Delaware, Newark 19716, DE, USA
e-mail: enickl@udel.edu

L. Schneider
Rutgers University, New Brunswick 08901, NJ, USA
e-mail: laschnei@rci.rutgers.edu

J. Rogan
Clark University, Worcester 01610, MA, USA
e-mail: jrogan@clarku.edu

D. Lawrence
University of Virginia, Charlottesville 22904, VA, USA
e-mail: lawrence@virginia.edu

W. Leal Filho et al. (eds.), *International Perspectives on Climate Change*,
Climate Change Management, DOI: 10.1007/978-3-319-04489-7_13,

Large-Scale Natural Disturbance Driven by Climate Change, this chapter outlines the spatial and temporal variability of precipitation across the Southern Yucatán Peninsula, Mexico, and examines Mayan farmers' adaptations to droughts and other climate perturbations. The authors analyzed precipitation trends and anomalies from 1953 to 2007, using linear regressions and the quintile method to classify meteorological droughts. Authors also conducted 150 household interviews across 10 communities to investigate Mayan farmers' adaptations to climate perturbations. Results demonstrate a significant decrease in annual and rainy season precipitation across much of the study area, coupled with an increased occurrence of droughts, especially since 1980. Interviewed subsistence maize farmers have adapted to decreasing and irregular precipitation by adjusting agricultural calendars, planting more maize varieties, increasing water storage, and diversifying their practices both within the agricultural system and beyond it. Through this research, the authors demonstrate the importance of incorporating farmers' local and traditional knowledge into prevention and mitigation policies of governmental and non-governmental institutions in the region.

Keywords Drought • Precipitation variability • Trend analysis • Adaptation Yucatán Peninsula • Mexico

Introduction

Due to its geographical location, Mexican farmers are extremely vulnerable to the disastrous impacts of droughts (Hernandez et al. 2007). Combined with the fact that irrigation infrastructure is only utilized across 23 % of the country's cultivated area (Burgin et al. 2009), Mexico is highly susceptible to climatic variations in water availability. Further, the Yucatán Peninsula is one of the most vulnerable areas to the effects of climate variability and climate change across the country, especially due to the occurrence of droughts and hurricanes.

In ancient times, a series of extreme droughts that occurred in approximately 760, 810, 860 and 910 CE seem to have greatly contributed to the collapse of the Mayan civilization. Thus, droughts have long affected society at different levels, but are especially impactful to farmers practicing rain-fed agriculture. Through the centuries, farmers in the Yucatán Peninsula have developed strategies to face changing environmental conditions (Netting 1993; Denevan 1980). The flexibility of these strategies is key to effectiveness: farmers are not only constantly adapting to climate changes, they also face the combined impacts of social, economic, and institutional changes–a "double exposure" (O'Brien and Leichenko 2000). Farmers' ability to cope with adverse impacts of climate change depends not only on agronomic knowledge, but also on market conditions, crop prices and production costs, and household economic stability (Eakin 2000).

Understanding how certain sectors of the population are affected by drought and how they respond to it are considered essential steps towards climate mitigation through new policy structures and disaster relief programs (Endfield and Nash 2002). The reconstruction of regional histories of climate impacts and social response to drought throughout history are of critical importance to understand and anticipate the impacts of future events.

Droughts and Agriculture

There are numerous definitions and methods to classify droughts. In this chapter, drought is defined as meteorological drought, which, according to the World Meteorological Organization (WMO), is the "prolonged absence or marked deficiency of precipitation." Meteorological droughts are defined as a meteorological phenomenon that occurs when rain or natural runoff in a given period is less than the normal value (see Russell et al. 1970; Hernández-Cerda et al. 2000).

Within the various types of climate anomalies (e.g., variable rainfall, hurricanes, droughts, heat waves, etc.), drought is globally considered the most worrisome (Sauchyn 2004; Hagman 1984). Historically, droughts have caused economic, social and environmental problems in many parts of the world because their spatial distribution is generally greater than the area damaged by other natural phenomena (Hernández-Cerda et al. 2007). Some examples of major droughts with devastating effects on the world include India, from 1790 to 1796, whose impacts were felt across the globe with civil unrest and riots, and The 'Great Drought' of the Victorian era, from 1876 to 1878, extending from China to India and Indonesia, causing famine that took the lives of about 30 million people.

According to the Severity Index (IS) of areas vulnerable to meteorological drought in Mexico (Hernández et al. 2000a), the Yucatán Peninsula, especially the Northeastern and Southern regions, is considered one of the top five areas of Mexico with severe drought incidence. Nearly all of the state of Quintana Roo (99.8 %) and most of the state of Campeche (75.2 %) are among the most vulnerable areas to meteorological drought. This situation is worsening (Hernández et al. 2000b). Additionally, the region is frequently hit by hurricanes (Orellana et al. 2009), with subsequent risks of fire events (Cheng et al. 2013).

Rain-fed agriculture is among the economic activities most vulnerable to climate change (Parry et al. 1988; Reilly et al. 1996), as climate strongly determines agricultural productivity (Adams et al. 1998). Studies on the interplay of climate change and agriculture increasingly rank agriculture in developing countries as extremely vulnerable because of its "double exposure." Agriculture is especially sensitive to price and market fluctuations within a globalized economy and climate variations.

Farmers respond to climate change and economic globalization depending on how they perceive changes. Ban and Hawkins (2000) define "perception" as the process by which we receive information or stimuli from our environment and transform it into psychological awareness. Knowledge, interest, culture, and other social processes influence individual and societal thoughts and shape behavior (Frank et al. 2011).

Methods

To address rainfall variability and farmers' adaptations and perceptions about that variability, this project included rainfall data analyses from 1953 to 2007 as well as surveys and interviews undertaken from May to August 2010.

Study Area

The study area is located in the south-central Yucatán Peninsula, Mexico, between two of the largest biosphere reserves in Mexico: Sian Ka'an and Calakmul. The region covers approximately 26,870 km^2 of Southwestern Quintana Roo and Southeast Campeche and is a biodiversity hotspot (Ramamoorthy et al. 1993). Ten *ejidos* with high levels of indigenous Mayan population, and rainfed agriculture as a main economic activity were chosen. *Ejidos* are a common property form implemented through the process of agrarian reform stemming from the Mexican Revolution and formalized in the Constitution of 1917.

Study *ejidos* practice predominantly slash and burn cultivation, locally known as *milpa,* with corn, beans, and squash as main staples (Klepeis et al. 2004). Ancient in origin, the *milpa* system refers to a complex combination of agronomic practices, crop associations, and rotation sequences that vary with cultural context and agro-environment. Traditional *milpa* cultivation involves cutting an area of forest, burning the area soon after, and planting maize mixed with squash and beans (Schmook et al. 2013). Chili peppers, cattle, timber harvesting, honey, and charcoal production are other important economic activities for local farmers (Steward 2009).

Drought and Rainfall Analysis

To address climate change–specifically rainfall variability and droughts—nine weather stations were chosen in the vicinity of the *ejidos* studied (Fig. 13.1).

To address rainfall variability and drought occurrence from 1953 to 2007, precipitation data from the Quintana Roo and Campeche weather station network (CONAGUA) were analyzed to identify increases or decreases in precipitation trends, using linear regression with both the annual and seasonal rainfall data. With these precipitation anomalies, the stability, deficit, or surplus of precipitation were analyzed for each year or season. The intensity of meteorological droughts was determined using the quantile method.

It should be noted that some stations did not have a complete data series for all dates from 1957 to 2007. Therefore, each analytical method utilized different time periods depending on the available data for each station. The anomaly detection and the quantile method to address drought severity both employed the

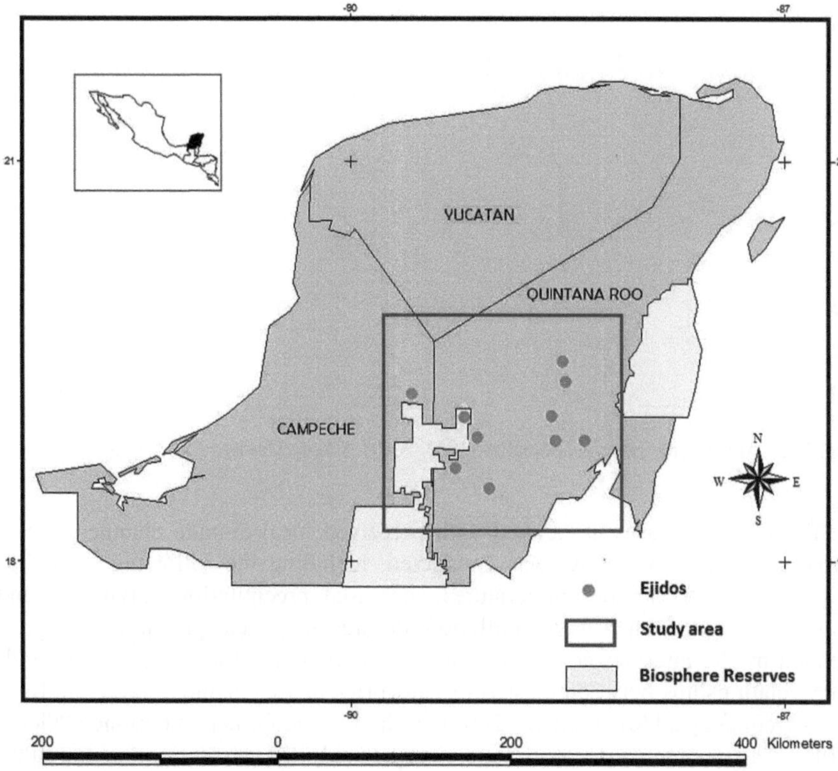

Fig. 13.1 Location of *"ejidos"* and study area

entire series from 1957 to 2007, because no comparisons were made between stations and gaps in the series would not affect the final result. The precipitation trend analysis and comparisons between stations used an abbreviated time series with complete data for all the stations, from 1973 to 2002.

Farmers' Perceptions and Adaptations

In addition to the climate analyses, the perceptions and coping strategies of Mayan farmers were investigated through interviews. One hundred fifty farmers were interviewed in the 10 *ejidos* from May to August 2010, the topics asked included the household's socioeconomic situation, degree of dependence on agricultural income, and individual perceptions of climate and climate change. Questions emphasized the occurrence of and adaptations to drought over the last two decades, and strategies for coping with these events. The farmers were asked if they thought climate had changed over the last 15 years and were then asked to consider the kind of climate changes they perceived and the causes of these changes.

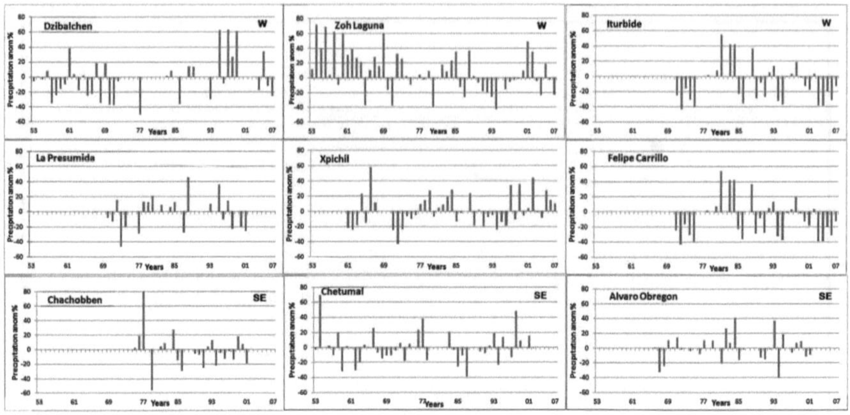

Fig. 13.2 Annual precipitation anomalies (1953–2007). *Source* Márdero (2011, p. 54)

To visualize causes associated with perceived local climate changes, a correspondence analysis (CA) was conducted, including the following variables: reduced rainfall, higher temperatures, increased precipitation variability, and associated causes, such as air pollution, deforestation, and global warming, as defined by the interviewee. CA is a statistical technique for displaying dependency relationships between classes or categories of two or more variables from cross-frequency tables (Figueras 2003). As half of the farmers mentioned "God's will" as a possible cause of climate changes, only the perceptions of the other half were included.

Finally, a list of possible adaptation strategies to cope with climate changes was compiled and all farmers were asked if they employed any of these strategies.

Results

Drought and Rainfall

Results of the precipitation data analysis reveal high spatial and temporal rainfall variability in the study region from 1957 to 2007. The highest annual rainfall variability was found at weather stations in the North. Stations with higher annual rainfall averages and positive anomalies were located near the coast, where trade winds carry moist, warm air masses in the summer. Overall, rainfall gradually decreased toward the central region. The driest zone was detected around the Zoh Laguna weather station, adjacent to the Calakmul Biosphere Reserve. Precipitation analysis from this station indicated the highest incidence of negative anomalies (up to 10 consecutive years with below average precipitation) and a clear decrease in positive anomalies of rainfall after the mid 1980s (see Fig. 13.2).

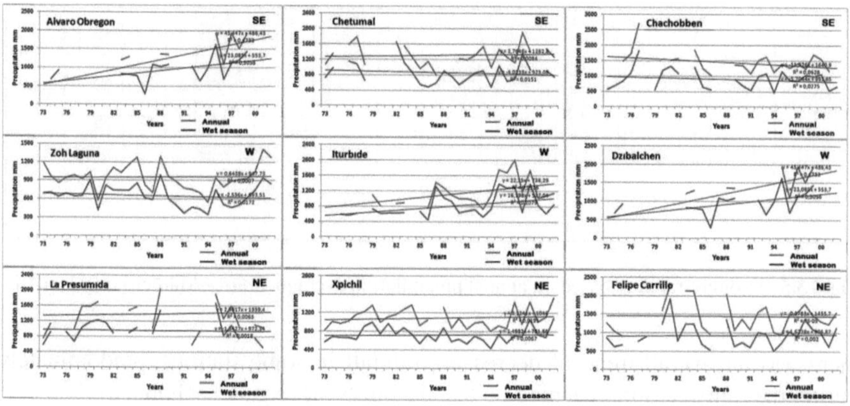

Fig. 13.3 Annual and rainy season (Jun–Oct) precipitation trends from 1973 to 2002. *Source* Márdero (2011, p. 56)

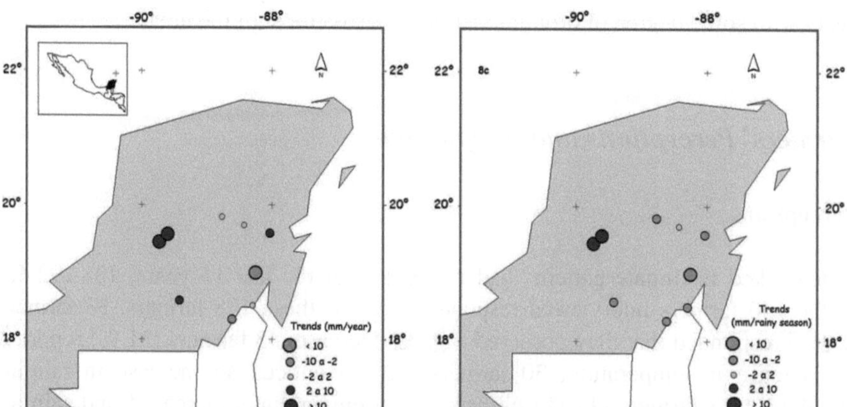

Fig. 13.4 1973–2002 **a** map of weather stations, **b** annual precipitation, **c** rainy season precipitation, **d** dry season precipitation. *Source* Márdero (2011, p. 68)

Figures 13.3 and 13.4 demonstrate an overall reduction in rainfall over the last thirty years, especially during the rainy season, with a decrease of up to −12 mm/year at the station of Chachobben. Only two stations (Iturbe and Dzibalchen), located in the West, indicate an increase in rainfall. However, both of these stations have gaps in the data series—some for consecutive years and others for a single year—which could influence the results.

Results of the rainfall anomaly analysis revealed two major trends. First, considerable precipitation variability was detected at several weather stations, indicating up to a 30 % increase in rainfall in one year followed by up to a 40 % decrease in rainfall the next year (e.g., in both Zoh Laguna and La Presumida). Second,

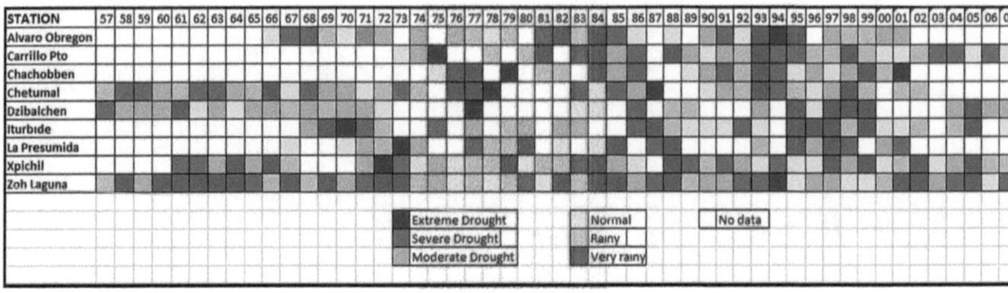

Fig. 13.5 Drought ranking based on annual precipitation average. *Source* Márdero (2011, p. 61)

there was an overall trend of decreasing rainfall in some stations, predominately during the rainy season, reaching up to −12 mm/year (e.g., Chachobben).

The quantile method shows that droughts were most severe in 1970–1973, 1985–1987 and 1994–1995 (Fig. 13.5); although the years classified as extreme droughts are not evenly spread.

After the mid-1980s, both rainy years and very rainy years became rare, and years with some degree of drought increased across most of the area.

Farmers' Perceptions and Adaptations

Perceptions

When asked if climate patterns had changed over the last 15 years, 138 (92 %) of the 150 farmers interviewed responded yes. Of these 138 farmers, 33 farmers (24 %) responded that they observed rainfall decrease; 43 farmers (31 %) reported an increase in temperature; 30 farmers (22 %) noticed an increase in rainfall variability; 23 farmers (17 %) observed that temperatures increased and rainfall decreased; 7 farmers (5 %) said temperature and rainfall variability increased; and just one farmer (< 1 %) reported increased rainfall.

According to the CA results, farmers link temperature increases (*+T*) with pollution (*cont*), and rainfall decrease (*−PP*) is seen as an effect of "global climate change" (*cg*). Some farmers claimed directly that "all these changes in rainfall are caused by the worldwide climate change." In addition, the combination of less rainfall and increased temperature (*+T−PP*) was attributed to deforestation (*def*) in the region, and some farmers did not attribute rainfall variability (*var*) to any specific cause (Fig. 13.6).

Adaptations

Common adaptations to climate anomalies such as rainfall variability include adjustments in the agricultural calendar, income diversification, and storage of water and maize. Other strategies mentioned to cope with climate variability,

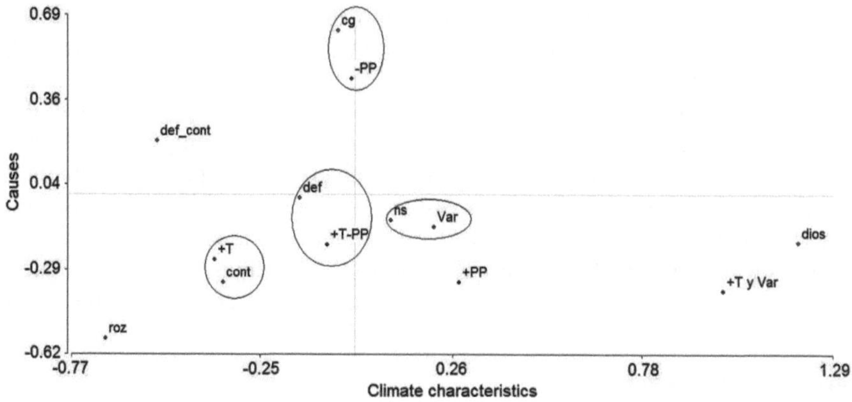

Fig. 13.6 Correspondence analysis of farmer´s perception about climate change and the possible causes. *Source* Márdero (2011, p. 72)

especially droughts, included planting and maintaining house gardens and changes in cultivation techniques. Another strategy was seeking out assistance from institutions dedicated to disaster relief and increased participation in government-sponsored conservation programs providing payments to participating farmers. The most frequent adaptation techniques are explored below.

Agricultural Adaptations

- **Adjustments in agricultural calendar** None of the studied communities have access to irrigation systems. Given the resulting dependence on rainfall, the timing of planting and type of maize planted are crucial for *milpa* production. Adjustments in the agricultural calendar may include delaying the start of burning and planting, planting twice in the same season, or adjusting planting dates according to farmers' weather observations and predictions. Delaying planting dates could lead to a short growing season, precluding the crop maturation of some maize varieties. It is important to avoid the *canícula*, or mid summer drought, during the plants' reproductive stage, with the highest water requirements. Eighty farmers (53 %) mentioned adjustments of their agricultural calendar in response to rainfall variability. Some farmers reported dividing their *milpas* into several plots, which they plant at different times depending on rainfall, to reduce climate risk.
- **Combining corn varieties and multiple planting.** Some farmers use corn varieties with less water requirements and a shorter cultivation period to avoid the height of the *canícula*. Fifteen farmers (10 %) indicated that, occasionally, they plant several times in one year and use different corn varieties with a shorter growing time.
- **Reducing working hours.** To reduce discomfort and fatigue caused by the sun and rising temperatures, farmers reduced the number of hours worked. Rising temperatures—according to the interviewees—have become common in recent

years. Eighty (53 %) out of 150 farmers interviewed reduced working hours in the *milpas* during the last 15 years. On average, they reduced working hours by two hours/day.

- **Storing water and maize.** About 56 farmers (37 %) employ water-storing mechanisms to cope with droughts. Water storage includes digging small ponds for cattle and/or storing water in plastic containers. Ninety percent of the interviewees store corn from every harvest, for the next growing season, because according to them, their seeds are better than commercial ones.
- **Moving to moister plots.** Only 28 farmers (19 %) moved their plots to areas with higher soil moisture or close to water bodies or to areas with more fertile soils.

Income diversification

- **Diversifying Income**. Surveys showed that only 22 farmers (15 %) relied on a single economic activity for income, while 85 % of all farmers engaged in several economic activities. Forty-five farmers (30 %) combined only on-farm activities such as agriculture, beekeeping, livestock, and charcoal production; fifteen of them (10 %) are involved just in off-farm activities such as *jornales*, or wage labor, masonry, merchant, and carpentry work, and most of them (51 %) prefer to combine on-farm and off-farm activities (76 farmers). Interviewees stated that combining on-farm activities with off-farm employment makes them less dependent on crop success and therefore less vulnerable to the uncertainties of climate change.
- **Government support.** Monetary support from government programs constitutes another pillar of local livelihoods. Ninety-seven percent of all interviewed farmers claimed to receive some kind of government support. The combination of changes in agricultural practices, income diversification, and governmental financial support programs are the most important strategies of local farmers to confront and cope with the challenges and adversities of climate variability and economic globalization. Despite these difficulties, 115 (77 %) of the 150 farmers interviewed claim that they will continue farming. Twenty interviewees of them (13 %) said that they want to quit agriculture, and 15 farmers (10 %) did not specify their future plans.

Discussion

Through the precipitation analyses, we observed considerable rainfall variability in the Southern Yucatán Peninsula. Measurement from some stations revealed marked fluctuation in the amount of rainfall between years. In some years rainfall was 30 % above average rainfall and in other years rainfall was 40 % below average, making agriculture an extremely unpredictable and risky business. A further difficulty for agriculture is an overall reduction in the quantity of rainfall, especially during the rainy season.

Explanations for decreased rainfall in the study area are varied. It may be triggered by anthropogenic climate change, especially since the Industrial Revolution (Wisner et al. 2007). Recent research indicates a link between anthropogenic activities and global climate change, with either a direct or inverse relationship with decreasing rainfall, depending on climatic and environmental characteristics of each region (Heller and Muthukumara 2002; Burgin et al. 2009). A link has also been speculated between global dimming, or the reduction of solar radiation reaching the surface due to the increase of aerosols in the atmosphere, and increased droughts in some regions (Stanhill and Moreshet 2004; Liepert 2002). Another possible cause of rainfall decline could be deforestation, which results in less evapotranspiration, thereby decreasing precipitation (Shukla et al. 2008; Kanae et al. 2001).

The noticeable decrease and variability of local rainfall patterns make practicing rain-fed agriculture challenging. The small-scale agricultural sector is extremely vulnerable to climate change because crop productivity is directly linked to the occurrence of rainfall. In addition to less reliable rainfall patterns and more frequent drought events, small scale farmers in Mexico face additional challenges from the lack of appropriate technologies and irrigation systems, cheap grain imports, and public policies that have not focused on mitigating negative effects of climate change by compensating farmers for their losses (Fernández-Turrent et al. 2013).

Farmers' income diversification emerged as an important strategy for drought adaptation and rainfall variability, though it would be inaccurate to state that income diversification is solely triggered by climate change. In addition, due to changes in Mexican economy and economic globalization, rain-fed farmers are not earning enough money to subsist and continue farming. Farmers diversify their livelihoods by engaging in several economic activities to manage the hardship imposed on them by economic globalization and poorly designed agricultural policies (O'Brien and Leichenko 2000).

Based on local adaptation strategies, public policies and governmental programs geared at improving adaptation to climate change in rural areas should consider preexistent coping mechanisms and incorporate them into their recommendations whenever possible. Recommendations should improve and introduce adapted technology for small-scale agriculture. Increased mechanization and the construction of irrigation systems could be an efficient mechanism to make smallholder agriculture less vulnerable to climate changes. Supporting the creation of alternative markets for crops, honey, and other agricultural products, as well as the creation of more programs focused on payment for environmental services and increasing employment opportunities outside the agricultural sector constitute other alternatives.

Conclusion

In Mexico, from the 1970s onward, maize production has been in relative decline. Since the 1980s, maize production demonstrates an absolute decrease (Lazos Chavero 1995), Nevertheless, according to authors like Fernández-Turrent et al. (2013)

Mexico can regain greater self sufficiency in maize production and reduce its import dependence and costs, suggesting that within 10–15 years Mexico could increase annual production from current lands to 33 million/mt; irrigation and infrastructure projects in the southern part of the country could add another 24 million mt/year. This would be more than enough to meet Mexico's growing demand for maize, estimated to reach 39 million mt/year by 2025 (FAPRI 2011). Additional research confirms the viability of these estimates. Either the deficit in the supply of maize is caused by the neoliberalization of the national and global economy or increased climate variability due to the effects of climate change, the small-scale agricultural sector is undergoing a serious crisis. Though farmers relayed how important it was to plant a cornfield each year, and 77 % stated that they would continue to do so in the future, younger generations are less interested in being farmers, claiming agriculture to be a high-risk activity requiring substantial investment and work and providing little in return. Younger generations may very well choose to work outside of their communities, migrate to another country, and lose the traditional knowledge associated with *milpa* cultivation.

References

Adams RM, Hurd BH, Lenhart S, Leary N (1998) Effects of global climate change on agriculture: an interpretative review. Clim Res 11:19–30

Van Den Ban AW, Hawkins HS (2000) Agricultural Extension, 2nd edn. Blackwell Science, UK

Burgin MJ, Berdegué J, Aguirre M, Medina R, Ramírez M (2009) Adaptación local al cambio climático y dinámicas territoriales rurales. (Local adaptation to climate change and rural adaptation dynamics.) Informe. Gobierno del Estado de Michoacán, SEDRU, p 62

Cheng D, Rogan J, Schneider L, Cochrane M (2013) Evaluating MODIS active fire products in subtropical Yucatán forest. Remote Sens Lett 4(5):455–464

Denevan W (1980) Swidden and cattle versus forest: the imminent demise of the Amazon rain forest reexamined. Stud Third World Soc 13:25–44

Durgin FH, DeWald D, Lechich S, Li Z, Ontiveros Z (2011) Action and motivation: Measuring perception or strategies? Psychon Bull Rev 18:1077–1082

Eakin H (2000) Smallholder maize production and climatic risk: a case study from central Mexico. Clim Change 45(1):19–36

Endfield GH, Nash DJ (2002) Drought, desiccation and discourse: Missionary correspondence and nineteenth-century climate change in Central Southern Africa. Geog J 168(1):33–47

Fernández-Turrent A, Wise AT, Garvey E (2013) Achieving Mexico's maize potential. In: Food sovereignty: a critical dialogue. Yale University, New Haven

Figueras SM (2003) Análisis de Correspondencias [en línea] Estadística. http://www.5campus.com/leccion/correspondencias [consulted on June, 2013]

Hagman G (1984) Prevention better than cure: report on human and natural disasters in the Third World. Swedish Red Cross, Estocolmo, Suecia

Heller SP, Muthukumara M (2002) La adaptación al cambio climático. Rev Finanzas Desarrollo, Fondo Monetario Internacional 39(1):29–31

Hernández-Cerda ME, Torres-Tapia LA, Valdez-Madero G (2000) Sequía meteorológica; Mexico: una visión hacia el siglo XXI. El cambio climático en Mexico, Instituto de Geografía, UNAM, Mexico

Hernández- Cerda ME, Carrasco-Anaya G, Sánchez GA (2007) Mitos y realidades de la sequía en México. Temas Selectos de Geografía de México (I.6.3) Instituto de Geografía, UNAM, México

Klepeis P, Vance C, Keys E, Macario-Mendoza P, Turner II BL (2004) Subsistence sustained: swidden or milpa cultivation. In: Turner II BL, Geoghegan J, Foster D (eds) Integrated land-change science and tropical deforestation in the Southern Yucatán: final frontiers. Clarendon Press, Oxford, pp 189–206

Kanae S, Oki T, Musiake K (2001) Impacts of deforestation on regional precipitation over the Indochina Peninsula. J Hydrometeorol Am Meteorol Soc 2(1):1–20

Lazos Chavero E (1995) De la candela al mercado: el papel de la mujer en la agricultura comercial del sur de Yucatán. Relaciones de género y transformaciones agrarias. El Colegio de Mexico, Mexico, pp 91–133

Liepert, B (2002) "Impacts of aerosols on the hydrological cycle in Echam4 experiments", Max-Planck Institute Visiting Scholar Fellowship by Max-Planck Society (PI). Hamburg, Germany

Márdero S (2011) Sequías y efectos en las prácticas agrícolas de familias campesinas del Sur de la Península de Yucatán. Unpublished Master's Thesis, El Colegio de la Frontera Sur

Netting RM (1993) Smallholders, householders: farm families and the ecology of intensive, sustainable agriculture. Stanford University Press, Stanford

O'Brien K, Leichenko R (2000) Double exposure: assessing the impacts of climate change within the context of economic globalization. University of Oslo and Center for Urban Policy Research, Rutgers University, CICERO

Orellana R, Espadas C, Conde C, Gay C (2009) Atlas Escenarios de Cambio Climático en la Península de Yucatán, Unidad de Recursos Naturales, Centro de Investigación Científica de Yucatán y Centro de Ciencias de la Atmósfera de la Universidad Nacional Autónoma de México, Mérida, Yucatán, Mexico

Parry ML, Carter TR, Konijn NT (1988) The impact of climate variation on agriculture, vols 1, 2. Kluwer Academic, Norwell

Ramamoorthy TP, Bye R, Lot A, Fa J (eds) (1993) Biological diversity of Mexico: origins and distribution. Oxford University Press, New York

Reilly JM, Baethgen W, Chege FE, Van de Geijn SC, Lin E, Iglesias A, Kenny A, Patterson D, Rogasik J, Roetter R, Rosenzweig C, Sombroek W, Westbrook J (1996) Agriculture in a changing climate: impacts and adaptation. In: Watson RT, Zinyowera MC, Moss RH (eds) Climate change 1995: impacts, adaptations and mitigation of climate change—scientific-technical analyses—contribution of working Group II to the second assessment report of the intergovernmental panel on climate change. Cambridge University Press, Cambridge, pp 427–467

Russell CS, Arey DG, Kates RW (1970) Drought and water supply. Johns Hopkins University Press, Baltimore, Maryland

Sauchyn DJ (2004) The paleoclimate record. Section 2.3. In: Barrow EB, Maxwell, Gachon P (eds) Climate variability and change in Canada: past, present and future. ACSD Science Assessment Series, Meteorological Service of Canada, Environment Canada, Toronto, Ontario 2: 7–9

Schmook B, Van Vliet N, Radel C, Manzón-Ché MJ, McCandless S (2013) Persistence of swidden cultivation in the face of globalization: a case study from communities in Calakmul Mexico. Hum Ecol 41(1):93–107

Shukla J, Nobre C, Sellers P (2008) Amazon deforestation and climate change. Science 247(4948):322–325

Stanhill G, Moreshet S (2004) Global radiation climate changes in Israel. Clim Change 22:121–138

Steward SA (2009) Livelihoods, coping strategies and processes of change following Hurricane Dean, Southern Yucatán, Mexico. Unpublished Master's Thesis, Clark University, U.S.A, pp 109

Wisner B, Fordham M, Kelman I, Johnston B, Simon D, Lavell A, Günter H, Oswald U, Wilches-Chaux G, Moench M, Weiner D (2007) Cambio Climático y Seguridad Humana. Red de estudios sociales en prevención de desastres en América Latina. http://www.desenredando.org/public/articulos/2007/clim-change/CCySH.pdf. Accessed 6 Feb 2011

Chapter 14
Making Science Count: Climate Change and the Science/Practice Interface

Harry Diaz and Margot Hurlbert

Abstract There is mounting evidence that global warming is producing variations in local weather patterns and water supplies, disturbing ecosystems and soil landscapes and impacting economic production and social conditions. Important changes in rainfall patterns, increases in temperatures and more intense and severe climate events have already been observed, with negative impacts for people's livelihoods. The multiple impacts of climate change could be a considerable source of risk, affecting patterns of development and local livelihoods, but they could also provide new opportunities (i.e., expansion of cultivated areas). In this perspective, climate change has been defined as a "wicked" problem for which there is no easy solutions and no simple approaches. It requires not only a multiplicity of perspectives to understand the phenomenon in all its complexity but also an increasing proximity between university researchers, policy-makers, industry, and citizen sector organizations to manage the risks and opportunities. A significant challenge, nevertheless, has been the limited integration between researchers and those government agencies that play a central role in the everyday management of development and natural resources. There are significant institutional and cultural barriers between researchers and policy-makers that hinder the transformation of scientific knowledge into plans and actions able to strengthen adaptive capacity. The chapter discusses climate governance and the science/practice interface in terms of the process of climate knowledge mobilization. Based on the dissemination experience of two multi-disciplinary policy-oriented projects, the presentation applies the lessons learned and compares these lessons to a variety of insights found in the related literature.

Keywords Climate change • Adaptation • Governance • Science/policy interface

H. Diaz (✉) · M. Hurlbert
Polo Diaz, Sociology and Social Studies, University of Regina, Regina, SK, Canada
e-mail: harry.diaz@uregina.ca

M. Hurlbert
e-mail: Margot.hurlbert@uregina.ca

W. Leal Filho et al. (eds.), *International Perspectives on Climate Change*, Climate Change Management, DOI: 10.1007/978-3-319-04489-7_14,

Introduction

Global warming is already producing variations in local weather patterns and water supplies, disturbing ecosystems and soil landscapes and impacting economic production and social conditions: intensive droughts in north-eastern Brazil, torrential rains in Colombia, tornados in North America, are a few examples of these disturbances. The future could definitely be worse to the extent that warming of the planet go unabated. Based on various emission scenarios, the Intergovernmental Panel on Climate Change (IPCC) estimates that by the year 2100, mean temperatures will rise between 2 and 6 °C in Latin America and North America (Field et al. 2007; Magrin et al. 2007). Based on the already established correlation between increases in the global average temperature and an increasing transformation of climate patterns (such as seasonal changes in rainfall and, more intense and severe climate events) and ecosystems (including the displacement of forests, degradation of agricultural soil and water resources, and others) we should expect increasing risks, with serious repercussions for the patterns of development and local livelihoods. There is also the possibility that climate change could also bring new opportunities (e.g., expansion of cultivated areas or new crops), but it is still uncertain what they are and how we could mobilize them in a perspective that is efficient, sustainable and just. What is absolutely certain is that an effective management of both threats and opportunities requires not only an expansion of our knowledge about climate change and its impacts but also an effective and robust integration of this knowledge into public policies and programs in order to increase the adaptive capacity of people, the state, and civil society.

In many American countries there is a deficit in the integration of scientific knowledge into climate policy-making. Some indicators of this deficit are the governments' failures in reaching an efficient international agreement around mitigation policies and their inability to mainstream climate into the public agenda, as well as a lack of attention among public agencies to anticipatory adaptive programs. The deficit resides not only in governments but also among scientists, who have a role that go beyond the simple task of disseminating the results of their work in society. In a world characterized by increasing risks scientists must also contribute to a more robust risk-related policy development by identifying obstacles, recognizing and understanding the needs and priorities of government agendas, and establishing effective communications channels that facilitate dialogue and mutual understandings to facilitate knowledge mobilization (Cash et al. 2003). In these terms, it is urgent to identify the specific obstacles that characterize the science/policy nexus and how they could be overcome, including the most efficient tools, strategies and methods that can be utilized to improve the flow of information between science and policy, to foster institutionalising evidence informed policy processes in the field of science, technology, innovation and climate change (Jones et al. 2008, p. vii). Only a robust, smooth interface between science and practice could "make research count" (Cardoen 2009) in a world where uncertainties and risks are becoming the norm. This chapter is part of an increasing collaboration between Canadian and Latin American researchers oriented to understand and research the interface in the area of global change in general and climate change in particular.

This chapter discusses the complexities of the science/policy interface in the area of climate change, an interface that in itself has its own complexities. Climate change is clearly one of those seemingly intractable (i.e. wicked) problems, which, it is argued, requires new ways of knowledge development and mobilization, including a more interdisciplinary and inclusive responsive science (Batie 2008; Brown et al 2010; Gibbons et al. 1994; Funtowicz and Ravetz 1993; Jasanoff 2005; Nowotny et al. 2001). Similarly scholars of policy processes emphasize the need for a new way of policy and management processes to deal with global change: not the exclusive activity of government bodies but principles of network steering whereby traditional intervention modes and policy instruments are replaced or supplemented by notions of reflexive governance (Voß et al. 2006), deliberative democracy (Hajer and Wagenaar 2003), transition management (Loorbach 2007; Rotmans et al. 2001), Mode-2 strategies (Regeer et al. 2009), and Q methodology (Brown 2006).

Taking into consideration these two assumptions, the need for an interdisciplinary science and a governance system, the chapter begins with a self-reflexive evaluation of two large research projects—both focused on the relationships between the environment and society—in terms of their ability to contribute to policy development. The projects were chosen due to their initial definition as policy-oriented projects and their strong emphasis on adopting an interdisciplinary perspective.

The chapter provides a short description of the two projects and of their organizational process. The final section of the chapter discusses some the insights generated by the evaluation of both projects a research framework able to generate a more fruitful understanding of the complexities of the interface.

Evaluation of the Two Projects

The two research projects that frame the discussion were focused on environmental issues, the first and the second dealing with the always difficult relationship between development and biodiversity. Both projects were considered scientific to the extent that their research components followed strict scientific procedures but they were also defined and organized as policy-oriented and, accordingly, they sought to influence decision policy making processes. The first of these projects—the Institutional Adaptation to Climate Change (IACC) project—dealt with the social dimensions of climate change and it involved an evaluation of the existing institutional capacity in Canada and Chile to reduce the climate vulnerability of rural stakeholders. The second project—the Great Sand Hills (GSH)—involved an assessment of human impacts on a very sensitive area of native prairie biodiversity and sand dunes in Western Canada. Table 14.1 provides some basic information on the two projects.

Table 14.1 The projects

Projects	Research teams	Stakeholders	Duration	Financial resources
IACC	30 researchers and young researchers from eleven disciplines	*Primary*: PFRA (Canada) & CONAMA (Chile) *Secondary*: Water and agricultural agencies, NGOs, local organizations, study communities	2004–2009	$ 2.5 million from the Social Sciences and Humanities Research Council of Canada
GSH	A "scientific advisory committee" (6 senior researchers) and 70 associate researchers (social and natural scientists)	*Primary*: Saskatchewan Environment and Regional Planning Committee of the GSH *Secondary*: Public agencies with an interest in the area, local governments and communities, first nation organizations	2005–2007	3.4 million from Saskatchewan Environment

The IACC was a unique inter-disciplinary and international project constituted by forty natural and social scientists focused on climate, water and institutions in Canada and Chile. The program was funded by the Major Collaborative Research Initiative (MCRI) Program of the Social Sciences and Humanities Research Council of Canada (SSHRC). Its goal was to develop a systematic and comprehensive understanding of the capacities of institutions to formulate and implement strategies of adaptation to climate change risks and to determine the projected impacts of climate change on the supply and management of water resources in the rural communities in two dry land river basins. The project research objectives were three: (1) to identify current and past vulnerabilities of rural communities to water and climate conditions; (2) to evaluate regional institutional capacities to reduce the future vulnerabilities of rural communities to climate change and its impact on the hydrological resources; and (3) to examine local vulnerabilities and institutional capacities in the context of potential future scenarios of climate change and potential risks. The project was conceived by a group of social and natural scientists concerned with the process of climate change and the lack of understanding of the role that society should play in order to reduce the risks associated with the transformation of climate. A small group of government scientists were integrated into the team and a scientific advisory committee was organized to provide permanent advice to the project.

The project was organized around the vulnerability approach, which emphasizes the need to analyze not only the future vulnerability of a system to climate change, but also its vulnerability in the context of past and current climate conditions. To maintain a coherent work it was decided to focalize on water issues given the dry climate of both basins. The empirical application of the approach requires both the development of a systematic understanding of both the current exposure of a system to climatic and other stresses as well as its adaptive capacity and the construction of future climate scenarios for the study areas; as well as an analysis of how existing vulnerabilities may be affected by future climatic conditions. The adoption of this vulnerability model facilitated an interdisciplinary approach informed by the work of social scientists, mainly focused on assessing present and past vulnerabilities, and the contributions of natural scientists in the areas of environmental vulnerabilities and building scenarios of future climate conditions.

The large research team—composed by 30 senior and junior researchers and research assistants—was formed during the design the proposal and involved a diversity of disciplines, including sociology, geography, psychology, history, political science, economy, engineering, mathematics, philosophy, agronomy, and biology. The research team has also included the active participation of governmental and non-governmental partners. This team was involved the development, implementation and integration of a large number of studies. They included an assessment of the current climate vulnerabilities of rural communities, an analysis of water conflicts, a historical study of institutional adaptation in periods characterized by water scarcities, studies of environmental vulnerabilities identified by stakeholders, and an assessment of the capacities of governance institutions to reduce the vulnerabilities of rural communities, and an assessment of the future climate scenarios for the two basins and their potential impacts. All these research activities and results are documented in the website of the project (http://www.parc.ca/research_projects-iacc.htm).

The GSH project, on the other side, was spatially and topically more focused. The project was supported by the Ministry of the Environment of Saskatchewan. The goal of the project was to develop and implement a strategic assessment of human activities that affect or impact the ecological integrity, as well as the sustainability of the Great Sand Hills area in the Western of Saskatchewan, one of the Canadian prairie provinces. The area, approximately 2,000 km^2, is a continuous patch of unconsolidated sands covered by native prairie and, accordingly, extremely sensitive to erosion if the native vegetation is disturbed. The area contains high levels of native biodiversity, including a number of rare and endangered species, and is considered to have a high degree of ecological integrity. It is mostly under the control and protection of the provincial government. The most important economic activity of the area is ranching. In recent years, however, there has been a significant development of the gas industry. Giving the significant deposits of gas in the area, and the potential conditions for conflicts between the gas industry and ranchers, the provincial government decided to carry out a large-scale assessment of the cumulative impacts of human activities over the GSHs in order to define future land uses and to delineate future areas for protection.

To carry out this assessment, an interdisciplinary Scientific Advisory Committee was established by the Ministry of Environment for the purpose of overseeing the design and implementation of the regional assessment and to formulate a set of recommendations to maintain the ecological integrity of the area. The committee was composed by a group of senior scientists from different disciplines, including a biologist from the Ministry, and worked with a small group of assessors from the Ministry, who provided administrative support, information, and contacts with several government agencies. This core was responsible for recruiting a number of researchers based on existing professional relationships and previous collaborations in order to carry the multiple studies required by the regional assessment, This arrangement allowed for a continuous two-way interaction between the scientists and the Ministry, an interaction that allowed to reduce many institutional obstacles.

The overall approach was based on a Strategic Environmental Assessment (SEA) methodology, which involved a systematic process for evaluating the environmental consequences of a diversity of actions, including policies, economic activities, programs and proposals. SEA is a proactive approach to planning and assessment, which anticipates future challenges and objectives and creates and examines alternatives that lead to a preferred sustainable development action(s). In this context studies were carried out in several area, including a climate profiling, surficial geology, gas reserves, water resources, biodiversity (fauna and vegetation), population and community, qualitative of life measures, governance, heritage resources, aboriginal use of the area, land use patterns, income and employment, and government revenues. A final report is available at the website of the Saskatchewan's Ministry of the Environment (www.environment.gov.sk.ca/GreatSandHillsRegionalEnvironmentalStudy).

These two projects had important successes, although with some significant differences. Both projects, as indicated before, were based on scientific research and oriented to inform policy, so they should be evaluated in terms of these two dimensions.

As scientific studies, a central goal of these projects was to improve our understanding of the areas that constituted their focus of attention. The IACC project was highly successful in these terms. The completion of the project allowed us to generate a deeper understanding of the capacity of governance to reduce the vulnerability of rural people. Moving beyond the idea that the simple presence of an "institutional capacity" and "well developed institutional systems" (IPCC 2001: 896–897) are conditions that ensure an adaptive capacity, the IACC project was able to demonstrate that even well established institutional systems have significant shortcomings and these are related not only to a lack of policy but also to the daily management practices of governments. The GHS project was no less successful to the extent that was able to generate significant empirical information about the impacts of specific human activities on biodiversity and integrate a multiple body of information into a range of sustainability scenarios. Both projects were also able to address the complexities of organizing and implementing an interdisciplinary approach, integrating different interdisciplinary interests and emphases into well coherent research results. In both cases the key for this interdisciplinary success was on an emphasis on a problem-driven approach, which channelled all efforts into a single perspective. In the IACC project the interdisciplinary effort was attained by establishing the concept of vulnerability as the central point of convergence for the multiple research programs. In the case of the GHS, the issue of regional sustainability was the lighthouse that provided the necessary direction and integration to the diversity of disciplinary studies. In both cases it was very helpful the early integration of scientists with a robust interdisciplinary experience able to foster the development of a combined and integrated perspective.

The two projects differed, however, in terms of the dissemination of the newly produced knowledge within the academic community. An intrinsic characteristic of any scientific project is the facilitating and enabling access to other scientists to the information that is being produced, so researchers always make a considerable effort to disseminating new findings in publications and scholarly meetings. In the case of the IACC project this task was an intrinsic part of the project. Many publications—books, journal articles, working papers, and graduate theses were produced in the 6 years of the project, as well as a larger number of conference papers were produced and delivered, increasing the profile of the project. As a result, spin-off outcomes of the project were several new international and national research proposals in the areas of vulnerability and adaptive capacity and the development and consolidation of an interdisciplinary research network in the area of climate change and adaptation, a network that has involved researchers from Canada, Colombia, Brazil, Argentina, Bolivia, and Chile. New projects and academic collaborations have also resulted in the training of new scholars and significant job opportunities for research assistants. The GSH project's academic dissemination results were very limited, with a very limited number of publications and public presentations. This shortcoming was very much the product of the conditions that framed the origin and development of the study. It was a government requested and financed study that was oriented to establish a program of regional development and, accordingly, a report with specific recommendations was expected from the Scientific Advisory Committee to inform the decision of the Minister regarding the future of the area. In these terms, there was a continuous process of negotiation about the use of the information for other purposes. This tension between the institutional needs of the Ministry of Environment and the process of knowledge mobilization was finally resolved by establishing rules that defined the conditions and control mechanisms for the dissemination of results, rules that certainly inhibited scholarly production.

Knowledge dissemination to the scholarly community is not the only form of mobilization that was relevant to both projects. As indicated before both projects were oriented to influence policy in their own areas and, accordingly, a systematic process of knowledge dissemination was implemented by both teams. In the case of the GSH project this process of informing policy and facilitating the dissemination of information among a diversity of stakeholder was highly successful. These stakeholders were involved throughout various stages of the baseline data collection process, particularly in terms of land use patterns and characteristics of community and quality of life studies. Moreover, they were offered constantly the opportunity to express opinions and make suggestions. The final report of the project, a document well over 200 dense pages accompanied by a CD containing a multiplicity of study reports, contained 62 recommendations. They ranged from issues related to biodiversity and conservation to governance the sustainability of regional communities. The Minister finally accepted most of these recommendations. In terms of a fruitful linkage between science and policy the GSH project was highly effective.

The story of knowledge mobilization within the IACC project has another narrative and, accordingly, different results. The IACC project, as in the case of the GSH, involved the organization and delivery of a multiplicity of activities oriented to inform local communities, practitioners, and the policy community. A large number of local, regional, and provincial workshops were organized to deliver the project's results and insight to the multiple stakeholders of the project. In addition, a Public Participation Geographic Information System approach was developed and implemented in all the rural communities involved in the project. Knowledge was also mobilized through a website that made all the documents produced by the project available to the public. In the case of the policy community, an especial effort was made to facilitate the transfer of knowledge from the project, an effort that included annual and final reports, booklets, participation in government workshops and conferences, and a special report for government agencies, which included ten general recommendations for changes to government programs and organizational patterns. The results of these multitude of dissemination activities were mixed.

On one side, it certainly contributed to increasing awareness of climate change within the rural community and practitioners. No less relevant, it allowed for establish solid linkages between some of the researchers and a group of professionals working for one of the agencies of the federal Ministry of Agriculture, a linkage that has become very fruitful for the development of new forms of research and working collaboration. On the other side, there has been no tangible results in terms of the incorporation of the scientific results and the assessments of risks to government policies and programs.

It is pertinent to say that there is indication that the government has redefined some of its programs in the perspective of the recommendations presented by the IACC project. As expected, establishing causal links between scientific results and these changes is difficult to prove. We would like to think however, that the IACC provided additional grist for the decision making mill and that the project was part of the wider conversation going on in the province regarding water management. While it is difficult to identify how much it influenced the process, one can nonetheless reasonably assert that it added a few additional grains of influence onto the balance that has led to changes.

A Discussion of the Barriers: Some Insights

There is an increasing number of publications that provide significant insights about the science policy interface and the barriers that impede a smooth flow of the information between scientists and policy makers. Some of these barriers seems to be related to the general dynamics that characterize each one of the spheres. In the policy area there are issues related to social and political contextual processes, such as the politicisation of science and reciprocal scientisation of policy (Hoppe 1999; Weingart 1999) as well as still existing tensions between economic development and environmental conditions. No less relevant are the tensions that emerge around epistemological issues, such as the tension between engaged citizen scientists and neutral scientists in relation to the role of scientists in the policy process (Higgins et al. 2006), the incorporation of local and indigenous knowledge to the knowledge production process (Liberatore and Funtowicz 2003), or the need for specialized expertise versus the democratization of knowledge (Cash et al. 2003).

Other contributions speak to us about differences between the realms of policy and science, which impose a cleavage difficult to bridge. Von Storch (2009), for example, argues that the "cultural construct" and the "scientific construct" offer competitive interpretations of the reality, a discrepancy where scientific assessments are only appreciated by policy-makers when they confirm prior knowledge established in their cultural constructs. In the same vein, Hegger et al. (2012) and Hegger and Dieperinck (2012) identified argue that science and public policy have their own timeframes, epistemologies, goals, reward structures, process-cycles and criteria for judging the quality of knowledge (see also Dilling and Lemos 2011; Kasperson 2011a). Kasperson argues in the same direction, indicating that the most serious barriers to a successful transfer of knowledge are constituted by knowledge inadequacies, conflicting conceptual frameworks, limited communication between science and practice, as well as social distrust and lack of credibility (2011b: 433). In this perspective, the rift between science and practice seems to be linked to existing different institutional nests—each one of them with its own horizon of meanings and definitions of what is necessary and relevant in relation to the generation and social use of knowledge.

No less relevant seems to be issues pertinent to the specific area that the science/policy interface intend to cover. There are certain areas where there is a social and political consensus of their relevance and of the need to develop evidence-based policy in order to confront their challenges. Other areas seems to be more problematic, such as climate change. Thus, climate science may be questioned, either because of the high level of uncertainty that characterize many areas of the scientific work, or because the findings demand important redefinitions of the ideas that sustain these cultural constructs (Choi et al. 2005; Sarewitz 2011). In this context, it is not strange that issues such as skepticism, misunderstanding, and limited communication emerge from these gaps in the science/practice interface.

A systematic analysis of the existing literature identifies three main areas relevant to a systematic analysis of the science/policy interface: the political context, the quality of the scientific evidence, and the nature of the links that exist between the policy and research communities (Jones et al. 2008; Court and Young 2003). Using these three areas as lens to analyze the two projects provides us with some relevant insights.

The dimension of evidence is related to the research process and has to do with issues of credibility, the methodological approach, and the degree of consensus arrived within research teams (Court and Young 2003: 16; see also Ingram et al. 2007; Brenner 2011). Both projects were very strong in relation to the quality of the evidence produced in the course of the research. The integration of well known scholars in both teams provided the necessary scientific credibility and legitimacy to the results. Moreover, the interdisciplinary nature of the teams and the high degree of integration reached in the projects facilitated a proper methodological approach and a comprehensive set of results.

Issues related to the political context and the nature of the links between science and policy were significantly more relevant in the both projects. As discussed previously, the GSH project was a project oriented to produce a "tailored assessment" and, consequently, its results were highly helpful to decision-making. The goal and objectives if the project were established by the user, the provincial ministry of the environment, leaving to the discretion of the scientific team the procedures to attain these objectives. In these terms issues relevant to the political context were irrelevant to the research team. In the same vein, the fact that the project was originally tailored to the need of the user facilitated the links between the scientists and the policy makers. As explained previously, a group of professional from the staff of the Ministry of the Environment participated as advisors in the Scientific Committee, facilitating information and contacts that otherwise would have been difficult to obtain or establish. In this context, this tailored project had not only the necessary scientific credibility but it also produced a knowledge that was practical, useful, accessible, and acceptable (Ingram et al. 2007). It imposed, however, some limits to the capacity of disseminate knowledge to other users, an aspect that it is important to register as an experience.

The experience of the IACC project was clearly different. It was a policy-oriented project designed and implemented by scientists with a somehow naïve understanding of the complexities of the policy-making process, assuming that the process of informing policy was a unidirectional and direct process. In spite of a significant effort to create and implement a strategy of knowledge mobilization, the project was unable to establish the required credibility to interest those involved in establishing a policy agenda. Part of the problem was related to the absence of mechanisms to monitor the impacts of knowledge mobilization efforts. Under the light of the insights developed by the literature we could argue, however, that the main problems were related to a lack of knowledge of the political context that the IACC project was trying to influence, as well as the absence of well forged links between the scientists and the policy community.

One of the most significant problems was a lack of recognition of how difficult is to influence the Canadian federal government and some of the provincial governments in the area of climate change. During the period in which the project took place both the federal and the two provincial governments relevant to the project, Saskatchewan and Alberta, were controlled by a group of conservative politicians who ignored the relevance of climate change in the public agenda. In these terms the high quality knowledge produced by the project never attracted the interests of key policy-makers. Mobilization. The lack of a systematic knowledge of government and its processes was not only related to the limited significance of climate change in the policy agenda but also about the government organizational structure. The IACC project was focused on the impacts of climate on rural water resources, which required a knowledge mobilization program directed to a multitude of government agencies operating at different levels of government. These organizational balkanization of water governance in Canada, characterized by limited levels of inter-agency coordination and communications, made the process of knowledge mobilization more difficult. The dissemination program was, accordingly, easily lost in the myriad of organizational interests. Our limited knowledge of the policy context and of the organizational characteristics clearly limited the initial organization of robust links between the project and the policy community. The project established excellent contacts with one of the federal agencies, contacts that have allowed us launch new projects with this agency as a central partner, but obviously they were not enough to consolidate a multiplicity of links that were required by the organizational diversity of governance.

Conclusions

The experience of the IACC project generated the need for a better understanding of the science/policy interface. It is clear that significant improvements in the areas of interdisciplinary research lead to a better knowledge of climate change, but this cannot become useful and practical scientific knowledge if we do not understand the political context and the complexities of establishing robust links with a multiplicity of actors, including not only professionals, but also high level managers, advisors, and legislators, as well as civil society stakeholders with the capacity to "reach" those who establish the research agenda.

An important step in building this new knowledge is to develop a better understanding of the interface by using those research capacities and tools that are familiar to us. Given the particularities of different government contexts and organizational patterns it is almost essential to carry out international comparative research as a way to have a better knowledge of the multiplicity of contexts and processes that characterize the science/policy. Some areas relevant to this purpose are the impact of ideologies—such as neoliberalism—on the degree of aperture of policy-makers to scientific evidence; the mechanism that allow for the credibility of scientists and research processes, and the roles of a diversity of intermediaries that could facilitate bridging the gap between science and the policy community. An interesting alternative—or complement—could be to allocate funds and resources to the creation of a group of science and policy intermediaries with the capacity to understand policy processes and answer key questions of policy makers, as well as represent the views of the scientists and policy makers in a facilitative, communicative, and conflict resolving (if necessary) manner. Building this concept into research institutions, research projects and government might reduce what has been the wicked problem of climate change to a structured problem of managing climate risk, mitigating climate change, and capitalizing on potential adaptations.

References

Batie S (2008) Wicked problems and applied economics. Am J Agric Econ 90(5):1176–1191

Brenner RM (2011) Knowledge to practice in the vulnerability, adaptation, and resilience literature: a propositional inventory. In: Kasperson R, Berberian M (eds) Integrating science and policy. Vulnerability and resiliency in global environmental change. Earthscan, London

Brown SR (2006) Q methodology and naturalistic subjectivity. In: Midgley B, Morris E (eds) Modern perspectives on J.R. Kantor and Interbehaviorism. Context Press, Reno, pp 251–268

Brown V, Harris J, Russell J (2010) Tackling wicked problems through the transdisciplinary action. Earthscan, London

Cardoen F (2009) Knowledge to policy. Making the most of development research, IDRC/Sage, Ottawa

Cash DW, Clar WC, Alcock F, Dickson NM, Eckley N, Guston DH, Jager J, Mitchell RB (2003) Knowledge systems for sustainable development. Proc Nat Acad Sci 100(14):8086–8091

Choi BCK, Pang T, Lin V, Puska P, Sherman G, Goddard M, Ackland MJ, Sainsbury P, Stachenko S, Morrison H, Clottey C (2005) Can scientists and policy-makers work together. J Epidemiol Commun Health 59(8):632–637

Court J, Young J (2003) Bridging research and policy: insights from 50 case studies. ODI working paper #213, ODI, London. Available at www.odi.org.uk/publications/219-bridging-research-policy-insights-50-case-studies

Dilling L, Lemos MC (2011) Creating usable science: opportunities and constraints for climate knowledge use and their implications for science policy. Glob Environ Change 21:680–689

Field CB, Mortsch LD, Brklacich M, Forbes DL, Kovacs P, Patz JA, Running SW, Scott MJ (2007) North America. In: Parry ML, Canziani OF, Palutikof JP, van der Linden PJ, Hanson CE (eds) Climate change 2007: impacts, adaptation and vulnerability. Contribution of working Group II to the fourth assessment report of the intergovernmental panel on climate change. Cambridge University Press, Cambridge

Funtowicz SO, Ravetz JR (1993) Science for the post-normal age. Futures 25(7):739–755

Gibbons M, Limoges C, Nowotny H, Schwartzman S, Scott P, Trow M (1994) The new production of knowledge: the dynamics of science and research in contemporary societies. SAGE, London
Hajer MA, Wagenaar H (eds) (2003) Deliberative policy analysis—understanding governance in the network society. Cambridge University Press, Cambridge, UK
Hegger DLT, Dieperink C (2012) Patterns in joint knowledge production projects for regional climate change adaptation—lessons for project and program design. Presented at the Lund earth systems governance conference, Lund, Sweden, April 2012
Hegger D, Lamers M, Van Zeijl-Rozema A, Dieperink C (2012) Conceptualising joint knowledge production in regional climate change adaptation projects: success conditions and levers for action. Environ Sci Policy 18:52–65
Higgins PAT, Chan KMA, Porder S (2006) Bridge over a philosophical divide. Evid Policy 2(2):249–255
Hoppe R (1999) Policy analysis, science and politics: from "speaking truth to power" to "making sense together". Sci Pub Policy 26(3):201–210
Ingram J, Stone J, Confalonieri U, Garvin T, Jutro P, Klink C, Lucjman B, Noellemeyer E, Toledo P (2007) Delivering global environmental change science to the policy process. In: Tiessen H, Brklacich M, Breulmann G, Menezes R (eds) Communicating global change science to society. An assessment and case studies. Editorial Island Press, Washington
IPCC (Intergovernmental Panel on Climate Change) (2001) Climate change 2001: impacts, adaptation, and vulnerability technical summary. A report of working Group II of the intergovernmental panel on climate change, Cambridge University Press, Cambridge
Jasanoff S (2005) Designs on nature: science and democracy in Europe and the United States. Princeton University Press, Princeton
Jones N, Jones H, Walsh C (2008) Political science? Strengthening science-policy dialogue in developing countries. Working paper 294, ODI (Overseas Development Institute, London
Kasperson R (2011a) Characterizing the science/practice gap. In: Kasperson R, Berberian M (eds) Integrating science and policy. Vulnerability and resiliency in global environmental change. Earthscan, London
Kasperson R (2011b) Directions for closing the scientific/practice gap. In: Kasperson R, Berberian M (eds) Integrating science and policy. Vulnerability and resiliency in global environmental change. Earthscan, London
Liberatore A, Funtowicz S (2003) "Democratising" expertise, "expertising" democracy: what does this mean, and why bother? Sci Pub Policy 30(3):146–150
Loorbach D (2007) Governance for sustainability. Sustain: Sci Pract Policy 3(2):1–4. Published online 11 Dec 2007 Available at http://www.bing.com/archives/vol3iss2/editorial.loorbach.html
Magrin G, Gay García C, Cruz Choque D, Giménez JC, Moreno AR, Nagy GJ, Nobre C, Villamizar A (2007) Latin America. In: Parry ML, Canziani OF, Palutikof JP, van der Linden PJ, Hanson CE (eds) Climate change 2007: impacts, adaptation and vulnerability. Contribution of working Group II to the fourth assessment report of the intergovernmental panel on climate change. Cambridge University Press, Cambridge, pp 581–615
Nowotny H, Scott P, Gibbons M (2001) Rethinking science: knowledge in an age of uncertainty. Polity, Cambridge
Regeer BJ, Hoes AC, van Amstel-van Saane M, Caron-Flinterman FF, Bunders JF (2009) Six guiding principles for evaluating mode-2 strategies for sustainable development. Am J Eval 30:515–537. http://aje.sagepub.com/content/30/4/515
Rotmans J, Kemp R, Van Asselt M (2001) More evolution than revolution: transition management in public policy. Foresight 3(1):15–31
Sarewitz D (2011) Does climate change knowledge really matter? Wiley Interdisciplinary Reviews: Climate Change 2(4):475–481
Von Storch H (2009) Climate research and policy advice: scientific and cultural constructions of knowledge. Environ Sci Policy 12:741–747
Voß JP, Bauknecht D, Kemp R (eds) (2006) Reflexive governance for sustainable development. Edward Elgar, Cheltenham
Weingart P (1999) Scientific expertise and political accountability: paradoxes of science in politics. Sci Pub Policy 26(3):151–161

Chapter 15
Island Erosion and Afflicted Population: Crisis and Policies to Handle Climate Change

Tuhin Ghosh, Rituparna Hajra and Anirban Mukhopadhyay

Abstract Environmental Refugees are unable to maintain a secure livelihood in their own habitat due to environmental hazards, with minor optimism to return. This category includes the people who are displaced due to the disruption in physical and/or social systems, and subsequent losses or degradation of ecosystem services. The impacts of sea level rise in combination with complex hydrodynamic conditions, have caused severe coastal erosion on islands of the Indian Sundarban. In a recent past, within the Hugli River (lower course of River Ganga) estuary, three islands namely Lohachahara, Suparibhanga and Bedford completely submerged and Ghoramara Island eroded significantly which resulted in a considerable population of environmental refugees. In 1991 there were 374 inhabitants in the Lohachara Island who became landless after submergence, and were compelled to move other places. Ghoramara Island is located between 21° 53′ 56″ N to 21° 55′ 37″ N latitude and 88° 06′ 59″ E to 88° 08′ 35″ E longitude within the Hugli estuary of western part of Indian Sundarban. The major occupation of local people is agriculture, fishing and prawn seed collection. Time series analyses using multi-temporal satellite imageries of 1975 and 2010 unfold the erosional pattern of this island. Some of the distinct villages of this island are already under water. Due to the displacement from their own habitat and also gradual loss in ecosystem services, increased rate of migration in this island has resulted. The poorer people who lost their homeland were compelled to move towards mainland areas like Kakdwip/Namkhana or comparatively stable islands like Sagar Island,

T. Ghosh (✉) · R. Hajra · A. Mukhopadhyay
School of Oceanographic Studies, Jadavpur University, Kolkata 700032, India
e-mail: tuhin_ghosh@yahoo.com

R. Hajra
e-mail: rituparnahajra2502@gmail.com

A. Mukhopadhyay
e-mail: anirban_iirs@yahoo.com

W. Leal Filho et al. (eds.), *International Perspectives on Climate Change*,
Climate Change Management, DOI: 10.1007/978-3-319-04489-7_15,

with little or without any token compensation from the Government. Some of the economically stable people migrated to their other properties in the central part of the Island, hence ensuring their wellbeing. This people are often still paying land tax for their lost land, with little hope of becoming compensated by either cash or land, in the near future.

Keywords Erosion • Refugee • Migration • Rehabilitation policy • Ghoramara Island

Introduction

In the context of climate change and aggravated natural hazards, environment related migration is evolving as a global crisis due to the displacement of larger sections of population from their own original areas. The concept of Environmental Refugee was first introduced by Lester Brown, from the World Watch Institute in the 1970s (Black 2001). Contemporary studies by El-Hinnawi (1985) and Jacobson (1988) on forced migration due to environmental degradation and natural hazards, further popularized the concept. This group of people are forced to leave their own habitat temporarily or permanently, due to the disruption in their supporting ecosystems or from environmental hazards (El-Hinnawi 1985). Jacobson (1988) expressed that the environmental refugees are evolving as a single largest cluster of displaced persons in the globe. The concept of environmental refugee has become much more popularized, and has been receiving proper attention during recent times (Mayer 2002).

The waves of environmental migrants that spill across borders can destabilize domestic law and order, and also international relations, associated with environmental degradation (Homer-Dixon 1991). Also, Myers (2002) described the environmental migrants often cannot have any hope to get secured livelihood in their homelands, and little or no chances to return. Here, the migrants are considered to experience the so-called trans-border migration. But, the migration may not be always trans-border. During decades, the close links between environmental degradation and forced migration has been observed (UNHCR 1993), and the social/economic/political reasons—rather than any single driving factor—have predominantly played a role. It is very difficult to distinguish the migrants driven by environmental or economic factors. Most often, poorer people driven by environmental degradation (Myers 1993), indicate that the definitions of such refugees are ambiguous and inconsistence. In fact, governments and international agencies find it difficult to recognise environmental migrants to, as they often do not have any legal basis.

The 1951 United Nation Convention on Refugees and the subsequent amendment in 1967 expressed the initial international law or standard criterion to define the term refugee and Bate (2002) showed that the compulsion can vary from moderate to intense. There are several conflicts on the definition of the refugees, on the basis of the 1951 UN Convention and a global debate is ongoing for its refinement (Bell 2004; Black 2001) as the Geneva treaty on International Refugee Legislation and United Nations High Commissioner of Refugees were established after World war II (Stojanov et al. 2008). This revision needs to clarify the new emerging groups of migrants like Internally Displaced people, Economic Refugee, Humanitarian Refugee and Environmental Refugee. As an alternative, the concept of internally displaced persons (IDPs) appears relevant to those suffering displacement by environmental change and the Guiding Principles on Internal Displacement, along with national implementation policies, indicate a developing and expanding regime potentially suitable for climate change migrants (Williams 2008). The UNHCR's Guiding Principles on Internal Displacement identifies rights and guarantees for the protection of those suffering forced displacement (UNHCR 1998). So far, the problem of environmental refugee is increasing. Based on this fact, organizations such as the International Organization for Migration, the UN Environment Programme and IPCC are taking initiatives on this matter (Black 2001).

Large scale migration is often taken as a 'cause' rather than the 'effect' of environmental degradation (Williamson 1996). Increasing flow of migrants put additional pressure on the resources of the destination areas, and threatens the livelihoods of present inhabitant (Panda 2010). In addition, movers may have to accept whatever opportunities come their way in the new location (Lonergan 1998). Complete resettlement of environmental migrants is difficult (Mayers 1993). Movement may lead to the substitution of one set of stresses (environmental) for another (economic, social, political and/or further environmental stresses) (Lonergan 1998). An appropriate plan is necessary to combat the negative impact of environmental migrants on the ecological sensitive area like Ghoramara Island (Hazra and Bakshi 2003).

Environmental Migrants from Vanishing Island

The IPCC (1990) estimated that climate change will aggravate the displacement of people and further migration due to various disruptions in their wellbeing. There is a projection of 150 million people by 2050 (Myers 2002), in which 50 million people will be displaced due to a 1 m sea level rise (Jacobson 1988). Almost 120 million people could be rendered homeless by 2,100 both in India and Bangladesh, due to sea level rise and given the proximity of Bangladesh to India much of the people will end up as migrants in Indian cities which are already facing resource scarcity (Rajan 2008).

Myers (2002) argues that climate migrants from Bangladesh alone might outnumber all current displaced population worldwide. Similarly, a large stretch of the Indian coast is under risk due to sea level rise, along with intensifying cyclone and larger storm surges. As India possess second largest population (63 million), 7,500 km coastline, 60 % people living in the low lying coastal zone and seventh in terms of area (82,000 sq km) (McGranahan et al. 2007). According to Kelman (2006) climate

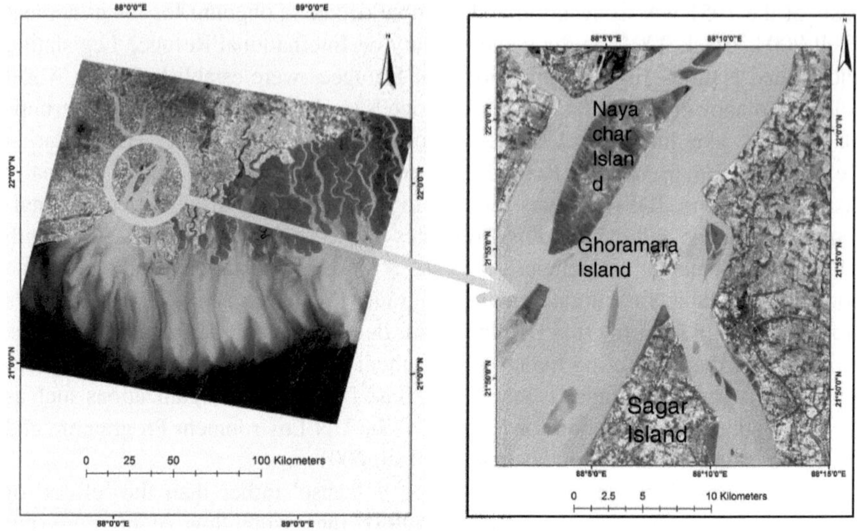

Fig. 15.1 Location map of Ghoramara Island

change often effects on freshwater quality and geomorphological changes which might make some Small Island Developing States uninhabited forcing migration. In the Hugli River estuary within the Sundarban Island system on Bay of Bengal there are evidences of submergence of islands, severe erosion and migration of island dwellers (Ghosh et al. 2003; Hazra et al. 2002; Hazra and Bakshi 2003). In recent past, three such islands namely Lohachara, Suparibhanga, and Bedford completely submerged (Ghosh et al. 2003) and another populated island Ghoramara experiencing severe erosion and aerial loss (Ghosh and Sengupta 1997; Ghosh et al. 2003). Ghoramara Island located between 21° 53′ 56″ N to 21° 55′ 37″ N latitude and 88° 06′ 59″ E to 88° 08′ 35″ E longitude within the Hugli estuary of western part of Indian Sundarban (Fig. 15.1). The major villages within the Ghoramar Island are Khasimara, Khasimara Char, Lakshmi Narayanpur, Bagpara, Baishnabpara, Hatkhola, Raipara, Mandirtala, and Chunpuri, of which first five villages are already under water (Jana et al. 2012; Ghosh and Sengupta 1997; Ghosh et al. 2003). The island is under threat of severe erosion, embankment failure, cyclone and storm surge (Hazra and Bakshi 2003; Ghosh et al. 2003). Largely, these are the factors for habitat loss, gradual loss in ecosystem services, depreciation in wellbeing and resultant migration of the people of this island.

Methodology

In this study, the use of primary and secondary data has been employed, as there is a serious lack of migration data with local administrations. Primary data of the study area was collected through field surveys with structured questionnaire

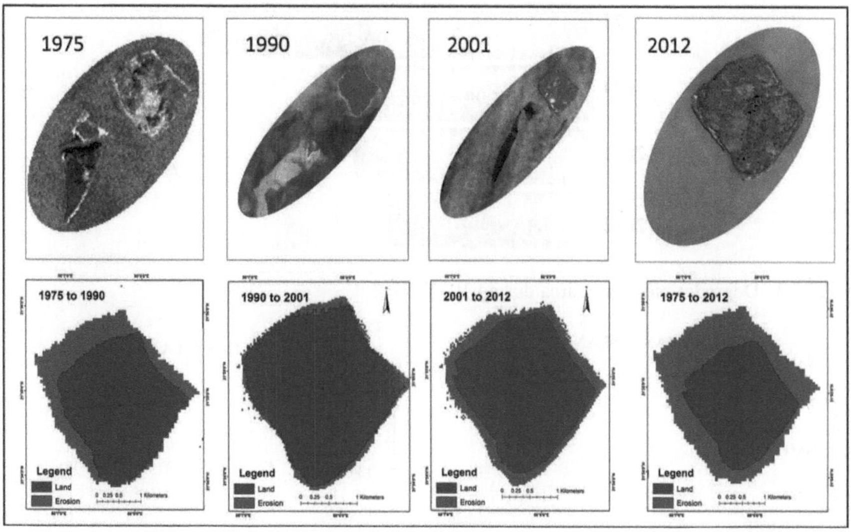

Fig. 15.2 Time series analysis to identify the extent of erosion in Ghoramara Island during 1975–2012

surveys involving the local inhabitants of Ghoramara and Sagar Islands, while satellite image analyses show the level of degradation on the basis of coastal erosion.

A questionnaire survey revealed the impact of coastal erosion on the population, their livelihood and the adaptive response as a form of accelerated migration. This has also unfold the perception of local people on erosion induced migration, managements etc. Time series satellite image analysis was done using the multitemporal satellite imageries of 1975, 1990, 2001, 2010 to understand the changing patterns of the islands. Secondary data from different years were collected from Census of India, and the population dynamics was studied.

Result and Discussion

The time series analysis is showing a considerably higher areal loss of Ghoramara Island. Satellite images of 1975, 1990, 2001 and 2010 have been used for this change detection study (Fig. 15.2).

The time series analysis shows that total area of Ghoramara was 8.51 sq km in 1975, decreased substantially to 4.43 sq km in 2012 which clearly indicates the severity of erosion (Fig. 15.3). Erosion was higher during 1975–1990 and in this period Lohachara, Suparibhanga and Bedford Islands become submerged along with the Khasimara, Khasimara Char, Lakshmi Narayanpur, Bagpara, Baishnabpara villages of Ghoramara.

Among these islands, Lohachara and Ghoramara were populated. Thus a gradual process of habitat loss is observed, which drives the urgency for migration.

Fig. 15.3 Degradation of land area during 1975–2012

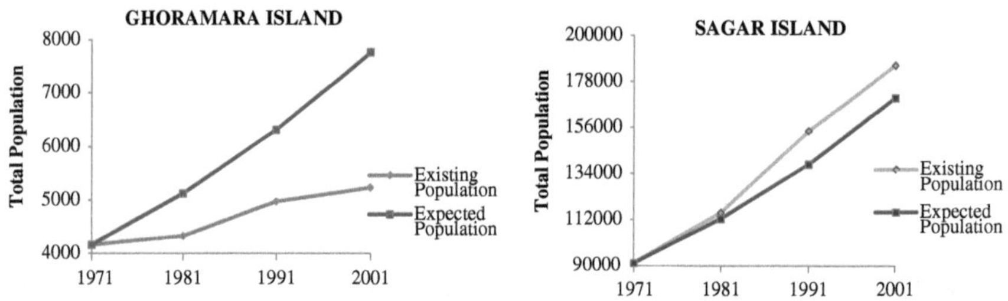

Fig. 15.4 Population growth trend in Ghoramara and Sagar Island from 1971 to 2001

The major event of migration occurred during 1975–1990, due to the massive erosion within that island cluster.

This populous island ecosystem is rapidly changing its morphology due to the exclusive erosion. Lohachara and Bedford Islands disappeared from its original location and triggered the migration of their local inhabitants. As a result, the rate of population growth in Ghoramara Island is 0.55 % while the overall growth the Sagar administrative block is much higher than that, about 2.1 % per annum. Only due to migration, the actual growth is much lower than the expected growth (as per the block level growth rate) (Fig. 15.4). The census of India shows very slow growth rate in Ghoramara from the year 1971, 1981, 1991, and 2001, in comparison to other stable islands nearby (Fig. 15.4) (Population Census of India; 1971, 1981, 1991, 2001). The displaced people mainly took shelter at nearby islands like Sagar Island. During 1981–1991 the actual population of Sagar Island became higher than the expected population (with usual 2.1 % growth rate), due to large number of migrants took shelter in Sagar Island for the submergence of Lohachara and five villages of Ghoramara. The population has been decreased from 5236 to 5193 during the year 2001–2011 which may be attributed to migration (Fig. 15.4), in spite of existing growth rate within the same administrative area.

It is hard to keep an eye on the migration flow as actual record of migrants is not available from the administration. The local people's perception in majority is around 4000 people migrated from Ghoramara island (Fig. 15.5). This quantification of migrated people has been inferred from the questionnaire survey, as the respondent estimated the number of the family migrated due to erosion.

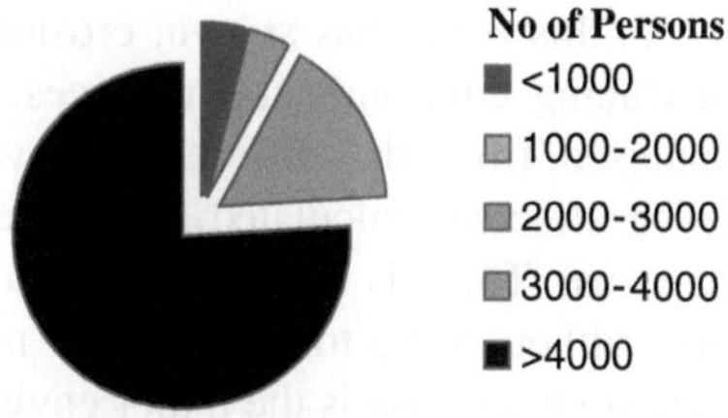

Fig. 15.5 Number of out migrants from Ghoramara Island (according to islanders' perception)

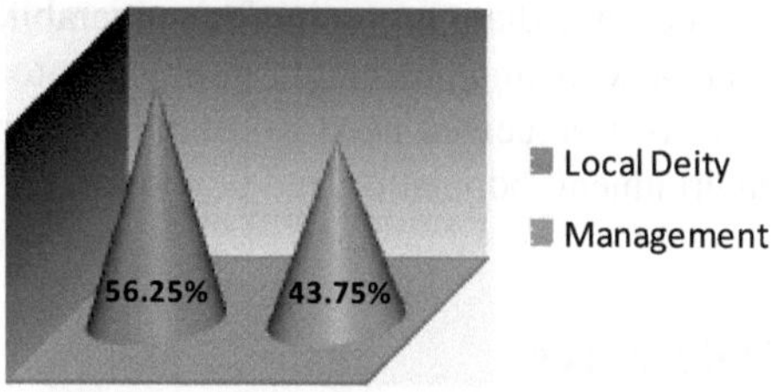

Fig. 15.6 Percentage of community trust on management and fatalism

As the migration is also related with human behavior with 'push' factors encourage migration out of their own decision. Gradually, with no or little effort for policy formulation to combat the worst situation was dramatically changed the mindset of the community. It is revealed from the survey that 56 % of the inhabitants of Ghoramara have faith on local deity rather than the management strategies (Fig. 15.6). It halts the implementation of preparedness efforts and the community becomes more vulnerable. This fatalism is often driven by their socio-economic condition, which mostly determined by the proximity to the hazard prone area, and becomes more 'unlucky' according to them. Absence of awareness and lack of capacity building is common among the inhabitants.

Conclusion

Worldwide coastal erosion is a common feature in any river estuary, often associated with huge and intensive sediment discharges. The estuary of the River Hugli is no exception, and the erosion within this estuarine island system is likely to displace thousands of vulnerable people in the future, compelling them to move away from their own home areas. According to international laws, these displaced people are yet to be recognized as environmental refugees. There seems to be no official data related to climate-related migration, which shows a deficiency in respect of research. The tracking of migrants from the sending areas and from the receiving areas is a critical task. In reality, the scope for the formulation of policy for their rehabilitation is still uncertain. The complexity of the natural systems and their functional relationship with society is a major hindrance for this kind of policy formulation.

As this paper has shown, erosion has already washed out several settlements, producing environmental refugees. The estimation of total migration has been analysed from the difference between the expected and estimated population to be affected, calculated using the population growth rate and population data (Census). There is a real time relationship between the coastal erosion and migration, although the total migration never complies with that. Even though, in this case study erosion is the major environmental driving factor, there are a number of other reasons and factors behind this migration.

The immediate necessity for vulnerability assessments to formulate a common agenda with implementable policy has to be considered in a regional scale, and there is a perceived need to establish a linkage between climate phenomena, the environment, population and poverty.

References

Bate DC (2002) Environmental refugees? Classifying human migrations caused by environmental change. Popul Environ 23(5):465–477
Black R (2001) Environmental refugees: myth or reality? Working paper No. 34, UNHCR, Mar 2001
El-Hinnawi E (1985) Environmental refugees. United Nations Environment Programme, Nairobi
Ghosh T, Bhandari G, Hazra S (2003) Application of a 'bio- engineering' technique to protect Ghoramara Island (Bay of Bengal) from severe erosion. J Coast Conserv 9:171–178
Ghosh T, Sengupta SK (1997) Morphological changes of Ghoramara Island, West Bengal: documentation. Indian J Geogr Environ 2:64–65
Hazra S, Bakshi A (2003) Environmental refugees from vanishing Islands. In: Bhattacharya P, Hazra S (ed) Environment and human security. Lancers' Books, pp 219–227
Hazra S, Ghosh T, Dasgupta R, Sen G (2002) Sea level and associated changes in the Sundarbans. Sci Cult 68(9–12):309–321
Homer-Dixon T (1991) On the threshold: environmental changes as causes of acute conflict. Int Secur 16(2):76–116
Jacobson J (1988) Environmental refugees: a yardstick of habitability. World watch paper, no. 8. World Watch Institute, Washington, DC, pp 46
Jana A, Sheena S, Biswas A (2012) Morphological change study of Ghoramara Island, Eastern India using multi temporal satellite data. Res J Recent Sci 1(10):72–81
Kellman I (2006) Island security and disaster diplomacy in the context of climate change. Les Cahiers de la Se′curit 63:61–94
Lonergan S (1998) The role of environmental degradation in population displacement. Environmental change and security project report, Issue 4 (Spring 1998), pp 5–15
McGranahan G, Balk D, Anderson B (2007) The rising tide: assessing the risks of climate change and human settlements in low elevation coastal zones. Environ Urbanisation 19(1):17–37
Myers N (1993) Environmental refugees in a globally warmed world. Bioscience 43(11):752–761
Myers N (2002) Environmental refugees: a growing phenomenon of 21st Century. Philos. Trans. R. Soc. Lond. B Biol. Sci. 357(1420):609–613
Panda A (2010) Climate refugees: implications for India. Econ Polit Weekly xlv(20):76–79. Available at: http://ssrn.com/abstract=1935922, 15 May 2010
Population census of India (1971) India Census Office, New Delhi
Population census of India (1981) India Census Office, New Delhi
Population census of India (1991) India Census Office, New Delhi
Population census of India (2001) India Census Office, New Delhi

Rajan SC (2008) Blue Alert. Greenpeace India society. Available at: www.greenpeaceindia.org
Stojanov R, Novosák J, Opiniano JM, Emenne F, Siwek T (2008) Development, environment and migration, analysis of linkages and consequences. Palacký University, Olomouc, p 222
United Nations High Commissioner for Refugees (1993) The state of the world's refugee: the challenge of protection. Switzerland, Geneva
United Nations Refugee Agency (UNHCR) (1998) Guiding principles on international displacement (E/CN.4/1998/53/Add.2)
Williamson C (1996) The immigration mystique. Basic Books, New York
Williams A (2008) Turning the tide: recognizing climate change refugees in international law. Law Policy 30(4):502–529

Chapter 16
Does Awareness Through Learning About Climate Change Enhance Farmers' Perception of and Adaptation to Climate Uncertainty?

Rosaine Nérice Yegbemey, Gauthier Biaou, Jacob Afouda Yabi and Sènakpon Eric Haroll Kokoye

Abstract Finding out ways to support farmers' decision to adapt to climate change is a challenge for agricultural policy. Hypothesizing that awareness through learning about climate change enhances farmers' perception of and adaptation to climatic uncertainty, this paper gives an empirical evidence of the relationships between learning about climate change and farmers' perception of, and adaptation to new climate conditions. The study was conducted in northern Benin by a survey method on 336 farmers. Talking about climate change, farmers mainly perceive it based on observed changes in their daily life environment, implying from own experience. About 56 % of the farmers were also aware through learning about climate change and the most predominant vectors of climate change information (although indigenous knowledge) are neighbour farmers or farmers associations and extension services. Using the Heckman probit model built on the assumption that the perception of is a precondition for the adaptation to, the results highlighted that the main driving forces

R. N. Yegbemey (✉)
Institute of Project and Regional Planning; Faculty of Agriculture, Nutrition, and Environmental Management, Justus-Liebig University of Giessen, Senckenbergstrasse 3, 35390 Gießen, Germany
e-mail: yrosaine@hotmail.fr; rosaine.n.yegbemey@agrar.uni-giessen.de

G. Biaou
Faculté des Sciences Agronomiques, Université d'Abomey-Calavi, Cotonou, Benin
e-mail: gbiaou@yahoo.fr

J. A. Yabi
Laboratoire d' Analyse et de Recherche en Economie et Développement (LARED), Département d' Economie et Sociologie Rurales, Faculté d'Agronomie, Université de Parakou, BP 123, Parakou, Benin
e-mail: ja_yabi@yahoo.com jacob.yabi@fa-up.bj

S. E. H. Kokoye
Department of Agricultural Economics and Rural Sociology, Auburn University, Auburn, USA
e-mail: kharsene@yahoo.fr

W. Leal Filho et al. (eds.), *International Perspectives on Climate Change*, Climate Change Management, DOI: 10.1007/978-3-319-04489-7_16,

of farmers' decision to adapt to climate change are level of education and organisation membership. On the other hand, the main determinants of farmers' perception are gender, contact with extension and awareness of climate change trends. Subsequently, enhancing farmers' awareness through learning about the predicted trends of climate change is likely to strengthen their perception of and furthermore encourage their adaptation to.

Keywords Climate change • Awareness • Perception • Adaptation • Heckman probit model

Introduction

Climate change is predicted to have harmful effects on agriculture which is mainly rain-fed in developing countries. In Benin (West Africa), agriculture contributes to about 39 % to the Gross Domestic Product (DSCRP 2007) and plays a major role in the livelihoods of the population, especially in rural areas. As a matter of fact, the whole livelihoods system is likely to be impacted by both climate variability and change.

Predictions of future climate conditions in Benin highlighted that rainfall, by 2100, will decrease about 0.2 % and 15 % in the southern and the northern parts of the country, respectively (MEHU 2011). Moreover, temperature is expected to increase between +2.6 and +3.2 °C by 2100 at the country level (MEHU 2011). In the recent years, the future climate conditions have become a concern as they are additional constraints on agriculture. According to Aho et al. (2006) indeed, the negative impacts of climate change are already perceptible in Benin, where farmers are experiencing yield reductions mainly due to high rainfall variability.

In this context, enhancing farmers' capacity to adapt to climate change is a priority for agricultural policy decision makers. In the frame of adaptation to new environmental change, it is now clear that perception of climate change shapes farmers' adaptation (Awoyé et al. 2012). If farmers learn gradually about the change in climate (Gbetibouo 2009), Maddison (2006) argued that they will also learn gradually about the best techniques and adaptation options available. Therefore, a sudden policy intervention (although designed with the best climate change adaptation strategies) might not meet the goal of building a sustainable farmers' resilience towards climate change. A simple way to go could be, in a first step, to raise farmers' awareness. In that prospect, enhancing farmers' knowledge about the predicted impacts of climate change and existing adaptation strategies might strengthen the awareness and furthermore lead to a better perception of and adaptation to. Accordingly, this study aims at providing empirical evidence of the impact of learning about climate change on farmers' perception of and adaptation to the climatic uncertainty.

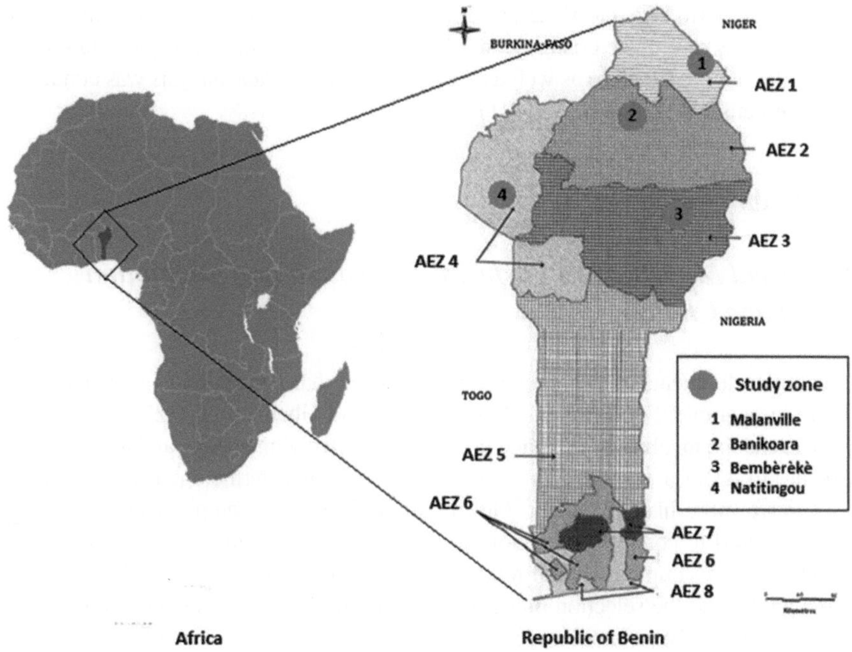

Fig. 16.1 Study area

Table 16.1 Sample structure

Agro-ecological zones	Municipalities	Villages	Sample size
1	Malanville	Koara-tédji	44
		Isséné	42
2	Banikoara	Bonhanrou	42
		Ounet	42
3	Bembèrèkè	Guéré	41
		Pédarou	42
4	Natitingou	Takonta	41
		Pam–Pam	42
Study zone			**336**

Materials

The study took place in northern Benin, covering four agro-ecological zones. The sampling took into account one municipality per agro-ecological zone (Fig. 16.1) and two villages per municipality. A total of 336 farmers were randomly selected for participation in a structured questionnaire containing questions about the farmers' socio-economic and demographic characteristics (gender, level of education, experience in agriculture, access to credit, organization membership, contact with extension service for instance, farm size, etc.), the perception of and adaptation to climate change and the awareness of climate change trend, impacts and adaptation strategies. Table 16.1 shows the sample structure as well as the sampling rates. Data analysis was conducted with the statistical software STATA 11.

Methods

Empirical Modelling of Farmers' Perception and Adaptation to Climate Change

Recent developments in the literature clearly stated that perception of and adaptation to are inseparable. Following Maddison (2006) and Gbetibouo (2009) adaptation to climate change involves a two-stage process: first perceiving change and then deciding whether to adopt a particular measure. This implies that farmers perceiving climate change are more likely to adapt. Given that, a Heckman's sample selectivity probit model is used to connect perception of and adaptation to climate change. Basically, the Heckman's sample selection model involved two sub-models. It is about the outcome model and the selection model referring to the adaptation and perception models, respectively. These models are based on two latent variables A and P defined as:

$$A = \beta Z + U_A \tag{16.1}$$

$$P = \beta' Z' + U_p \tag{16.2}$$

where, A refers to the farmers' decision to adapt while P refers to the farmers' perception. Z is a k-vector of regressors and Z' is a k'-vector of regressors. U_A and U_P are the error terms assumed to be jointly normally distributed, independently of Z and Z', with zero expectations. The empirical model is:

$$a_i = \beta_0 + \sum\nolimits_k \beta_k z_{ki} + u_{Ai} \tag{16.3}$$

$$p_i = \beta'_0 + \sum\nolimits_{k'} \beta'_{k'} z'_{k'i} + u_{Pi} \tag{16.4}$$

where, β_0 and β'_0 are constant terms, β_k and $\beta'_{k'}$ are parameters to be estimated. Coming back to the preliminary statements, the latent variable A is likely to be observed if $P > 0$. Thus, the actual dependent variable (adaptation to climate change) is defined as:

$$\begin{cases} Y = A & \text{if } P > 0 \\ Y \text{ is missing value} & \text{if } P < 0 \end{cases} \tag{16.5}$$

Let us impose A and P to be dummy variables defined as:

$$\begin{cases} A = 1 \text{ if farmers' adapt to climate change} \\ A = 0 \text{ otherwise} \end{cases} \tag{16.6}$$

$$\begin{cases} P = 1 \text{ if farmers' perceive to climate change} \\ P = 0 \text{ otherwise} \end{cases} \tag{16.7}$$

Given that, Eq. (16.5) becomes:

$$\begin{cases} Y = A = (0, 1) \quad \text{if } P > 0 \\ Y \text{ is missing value if } P > 0 \end{cases} \tag{16.8}$$

Fixing Eqs. (16.3) and (16.4) in Eq. (16.8), it comes out the following system:

$$\begin{cases} y_i = \beta_0 + \sum_k \beta_k z_{ki} + u_{Ai} = (0, 1) & \text{if } \beta'_0 + \sum_{k'} \beta'_{k'} z'_{k'i} + u_{Pi} > 0 \\ y_i \text{ is missing value} & \text{if } \beta'_0 + \sum_{k'} \beta'_{k'} z'_{k'i} + u_{Pi} \leq 0 \end{cases} \tag{16.9}$$

Interrelating Awareness of Climate Change and Farmers' Perception of and Adaptation to Climatic Uncertainty

In order to link awareness of climate change and the previous sub-models of farmers' adaptation and farmers' perception, we hypothesized that learning about climate change in terms of trends, impacts and adaptation strategies could enhance both perception of and adaptation to climatic uncertainty. Thus we set the awareness (W) into three latent variables: the awareness of climate change trends (T), the awareness of climate change impacts (I), and the awareness of climate change adaptation strategies (S). Moreover, we hypothesized that T and I are more likely to impact on farmers' perception and S on farmers' decision to adapt to climate change. Thus, by replacing W in Eq. (16.9), it comes out:

$$\begin{cases} y_i = \beta_0 + \sum_k \beta_k z_{ki} + \alpha s_i + u_{Ai} = (0, 1) & \text{if } \beta'_0 + \sum_{k'} \beta'_{k'} z'_{k'i} + \delta t_i + \theta i_i + u_P > 0 \\ Y \text{ is missing value} & \text{if } \beta'_0 + \sum_{k} \beta'_{k'} z'_{k'i} + \delta t_i + \theta i_i + u_P \leq 0 \end{cases} \tag{16.10}$$

where, β_0 and β'_0 are constant terms, β_k, α, $\beta_{k'}$, δ and θ are parameters to be estimated, and u and v error terms. The coefficients were estimated using the Heckman sample selectivity probit model. From the signs of the estimated coefficients and their magnitudes of significance, the factors affecting the farmers' perception of and adaptation to climate change were derived.

Selection of Exogenous Variables and Hypotheses to be Tested

The choice of exogenous variables presented in Table 16.2 was grounded in the literature and personal observations in the field.

- *Gender*: Agriculture in northern Benin men are more engaged in agriculture than their counterpart women (Paraïso et al. 2011, 2012). Men have more land access and more production assets than women. Thus, we hypothesized that men are more likely to perceive and adapt to climate change.

Table 16.2 Explanatory variables considered in the model

Variables	Modalities	Expected signs on A	Expected signs on P
Gender[d]	0 = Woman, 1 = Man	+	+
Level of education[c] (years)	–	+	+
Experience in agriculture[c] (years)	–	+	+
Access to credit[d]	No = 0; Yes = 1	+	
Organisation membership[d]	No = 0; Yes = 1	+	
Contact with extension[d]	No = 0; Yes = 1		+
Farms size[c] (hectare)	–	+	
Number of workers in agriculture[c] (person)	–	+	
Awareness of climate change trends[d]	No = 0; Yes = 1		+
Awareness of climate change impacts[d]	No = 0; Yes = 1		+
Awareness of climate change adaptation[d]	No = 0; Yes = 1	+	

Note [d]Discontinuous variable; [c]Continuous variable; A = Adaptation; P = Perception

- *Education level*: Following Maddison (2006), we assumed that educated farmers are more likely to respond to climate change by making at least one adaptation. Moreover, they might also be likely to have an easier understanding of climate change predictions (trends and impacts) and adaptation strategies. Therefore, the education level is expected to have a positive effect on both farmers' perception of and adaptation to climate change.
- *Experience*: From their experience farmers learn to know their environment and even anticipate on events such as rains, beginning of seasons, etc. We hypothesized that the more farmers have experience in agriculture, the more they are likely to perceive and adapt to climate change.
- *Access to credit*: We assumed that access to credit might enhance farmers' financial capital and make them more likely to adapt to climate change, especially when adaptation requires additional investments.
- *Organization membership*: Given that through organisations, producers might share farming experiences (i.e. successful strategies towards specific issues), we expected a positive relationship between farmers organisation membership and climate change adaptation.
- *Contact with extension services*: According to Maddison (2006), farmers who have benefited free extension advice are likely to adapt to climate change. Since extension officers cannot impose farmers farming practices to use, they mainly work on raising farmers' awareness. Thus, we expected a positive correlation between contact with extension services and perception of climate change.
- *Farm size*: Following Yegbemey et al. (2013), we assumed that the more the farm size increase, the riskier the production is likely to be under climate change conditions. Since farmers are not likely to be neutral to risk and actually tend to be risk averse agents (Serra et al. 2006; Yesuf and Bluffstone 2007), we hypothesized that larger farms are more likely to adapt to climate change.

Table 16.3 Socio-economic characteristics

Qualitative variables	Frequency	Percentage
Women	25	07.44
Men	311	92.56
Access to credit	71	21.13
Organisation membership	230	68.45
Contact with extension	184	54.76
Quantitative variables	*Mean*	*Standard deviation*
Level of education	3.06	3.99
Experience in agriculture	21.41	11.68
Farms size	4.87	6.75
Number of workers in agriculture	7.38	5.55

- *Number of household workers in agriculture*: We assumed that bigger households have more labor available for performing agricultural activities. So, we hypothesized that bigger households have more labor resource for adapting to climate change.
- *Awareness of climate change*: As stated previously from empirical researches, we hypothesized that knowledge about climate change in terms of trends, impacts and adaptation strategies could enhance both perception of and adaptation to climatic uncertainty.

Results

Descriptive Statistics

Table 16.3 shows the descriptive statistics of farmers' socio-economic characteristics. From this table, it comes out that most of the respondents were men. About 21 %, 68 % and 55 % of them had access to credit, belonged to a farmers association and had contact with extension services, respectively. The level of education was low (3 years on average) while the average experience in agriculture was about 22 years. The respondents were smallholder farmers (5 ha on average). On average the households had about 8 people working in agriculture.

Farmers' Perception of and Adaptation to Climate Change

About 95 % and 91 % of the respondents have perceived climate change and adapt to it, respectively. Farmers' perception of climate change was mainly about changes in rainfall and temperature. Talking about climate change adaptation, the most used adaptation strategies were crops diversification, changing farming practices (use of legumes for binding nitrogen, irrigation and change of fertilizers

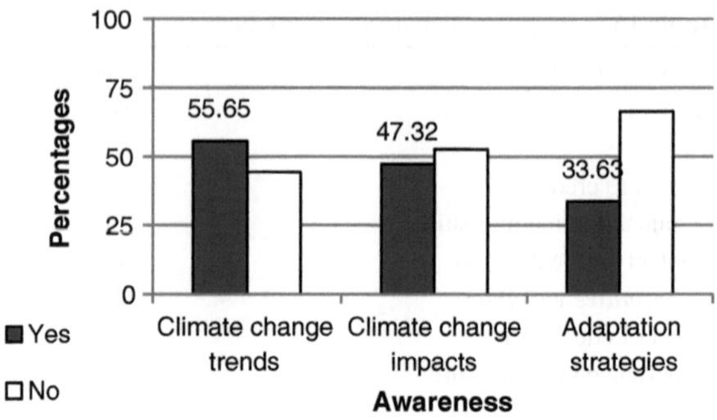

Fig. 16.2 Farmers' awareness through learning about climate change

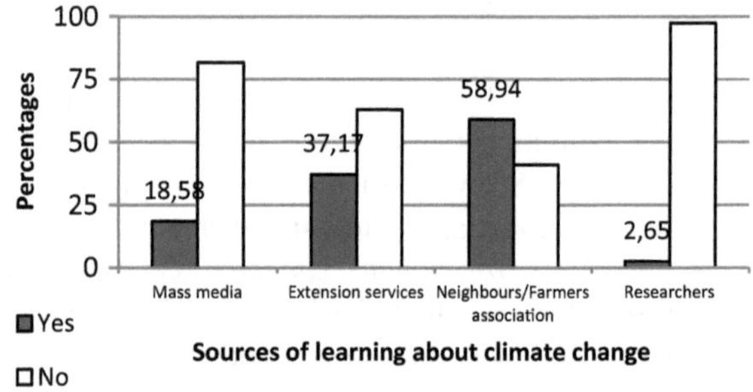

Fig. 16.3 Sources of learning about climate change

doses for instance) or the time of cultivation, land use strategies (change in the amount of land allocated to each crop and change of field location for instance), financial strategies (credit) and religious strategies such as prayers.

Awareness Through Learning About Climate Change

About 56 % of the respondents have learnt something about climate change. Farmers have information mainly about climate change trends. Then come climate change impact information and the climate change adaptation (Fig. 16.2). Farmers learn about climate change through various ways such as the mass media (local radios for instance), the extension services, the neighbour farmers or farmers associations and researchers. Among these ways, the most predominant vectors of climate change information are the neighbour farmers or farmers associations and the extension service officers (Fig. 16.3).

Determinants of Farmers' Perception of and Adaptation to Climate Change

The results of the Heckman probit model (Table 16.4) show that the model was highly significant ($P < 0.01$). Moreover, the Wald test of independent equations ($rho = 0$) indicates the correlations between adaptation and perception to be

Table 16.4 Results of the Heckman probit models with sample selection

Variables	Outcome (Adaptation) model			Selection (Perception) model		
	Coefficients	Standard errors	P > z	Coefficients	Standard errors	P > z
Gender	0.404	0.500	**0.419**	0.215**	0.106	**0.043**
Level of education	0.105**	0.041	**0.012**	0.140	0.116	**0.225**
Experience in agriculture	0.010	0.014	**0.492**	−0.016	0.025	**0.513**
Access to credit	0.539	0.509	**0.290**	–	–	**–**
Organisation membership	1.409***	0.287	**0.000**	–	–	**–**
Contact with extension	–	–	–	5.999***	0.904	**0.000**
Farms size	0.032	0.040	**0.426**	–	–	**–**
Number of workers in agriculture	0.028	0.035	**0.427**	–	–	**–**
Awareness through learning about climate change						
Awareness of climate change trends	–	–	–	6.301***	0.410	**0.000**
Awareness of climate change impacts	–	–	–	−0.076	0.748	**0.918**
Awareness of climate change adaptation	0.109	0.302	**0.716**	–	–	**–**
Constant	−0.061	0.553	**0.912**	−6.385	3.360	**0.057**
Model summary	/athrho coefficients	=	14.419	Rho coefficients	=	**1**
	Standard errors	=	7.280	Standard errors	=	**8.70e-12**
	P > z	=	0.048			
	Number of observations	=	**234**			
	Log pseudo likelihood	=	**−42.861**			
	Wald chi2(8)	=	**72.19**			
	Prob > chi2	=	**0.000**			
	Wald test of indep. eqns. (rho = 0): chi2(1) = 3.92 Prob > chi2 = 0.0476					

Notes *, **, *** significant at 10 % (0.05 < P < 0.10), 5 % (0.01 < P < 0.05), and 1 % (P < 0.01), respectively

statistically significant ($0.01 < P < 0.05$). Hence, the use of the Heckman model (i.e. based on the assumption that climate change perception is a prerequisite for adaptation) is relevant.

On the one hand, the main determinants of the farmers' decision to adapt to climate change were the level of education ($0.01 < P < 0.05$) and the organisation membership ($P < 0.01$). As expected, gender, experience in agriculture, access to credit, number of workers in agriculture, farms size, and awareness of climate change strategies were correlated with farmers' decision to adapt to climate change. However, these correlations were not significant ($P > 0.10$). On the other hand, the main determinants of farmers' perception are gender ($0.01 < P < 0.05$), contact with extension ($P < 0.01$) and awareness of climate change trends ($P < 0.01$). Level of education had a positive but not significant ($P > 0.10$) effect on farmers' perception of climate change while experience in agriculture and awareness of climate change impacts show unexpected negative effects on the same perception.

Discussion

The predominance of rain-fed agriculture in Sub-Saharan African results in food systems that are highly sensitive to rainfall variability (Wlokas 2008; Jones and Thornton 2009). Facing with the climatic uncertainty, farmers developed various strategies. The most common reported in literature are among others: use of new crop varieties (e.g. more suited to drier conditions), irrigation, crop diversification, mixed crop and livestock farming systems, changes of planting dates, diversifying from farm to non-farm activity, increased use of water and soil conservation techniques, change of capital and labour allocation (Maddison 2006; Nhemachena and Hassan 2007; Deressa et al. 2009; Till et al. 2010; Hisali et al. 2011; Yegbemey et al. 2013). To some extent, these adaptations strategies are responses to how farmers perceive climate change. Likewise, de Graft (2011) stated that farmer's ability to perceive climate change is a key precondition for their choice of adaptation. Understanding the connexion between climate change perception and climate change adaptation would be of a great interest for enhancing farmers' capacities of adaptation. Moreover, enhancing farmers' perception by improving their awareness of climate change will have further positive impacts towards the adaptation to climate change.

According to Hassan and Nhemachena (2008), awareness of a problem and potential benefits of taking action is another important determinant of adoption of agricultural technologies. Several studies reported that farmers' awareness and perceptions of soil erosion problems positively and significantly affected their decisions to adopt soil conservation measures (Traoré et al. 1998; Anim 1999; Araya and Adjaye 2001). Talking about farmers' awareness of climate change, Maddison (2006) found that farmers' awareness of changes in climate attributes (i.e. temperature and precipitation) is important for adaptation decision making. Farmers' awareness of climate change trends for instance is positively correlated with the perception of climate change. However, the awareness of climate change impacts has a negative but not significant effect on farmers' perception. Indeed, when farmers are aware of climate change they are easily able to establish by themselves the likely impacts of such changes on agriculture. Likewise, the awareness of climate change adaptation strategies also has a positive but unexpected not significant effect on farmers' decision to adapt to climate change.

At the first sight, this result can be explained by the fact that farmers are experienced in agriculture and therefore, are able to adapt from their own experiences. According Ofuoku (2011) indeed, farmers who have many years of farming experience have interacted much with the climate in relation to their farming activities and have good knowledge of how to adapt to changing. But one more time, as unexpected, experience in agriculture has a positive but not significant effect on farmers' decision to adapt to climate change. As a matter of fact the non-significance of the awareness of climate change adaptation might be due to the non-significance of the experience in agriculture. Gbetibouo (2009) also found out that the views between experienced and inexperienced farmers are statistically not significant. This latest result can be attributed to the fact that farmers learn about the best adaptation options through three ways: learning by doing, learning by copying and learning from instruction. Obviously, the results tell that quite all farmers learn adaptation options by doing (as a basis situation) whereas some of them learn by copying. The high significance of farmers association membership supports this argument. Finally, gender, level of education and contact with extension behave as expected.

Conclusion

Learning about climate change in general and climate change trends in particular is very likely to raise farmers' awareness of climate uncertainty. This awareness results in a better perception of climate change. As well, because farmers perceive climate change, they are more likely to adapt to it. Thus, as an intervention tool, agricultural policy decision makers should encourage farmers' learning about climate change through extension services for instance. Moreover, research outcomes related to climate change trends, impacts and existing adaptation strategies should be made available to farmers.

Acknowledgments The authors are grateful to the German Academic Exchange Service (DAAD) for financing this research. Thanks also to Ghislain Aïhouton and Michel Atchikpa for helping in data collection.

References

Aho N, Boko M, Afouda A (2006) Evaluation concertée de la vulnérabilité aux variations actuelles du climat et aux phénomènes météorologiques extrêmes. PANA/Bénin, p 93

Anim FDK (1999) A note on the adoption of soil conservation measures in the Northern Province of South Africa. J Agric Econ 50:45–336

Araya B, Adjaye JA (2001) Adoption of farm level soil conservation practices in Eritrea. Indian J Agric Econ 56:52–239

Awoye OHR, Agbossou EK, Yegbemey RN, Kokoye SEH (2012) Scientific and local knowledge on climate change in West Tropical Africa: do farmers' perceptions fit the measured changes? Resilience of agricultural systems against crises, Book of Abstracts, Tropentag, ISBN: 978-3-95404-215-9, p 423

de Graft AH (2011) Farmers' perception and adaptation to climate change: a willingness to pay analysis. J Sustain Dev Afr 13(5):150–161

Deressa TT, Hassan RM, Ringler C, Alemu T, Yesuf M (2009) Determinants of farmers' choice of adaptation methods to climate change in the Nile Basin of Ethiopia. Glob Environ Change 19(2009):248–255

DSCRP (2007) Document de Stratégies de Croissance pour la Réduction de la Pauvreté (DSCRP); Gouvernement du Bénin. Cotonou, Bénin, p 117

Gbetibouo GA (2009) Understanding farmers' perceptions and adaptations to climate change and variability: the case of the Limpopo Basin, South Africa. IFPRI discussion paper no. 00849, Environment and Production Technology Division, p 52

Hisali E, Birungi P, Buyinza F (2011) Adaptation to climate change in Uganda: evidence from micro level data. Glob Environ Change 21(2011):1245–1261

Jones PG, Thornton PK (2009) Croppers to livestock keepers: livelihood transitions to 2050 in Africa due to climate change. Environ Sci Policy 2009(12):427–437

Maddison D (2006) The perception and adaptation to climate change in Africa. CEEPA discussion paper no. 10, Centre for Environmental Economics and Policy in Africa, University of Pretoria, South Africa

MEHU (2011) Deuxième Communication Nationale de la République du Bénin sur les Changements Climatique. Bénin, MEHU, Cotonou, p 168

Nhemachena C, Hassan R (2007) Micro-level analysis of farmers' adaptation to climate change in Southern Africa. IFPRI discussion paper no. 00714, International Food Policy Research Institute, Washington DC

Ofuoku AU (2011) Rural farmers' perception of climate change in central agricultural zone of Delta state. Niger Indonesian J Agric Sci 12(2):63–69

Paraïso A, Yabi AJ, Sossou A, Zoumarou-Wallis N, Yegbemey RN (2012) Rentabilité économique et financière de la production cotonnière à ouaké au nord–ouest du Bénin. Annales des Sciences Agronomiques 16(1):91–105

Paraïso A, Sossou ACG, Yegbemey R, Biaou G (2011) Analyse de la rentabilite de la production du fonio (Digitaria Exilis S.) Dans la commune de Boukombe au Benin. Rech. Sci. Univ. Lomé (Togo) Série A 13(1):27–37

Serra T, Zilberman D, Goodwin BK, Featherstone AM (2006) Effects of decoupling on the mean and variability of output. Eur Rev Agric Econ 33:269–288

Till B, Artner A, Siebert R, Sieber S (2010) Micro-level practices to adapt to climate change for African small-scale farmers. A review of selected literature, IFPRI discussion paper no. 00953, Environment and Production Technology Division, p 28

Traoré N, Landry R, Amara N (1998) On-farm adoption of conservation practices: the role of farm and farmer characteristics, perceptions, and health hazards. Land Econ 74:27–114

Wlokas HL (2008) The impacts of climate change on food security and health in Southern Africa. J Energy S Afr 2008(19):12–20

Yegbemey RN, Yabi JA, Tovignan DS, Gantoli G, Kokoye SEH (2013) Farmers' decisions to adapt to climate change under various property rights: a case study of maize farming in Northern Benin (West Africa). Land Use Policy 34(2013):168–175. doi:10.1016/j.landusepol.2013.03.001

Yesuf M, Bluffstone R (2007) Risk aversion in low income countries, experimental evidence from Ethiopia. IFPRI discussion paper no. 00715

Chapter 17
Local Knowledge and Participatory Climate Change Planning in the Northeastern U.S

Lauren C. Heberle, Sam Merrill, Chris H. Keeley and Sadie Lloyd

Abstract This chapter presents examples of U.S. local climate change planning in Connecticut, Massachusetts, New Hampshire, and Maine. The processes initiated storm surge and sea level rise mitigation, adaptation and resilience plans in the context of national environmental conflict regarding climate change. Here, we highlight the constructive role local knowledge and experience can have in deliberative policy making and adaptive governance processes. These case studies utilized multiple methods of data collection including participant observation, semi-structured interviews, and content analysis. The elements that inspired communities to pursue adaptation were consistent across the case studies. The projects were framed using local values and did not directly challenge world views of participants. All had recent experience with extreme weather. Each benefitted from local "climate champions". Finally, each had access to technical assistance providers who engaged in collaborative learning. Local participatory planning processes that use tools like COAST and the NOAA Roadmap connect technical tools with social, political, and economic realities. They provide a container within which communities are empowered

L. C. Heberle (✉)
Center for Environmental Policy and Management, University of Louisville,
Lutz Hall 103, Louisville, KY 40292, USA
e-mail: Lauren.Heberle@louisville.edu

S. Merrill
Catalysis Adaptation Partners, LLC, P.O. Box 531, Scarborough, ME 04074, USA
e-mail: smerrill@catalysisadaptation.com

C. H. Keeley
NH Sea Grant, UNH Cooperative Extension, Great Bay National Estuarine Research Reserve, 122 Mast Road, Lee, NH 03861, USA
e-mail: Chris.Keeley@unh.edu

S. Lloyd
Efficiency Maine, 151 Capitol Street, Suite 1, Augusta, ME 04330-6262, USA
e-mail: sadie.lloyd@maine.edu

W. Leal Filho et al. (eds.), *International Perspectives on Climate Change*,
Climate Change Management, DOI: 10.1007/978-3-319-04489-7_17,

to have difficult conversations about needed responses to sea level rise and storm surge. Novel in the COAST approach is the integration of visual 3D graphics with benefit-cost analysis, entirely stakeholder driven model parameters, and the focus on stakeholders' identified values as resources that inspire action.

Keywords Climate change • Resiliency • Community planning • Deliberative policy • Adaptive governance

Introduction

Municipal governments in the United States began taking steps to address climate change before state and federal agencies, taking lead in the development of U.S. policy debates and dialogue. This happened across jurisdictions through a variety of organizations like the U.S. Conference of Mayors, cap and trade programs such as the Regional Greenhouse Gas Initiative and the Western Climate Initiative, and within jurisdictions through creation of municipal climate action plans and sustainability plans. Thus, while some jurisdictions use others as models for best practice, policies and plans remain widely varied (Meyer and Heberle 2010; Heberle and Christensen 2011). They also vary in types of impacts addressed that depend on local physical, economic and social circumstances. While this variation can be characterized as a major obstacle to making progress in addressing climate change policy in the U.S., it has nonetheless allowed for the emergence of locally-based initiatives that address relevant concerns within very diverse economic, social, and physical environments. Many of these initiatives emerged in various forms of deliberative policy making such as municipal and regional community-based planning task forces and committees. These varied efforts in which community members and stakeholders develop plans to address the vast array of issues related to climate change provides an arena to examine shifts towards deliberative governance practices within network societies that challenge traditional policy making and policy analysis (Hajer and Wagenaar 2003). They also provide opportunities to discuss the role that local knowledge about climate change plays in these local planning processes.

U.S. policy makers at every level of government, who engage in climate change initiatives, know they need stakeholder and community buy-into implement programs successfully because implementation requires changes in everyday behavior and shifts in deep seated cultural beliefs. They have had to address climate change skeptics that have aimed to inhibit meaningful dialogue in the U.S. about climate change at all levels of institutional structures (Dunlap 2013). At local levels they have had to work in a context where trust in scientists and belief in anthropogenic climate change has declined (Leiserowitz et al. 2013). Furthermore, because developing plans regarding climate change involves risk perception and uncertainty, those who make policy in a rational, scientific framework and require costs and benefits to be laid out in a linear path have difficulty incorporating uncertainty and risk into discussion about what action to take (Hajer and Wagenaar 2003; Dunlap 2013).

Adaptive governance and deliberative democracy are forms of policy making that have emerged to develop solutions to contested, complex, non-linear problems such as climate change. Characterized as more reflexive, offering more dialogue space for collaborative learning, these types of policy making processes serve as alternatives to hierarchical decision making prevalent in the U.S. and elsewhere. With shifts toward adaptive governance and deliberative democracy models to solve environmental problems, we also see shifts in opportunities for participatory governance supported by U.S. policy through federal regulation, suggested best practices, and grant requirements for applicants to include some level of community and stakeholder participation in decision making processes. Finally, new technical tools and methods for engaging in meaningful dialogue about planning for climate change and its impacts are emerging to compliment processes of deliberative democracy and adaptive governance.

The NOAA Roadmap for Adapting to Coastal Risks ("NOAA Roadmap") and the Coastal Adaptation to Sea-Level Rise Tool (COAST) are two methods being used in combination with evolving forms of policy making. COAST is a tool that facilitators use to map lost real estate value from sea level rise (SLR) and storm surge (SS) over a specified period of time. COAST uses local data such as assessor databases and parcel maps, local flood data, Light Detection and Ranging (LiDAR), the U.S. Army Core of Engineer's depth-damage functions, and NOAA hurricane models. This information is shown in 3D extrusions of lost real estate value. These are compared to lost real estate value if a set of specified adaptation strategies are implemented within that same time period. Inputs for the model are created with base layer data and property values that the facilitators gather from the local community. These inputs can be altered based on community knowledge and input. The suggested strategies are also flexible and can be tailored to community values and capacity.

The "NOAA Roadmap" is a vulnerability assessment framework that is largely qualitative, and emphasizes anecdotes of the community's experiences with extreme weather. The "NOAA Roadmap" is also a stakeholder-driven, participatory process commonly conducted over several workshops with community members, regional entities and relevant state or federal agency staff. The process often begins with defining community goals, building the group of stakeholders, using hazard profiles, and identifying community strengths and vulnerabilities within infrastructure, ecosystems, and people. A final step guides participants through risk reduction strategies that are appropriate to the community in question (NOAA n.d.). Scientists and experts engaged in assisting community-based environmental decision making use these tools to facilitate decision making based on scenario planning processes that do not predetermine outcomes. This shifts the role of the scientist and expert from arbiter of the "truth" that will solve problems to facilitators of a variety of kinds of knowledge that informs community-based decisions (Ozawa 1996).

Analyzing efforts that fall within these categories of policy making require attention to the discourse within the processes and dialogue between decision makers. Deliberative and interactive policy making involve shared and contested knowledge of participants. How climate change is presented to and experienced by those engaged in the decision process matters. Direct experience with extreme weather events is one example of local knowledge that shapes how communities engage in climate change adaptation and mitigation. Using local knowledge to show how resources the community values may be impacted by climate change makes planning for impacts more relevant to community decision makers (Merrill et al. 2010; IPCC 2012). Stakeholder understanding of the complexities of climate change also shapes if, when, and how they engage in processes aimed at developing solutions. Those in opposition to planning for impacts from climate change can be difficult to engage in deliberative planning processes. Community organizers and social change advocates of all types use the "meet them where they are at" approach to describe how to engage in dialogue with individuals whose worldviews must shift for them to support the advocates' efforts or act in a way that is new to them. Dialogue around climate change is embedded in particular worldviews, and planning for mitigation, adaptation and resiliency ultimately means acknowledging, and at times challenging, those worldviews. The fact that communities of different cultural backgrounds respond to risk in different ways needs to be addressed in the process of adaptation planning (Daniels and Walker 2001). "Where" communities are "at" in regard to their understanding and experience of climate change, and the appropriate community responses available shapes the ensuing dialogue. Using local knowledge and direct experience are methods that "meet communities where they are at" and can be integrated into deliberative policy making processes and adaptive governance structures.

Here, we draw on four examples of diverse local climate change planning processes in Connecticut, Massachusetts, New Hampshire, and Maine in the northeastern U.S. which occurred from 2008 to 2012. The cases discussed here include: Bridgeport, Connecticut where Clean Air-Cool Planet and The Nature Conservancy partnered with the Greater Bridgeport Regional Council to convene and facilitate the city's adaptation process in a modified version of the "NOAA Roadmap"; the Hampton-Seabrook Estuary, where the New Hampshire Coastal Adaptation Workgroup used COAST to assist three New Hampshire towns in the region's first economic impact analysis of sea-level rise and storm surge; Cape Cod, Massachusetts where fifteen communities of Barnstable County updated the county-wide Multi-Hazard Mitigation Plan using COAST in the process; and a planning process in Portland, Maine where COAST was introduced to structure early stages of the process (Fig. 17.1).

The case studies employed a combination of participant observation, semi-structured interviews, and review of process documents (plans, news articles, meeting agendas and notes). The planning process in each centered on developing locally based and supported actions toward climate change mitigation and adaptation. Examples from each case are used to highlight the emergence of deliberative policy practice and adaptive governance in the U.S., the value of local knowledge and experience to climate change planning, and the importance of trusted social and professional networks for such processes. Each of these elements has implications for addressing a community's sense of uncertainty and disempowerment regarding climate change.

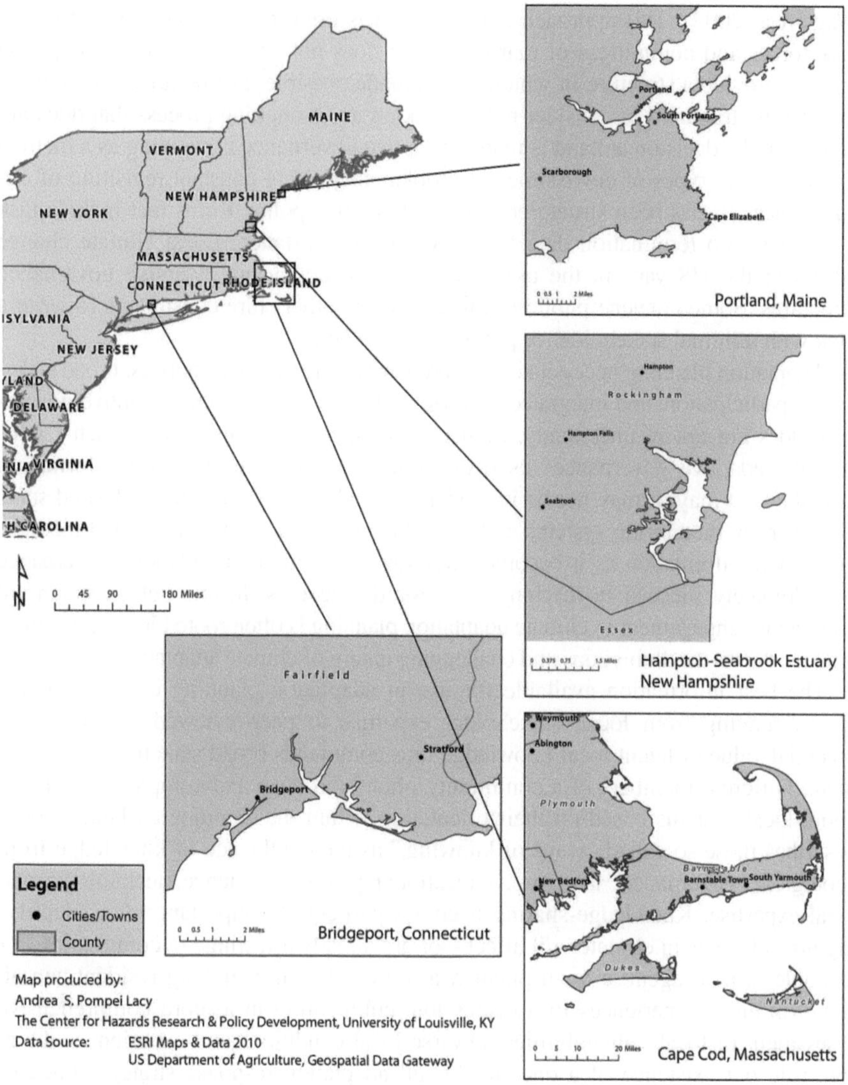

Fig. 17.1 Northeast region United States of America

Deliberative Policy, Adaptive Governance, and Climate Change

Deliberative policy making describes forms of decision making that encourage stakeholder participation in dialogue and ongoing collaborative problem solving. It suggests a different role for experts and analysts where they serve a facilitating role for the discussion of extant knowledge of a particular issue (Hajer and

Wagemaar 2003). This approach is not new; it is used to guide the work of policy task forces and committees of many types. It does not necessitate lay participation but can provide a structure in which that is made possible. Deliberative policy making also is structured to consider policy making as an ongoing process that does not end when the decision at hand is made. Adaptive governance is evolving as a method to solve many types of environmental problems requiring constant revisiting of circumstances. It has been structured using deliberative policy forms that include task forces with no termination date for convening for instance. Local climate change efforts in the US vary in the use of deliberative policy and adaptive governance structures. Some convene public taskforces, while others hire consultants to write a plan with minimal stakeholder or public involvement.

Adaptation planning necessitates comprehensive and iterative analysis, broad stakeholder participation, and integrated approaches to risk management. Adaptive governance does not ignore uncertainty. Rather, it embraces it. Brunner and Lynch (2010) describe adaptive governances as decision-making that uses the best information available and may or may not be immediately conducive to achieving a desired state. However, by monitoring systems and re-evaluating decisions over time, and incorporating new information as it becomes available, adaptation and adaptive governance can ultimately succeed in reaching the desired outcomes. In this light, meaningful community engagement in climate adaptation planning is often rooted in adaptive governance due to the ambiguous and challenging nature of climate adaptation.

The best information available for use in adaptation planning comes in many forms, ranging from local, stakeholder expertise to peer-reviewed research. The potential value of latent local knowledge thus commands broad stakeholder engagement. Different members of a community often hold deep and complex knowledge about local systems based on their orientation within the community. Feurt (2008) describes these so called "ways of knowing," as the confluence of knowledge from ecological, governance, land use, educational practices, science, technology, and local expertise. Knowledge-sharing becomes particularly important when identifying how changes in climate will affect—or are already affecting—a community. For example, a road agent, a town planner, a fire chief, and a lifelong resident can all bring different experiences to a discussion, culminating in a more comprehensive assessment of local vulnerabilities. Diverse local conditions and experiences therefore rule out existence of a one-size-fits-all adaptation response strategy. Each of these lessons is not unique to climate change. Rather, they are inherent to the principles of deliberative democracy (Dryzek 2001; Hajak and Wagenaar 2003).

Some studies are skeptical of the ultimate value of public participation and deliberation in climate adaptation planning processes. See for example, Milligan et al. (2009) or Few et al. (2007) on coastal management in the UK, Fast (2013) on renewable energy in rural east-central Canada, or Hobson and Niemeyer (2011) on the impact of deliberative processes on public participants' attitudes toward adaptation capacity. Concern in these studies revolve around identifying resolution of conflicting values, changes in public perceptions about climate change, or whether the deliberative process can ultimately replace strong governing leadership rather than more simply whether an agreed on path toward adaptation or mitigation was identified in the process irrespective of lingering value conflicts.

The planning processes in the U.S. case studies discussed here were structured to engage participants in consideration of adaptation strategies to sea level rise and storm surge events and to eventually develop action plans. The process of bringing together diverse sets of stakeholders varied. For example, in the multi-jurisdictional case in New Hampshire, members of Coastal Adaptation Workgroup (CAW) first met with various town boards and commissions to introduce the project and the benefits to obtain their support and assistance in inviting participants. In the Portland case, city officials first approved the process and then provided their logo as a stamp of approval to flyers and invitations sent out across the city thus providing legitimacy and support for wide spread participation. All were structured to elicit information about what local participants wanted to do about protecting their resources in the event of future severe events and encroaching sea levels. Each case incorporated deliberation and dialogue into the decision process.

The science of climate change and its impacts is complex and involves uncertain causal relationships across networks of social and natural systems. Non-experts and experts create narratives around cause and effect that are meant to resonate with individual social actors in a way that inspires action to engage in solving the problem. How that narrative is constructed, what tools are used to convey information, the content of information conveyed, and a community's sense of ability to affect change, impact whether planning happens and what type of plan emerges.

Tools and processes used in adaptive planning activities therefore need to afford an opportunity to incorporate local knowledge and experiences. In three of the four cases, facilitators used COAST to assist the participants in organizing expert and local knowledge and guide decision making. Economic impact modeling behind the COAST approach had already been published for eight towns in coastal York County, Maine (Colgan and Merrill 2008). Algorithms central to the software's calculations have also been published (Kirshen et al. 2012). The COAST software shell is available as a free download through the New England Environmental Finance Center. As a framework, the "NOAA Roadmap" has also been used widely and is available to facilitators on the NOAA website. Its process is made transparent and can be adapted as it was in the fourth case included here of Bridgeport, CT. Both approaches are structured to incorporate local expertise. Both have the capacity to transform local and expert knowledge into accessible visual representations (Burch et al. 2010). They also have the flexibility to include changes in participant inputs after they have a better understanding of the iterative process of adaptation and the potential economic, environmental, and social risks. The accessible, public, flexible features of these approaches to adaptation make them suitable for iterative, participatory decision making.

Visual representation of financial costs of community-identified resources elicits a community's anticipated risk as opposed to perceived risk, making solutions easier to imagine (McNall 2011). The "NOAA Roadmap" process used in the Bridgeport case was modified to include a participatory mapping approach that linked identified vulnerabilities and strengths across planning sectors with prominent hazards. COAST also employed community maps to visually represent predicted risks to participant identified resources including cultural and financial resources. In the Hampton-Seabrook Estuary communities, stakeholders placed value in a community as a unit rather than by parcel (they identified a collection of beach precinct real estate as a key focus area). They also requested that damages attached to individual homes be depicted in aggregate at the community scale rather than attached to each parcel.

Shifting from attempting to base action on perceived risk to anticipated risk occurred in these cases. COAST showed stakeholders the cost in damage if they did nothing compared to the cost of doing something now. Participants also discovered through dialogue that a combination of near-term and long-term actions was possible; not all adaptation must take place immediately. Participants were shown damages caused by storm surge versus sea-level rise. For coastal communities this helps them decide if immediate action is needed since storm surge is sudden and immediate while sea-level rise is slower but contributes to higher storm surge. COAST permitted stakeholders to visualize the damage and understand that unless they take adaptive actions, there could be heavy financial consequences both in the near and far term. One rarely sees how individuals implement what they learn in short-term behavior change. These changes in knowledge and attitudes are more readily identifiable in discussions of proposed adaptation strategies. For example, during one report out in the Bridgeport case, a real estate broker reported reconsidering plans to market properties at low elevations after seeing the predicted impacts of sea-level rise. Again, these cases showed that participants shifted from basing proposed action on perceived risk to anticipated risk through visualizing risks to resources they identified and in a predictive process that included their input.

Co-learning through an interactive dialogue can produce more robust solutions. Engaging stakeholders in determining model inputs and having the opportunity to question the modeling process meant that they had a more immediate understanding of how losses were being predicted and had influence in which community assets would be included. The ability to include this kind of community knowledge into the planning process for adaptation creates a platform of co-learning among technical assistance providers and local stakeholders. This dialectic increases the experts' knowledge of the community and the community's knowledge of the technical aspects they may not have understood previously. For example, throughout the process in the Seabrook case, participants asked increasingly informed questions that revealed growing awareness of the limitations of the modeling and the need for further analysis. One participant observed that the modeling did not include the costs of losing supporting services to vulnerable assets (e.g., sewer, roads, and electricity). Thus, one major outcome from this participatory meeting and the process as a whole was that stakeholders identified subsequent information needs and the facilitators now know specific data that need to be added to the model. Participants felt they shared knowledge with the facilitators that would be useful. Facilitators in some cases learned that participants needed the technical information presented in different sequences. They often needed the visual results first, then the numerical analysis or to see an example of the impact first. In one instance, participants suggested that to more effectively communicate, results be shown by a community member rather than an outside expert.

Valuing Local Knowledge and Experience

Deliberative, participatory and adaptive frameworks make space for the inclusion of local knowledge and co-learning but they do not pre-determine that inclusion. Understanding and incorporating local knowledge in climate adaptation is viewed as crucial to successful generation and implementation of solutions (IPCC 2012; Merrill et al. 2008). In the cases reviewed here, knowledge and experience at the local level intersect to provide a framework upon which future action can be built. This includes a community's set of local values and concerns, direct experience of climate change impacts such as increased severe weather events, and knowledge of solutions that are locally acceptable.

Framing the process using local knowledge and understandings, "meeting them where they are at", brought participants to the process. In all the cases, identifying shared values of each community or group of communities which would then inform decision making was the first step in the process. Doing this allowed facilitators to frame the discussion in a way that was meaningful to participants. In the case of Portland, Maine, the one day intensive session was structured intentionally to elicit what participants thought should be done to plan for storm surge and sea level rise, and participants were explicitly told that this was not a debate about climate change. The facilitator could then extract shared ideas for feasible solutions based on local knowledge and experience. In the Hampton-Seabrook case, facilitators intentionally framed the process as a public safety or economic issue that would result in concrete and immediate action items. This empowered stakeholders and helped them see relevance to what they do in their traditional municipal roles. Participants reported that if the project had been framed solely around protection of the estuary, those agencies tasked with protecting communities might not have participated. Community leaders in the Hampton-Seabrook case needed to be able to connect the process to their municipal leadership roles. The fifteen communities of Cape Cod came to the discussion of climate change through a regional hazard mitigation planning process after the Massachusetts Emergency Management Agency (MEMA) provided a grant to update hazard mitigation plans and communities were encouraged to include climate change as a hazard.

Local knowledge and experiences related to extreme weather impacts can provide a strong motivation for reducing vulnerabilities. There is generally immediate agreement on the need to reduce current vulnerabilities, which in turn better prepare communities for the future. All four areas in the case studies experienced extreme weather events in recent years, including multiple 100-year floods, tropical storms, and severe erosion. The local knowledge of these events provided tangible examples of climate impacts that community members rallied around. Informants in the Bridgeport case reported experience from the 2010 tornado, Hurricane/Tropical Storm Irene in 2011, and a snowstorm on Halloween in 2011 as motivating events. In the Hampton-Seabrook Estuary, the participants' direct experience with increased storm surge as an immediate risk set the stage to include discussions of sea level rise. In the Cape Code case, the experience of extreme weather impacts of those involved was more fluid where the participants are accustomed to changing landscapes caused by severe weather and thus planning for even more severe consequences was logical. These experiences helped to identify current vulnerabilities and seed discussions about the consequences of more severe impacts.

Identifying local experiences and attitudes toward solutions that are locally-acceptable provide a framework for discussions that do not automatically challenge cultural values. The Portland session asked participants to examine four adaptation approaches to protect real estate from storm surge and sea level rise (fortify assets, accommodate more water, relocate assets, or do nothing). They were then asked to apply the approaches to a specific area of privately held property and identify who would be responsible for implementation and how should the response be implemented. This strategy thus revealed solutions participants considered appropriate in their community. The facilitators did not make assumptions about the value placed on the properties nor about philosophical ideas about public or private responsibility.

Trust in Pre-existing and Emerging Leadership and Social and Professional Networks

Hajer and Wagenaar (2003) suggest that network societies need deliberative policy making structures to tap into information networks. Addressing local impacts of climate change is a relatively novel challenge for communities in the U.S. It draws on unconventional planning horizons and unfamiliar technical information. Since many municipalities in the U.S operate with volunteer boards and task forces, local capacity is limited both in regard to financial resources and knowledge resources. Access to social and professional networks of individuals and agencies who posses needed expertise within and across communities eases participants' uncertainty about this complex challenge.

In the cases reviewed here, participants asked to identify what made the processes succeed, they reported having community leaders with experience in and who continue to support climate action planning, a diverse network of participants willing to share their knowledge, and technical assistance providers who have previous experience with the area and are therefore trusted and credible. Cape Cod's climate change initiative emerged through several iterations. Attempts for adaptation planning were made over the period of several years, but the timing was not right for adaptation to take hold. The fifteen communities did not come together until the necessary local and regional leadership was in place, decision-makers' awareness of climate change reached a point conducive to taking action, several major storms altered the Cape Cod coastline, and a regional planning effort transpired. In the Hampton-Seabrook Estuary, it helped that the Town of Seabrook had recently worked with the Rockingham Planning Commission to investigate options for addressing sea-level rise in the town's hazard mitigation plan.

Because stakeholder engagement in the Hampton-Seabrook Estuary was multi-jurisdictional, the process was a unique collaboration for the participating communities and provided a learning opportunity about shared resources and distribution of costs and burdens. The positive outcome of this discussion was development of trust among participants that led to a reported shared desire to continue the collaboration. Learning together about the adverse impacts of climate change demonstrated the need to share knowledge and resources, and to develop responses that strengthen the region, not just one bounded jurisdiction. Other jurisdictions are taking notice and the participants are taking lessons learned in this process to a variety of other localities. There is a sense of momentum and accumulation of knowledge about responses to climate change that participants want to share across their region.

New Hampshire participants also acknowledged a critical need to engage with community members not directly involved in the process. As a next step, the three towns discussed the need for broader community engagement about their new understandings of storm surge and sea-level rise impacts. Participants reported that only then, with community support, will they have the ability to implement solutions—but importantly, engagement with a political process had emerged and would continue where formerly there had been neither engagement nor a process for implementation of an adaptation vision. Deepened trust among stakeholders was a stated positive outcome in all cases; it was established by sharing stories, experiences, and expertise. This is relevant because supporting implementation of plans can become contentious and difficult. Having the platform for stakeholders to share knowledge creates the opportunity for them to establish and build upon relationships that can facilitate better plans.

Finally, trusted technical assistance providers bring a strong understanding of a community's history, culture, and current planning context. They can help stakeholders to bridge the connection between their local or institutional knowledge and how it applies to adaptation planning. Access to trusted technical assistance providers was reported and highly visible in each case. Of special note is the role of networks of climate adaptation practitioners. For instance, New Hampshire's Coastal Adaptation Workgroup provides an example of service providers working collaboratively with communities to leverage diverse expertise, skills, and relationships with community members. One participant succinctly described the value of working within a network: "Key partnerships can help to stimulate action… and bring varying levels of expertise to a planning process." She continued, "regional entities can build off the work each other is already doing to make progress in climate adaptation planning, especially because it is such a cross-cutting issue."

Conclusions and Implications

Local responses to the threats posed by climate change fall into three broad, sometimes overlapping categories: do nothing, mitigation, and adaption. All three entail assumptions about risk of potential harm and human ability to reduce that risk. The deliberative policy process that includes community members in decision making has the capacity to produce a network of individuals and agencies that share an understanding of which responses are appropriate for their community. When community members participate together in learning about predicted losses of resources shared by that community or a network of communities, the potential for loss becomes more relevant to them and moves the risk from perceived to anticipated.

The case studies provide examples of a variety of planning processes centered on developing storm surge and sea level rise mitigation, adaptation and resilience plans in the context of national environmental conflict regarding the existence and causes of climate change. The cases highlight the constructive role local knowledge and experience can have in addressing climate change impacts. Three elements associated with local knowledge inspired communities to pursue adaptation and were consistent across the case studies. The projects were framed in ways that were derived from local values and experience. They were not framed to directly challenge world views of participants. The communities all had recent experience with extreme weather providing a common concern. Each benefitted from local "climate champions". Finally, each had access to technical assistance providers who engaged in collaborative learning with the participants and helped them identify resources to plan and implement decisions within their own community networks. These deliberative policy processes increased knowledge of both "lay" participants and "experts", improved social networks, and community support for continued climate change planning and implementation.

Local government processes observed in these examples suggest that the value in approaches like the NOAA Roadmap and COAST is that they connect technical tools with social, political, and economic realities. They provide a container within which communities are empowered to have difficult conversations about needed responses to sea level rise and storm surge. Because participant responses often invoke issues that are heavily values-laden, the nature of the public conversation is as important as the technical tools employed. This implies the need for care in structuring and facilitating public meetings so that the widest possible range of local political sensitivities can be encompassed and supported.

The decision making processes and tools discussed here have yet to be proven universally effective in addressing all complex, non-linear problems at a larger national or global level such as climate change. Nor do they predetermine positive local outcomes. These cases do provide evidence that adaptive and participatory planning processes that incorporate local knowledge and experience to address climate change impacts such as extreme flooding and storm surge can move communities to have a vested interest in adaptation planning processes, whether or not the planning process is explicitly framed around climate change. The conversations that need to be had about changing community practices are thus made possible. When local knowledge and experience is supplemented with "expert" risk analysis and high-quality climate change information, communities are better positioned to prepare for a changing climate.

Acknowledgments The case studies and portions of this chapter are drawn from research conducted and developed fully by Keeley (2012) and (Lloyd 2012) for their master's thesis and capstone project respectively. Funding was provided by the US EPA for development of the COAST software shell and for COAST iterations in ME and NH, with support from both the Casco Bay Estuary Partnership and the Piscataqua Region Estuaries Partnership. Planning process and community engagement support was provided by the EPA in ME and by members of the NH Coastal Adaptation Workgroup in NH.

References

Brunner R, Lynch A (2010) Learning from climate variability: adaptive governance and the Pacific ENSO Applications Center. Weather Clim Soc 2(4):311–319

Burch S, Sheppard SRJ, Shaw A, Flanders D (2010) Planning for climate change in a flood-prone community: municipal barriers to policy action and the use of visualizations as decision-support tools. J Flood Risk Manage 3:126–139

Colgan C, Merrill S (2008) The effects of climate change on economic activity in Maine: Coastal York County case study. Maine Policy Rev 17(2):66–79. Available at http://digitalcommons.library.umaine.edu/mpr/vol17/iss2/10. Accessed 27 Mar 2013

Daniels SE, Walker GB (2001) Working through environmental conflict: the collaborative learning approach. Praeger Publishers, Westport

Dryzek J (2001) Legitimacy and economy in deliberative democracy. Polit Theory 29(5):651–669

Dunlap RE (2013) Climate change skepticism and denial: an introduction. Am Behav Sci 57(6):691–698

Fast S (2013) A Habermasian analysis of local renewable energy deliberations. J Rural Stud 30:86–98

Feurt C (2008) Collaborative learning guide. Wells National Estuarine Research Reserve. Available at http://www.wellsreserve.org/sup/downloads/collaborative_learning_guide.pdf. Accessed 26 Mar 2013

Few R, Brown K, Tompkins EL (2007) Public participation and climate change adaptation: avoiding the illusion of inclusion. Clim Policy 7(1):46–59

Hajer MA, Wagenaar H (eds) (2003) Deliberative policy analysis: understanding governance in the network society. Cambridge University Press, Cambridge

Heberle LC, Christensen IM (2011) U.S. environmental governance and local climate change mitigation policies: California's story. Manage Environ Qual: Int J 22(3):317–329

Hobson K, Niemeyer SJ (2011) Public responses to climate change: the role of deliberation in building capacity for adaptive action. Glob Environ Change 21(3):957–971

IPCC (2012) Managing the risks of extreme events and disasters to advance climate change adaptation. A special report of working groups I and II of the intergovernmental panel on climate change. Barros CBV, Stocker TF, Qin D, Dokken DJ, Ebi KL, Mastrandrea MD, Mach KJ, Plattner G-K, Allen SK, Tignor M, Midgley PM (eds) Cambridge University Press, Cambridge, UK and New York, USA. Available at https://www.ipcc.ch/pdf/special-reports/srex/SREX_Full_Report.pdf. Accessed 27 Mar 2013

Keeley C (2012) Preparing for climate change in three New England coastal communities: lessons on motivations, approaches and outcomes. Submitted to the University of New Hampshire in partial fulfillment of the requirements for the degree of Master of Science in natural resources, Dec 2012

Kirshen P, Merrill S, Slovinsky P, Richardson N (2012) Simplified method for scenario-based risk assessment adaptation planning in the coastal zone. Clim Change 113:919–931

Leiserowitz AA, Maibach EW, Roser-Renouf C, Smith N, Dawson E (2013) Climategate, public opinion, and the loss of trust. Am Behav Sci 57(6):818–837

Lloyd S (2012) Planning for sea level rise in Portland, Maine using robust decision making as a guide. Unpublished capstone project for Masters of Science in Muskie School of Public Service at the University of Southern Maine, Fall 2012

McNall SG (2011) Rapid climate change: causes, consequences, and solutions. Routlege, New York and London

Merrill S, Sanford R, Lapping M (2008) Planners and climate change: an approach for communities. Maine Policy Rev 17(2):142–152

Merrill S, Yakovleff D, Holman D, Cooper J, Kirshen P (2010) Valuing mitigation strategies: a GIS-based approach for climate adaptation analysis. ArcUser Mag 13(4):28–29

Meyer PB, Heberle LC (2010) Local climate change initiatives in the United States: the primacy of short-term economic returns. In: van Staden M, Musco F (eds) Local governments and climate change: sustainable energy planning and implementation in small and medium sized communities. Advances in global change research, vol 39. Springer, Dordrecht, pp 181–190

Milligan J, O'Riordan T, Nicholson-Cole S, Watkinson A (2009) Nature conservation for future sustainable shorelines: lessons from seeking to involve the public. Land Use Policy 26:203–213

NOAA (n.d.) Approach—roadmap for adapting to coastal risk. NOAA Coastal Services Center http://www.csc.noaa.gov/digitalcoast/training/roadmap/approach. Accessed 27 Mar 2013

Ozawa CP (1996) Science in environmental conflicts. Sociol Perspect 39(2):219–230

Chapter 18
Evaluating of Post-2012 Carbon Policies

Olga Diukanova

Abstract A dynamic multi-regional computable general equilibrium model has been employed to evaluate the alternative post-2012 targets and policies for CO_2 reduction in Annex I and non-Annex I parties to the United Nations Framework Convention on Climate Change. Based on an analysis of conflicting strategies that different countries were pursuing during the recent international negotiations on climate change, this paper presents some ideas for an effective climate policy. The low-end pledges confirmed at the COP 16 negotiations on climate change were evaluated for the alternative policy settings, which include international carbon trading, domestic actions, and linking of the EU ETS with Russian and Ukrainian carbon markets. The study has a special focus on the implications of trading of surplus emission permits, or so-called hot air that occurs when a country's carbon abatement target is below its projected emissions. Results of computer simulations reveal that by 2020 global carbon emissions will exceed their 1990 levels by 71–96 %, depending on the model scenario. Although multilateral implementation results in the lowest emission levels, it overshoots the levels of CO_2 sufficient to stabilise GHG (greenhouse gas) concentrations at 450 ppm in 2100 by 5 Gt. Only international emission trading could motivate Russia and Ukraine to abate emissions. Such trade is profitable for the EU as it halves the carbon price compared with when the EU abates emissions domestically. Countries with non-binding abatement targets could be affected by the terms of trade effects resulting from emission reduction in other countries. Indeed, the possibility to trade emissions internationally provides the climate free-riders with an incentive for emission abatement. However, at the global scale, such policy lowers the cost of carbon abatement without meeting the GHG stabilisation goal. Therefore

O. Diukanova (✉)
Foundation for the Development of Environmental and Energy Markets,
21 Mayborody Street, Suite 34, Kyiv 04050, Ukraine
e-mail: odiukanova@gmail.com

W. Leal Filho et al. (eds.), *International Perspectives on Climate Change*,
Climate Change Management, DOI: 10.1007/978-3-319-04489-7_18,

ambitious abatement targets, along with a large emissions market, are necessary to make climate policy effective.

Keywords Carbon abatement • Emission trading • Computable general equilibrium model

Introduction

In order to restrict global warming to within 2 °C above the pre-industrial level by 2100, atmospheric GHG concentrations have to be stabilised at 450 ppm CO_2-eq by the end of the century. However, in 2012 GHG concentrations reached 400 ppm (EEA 2013). It is estimated that global GHG emission levels of 39–44 Gt of CO_2-eq in 2020 and below 21 Gt in 2050 would be consistent with a "likely" chance of limiting global warming to 2 °C (den Elzen and Höhne 2010; UNEP 2011). The International Energy Agency suggested a scenario of a 450 ppm target when energy-related CO_2 emissions peak at 30.9 Gt by 2020 and then steep decline to 15 Gt in 2050, with non-CO_2 greenhouse gases reduced proportionally (IEA 2010, 2011).

The COP 16 pledges of Annex I and non-Annex I parties to the United Nations Framework Convention on Climate Change (UNFCCC) (2010, 2011) were widely recognised as insufficient to meet the 2 °C target (Kartha and Erickson 2011; CAT 2012; UNEP 2010; Rogelj et al. 2010). The low end of countries' COP 16 pledges would leave a gap of 9 Gt of CO_2-eq in 2020; and if global emissions follow their baseline projections this gap could reach 12 Gt of CO_2-eq (UNEP 2011).

A large portion of the GHGs' gap is formed of surplus AAUs or so-called hot air. Russia and Ukraine are the main owners of excess AAUs, as in these countries the projected emissions are far below the abatement targets. During 2008–2012 Russian "hot air" was estimated at 6.2 Gt of CO_2-eq, and Ukrainian "hot air" at 2.4 Gt of CO_2-eq (European Commission 2009). For 2013–2020, the Russian and Ukrainian surpluses of AAUs are foreseen to be 2.2 Gt and 1.7 Gt of CO_2-eq, correspondingly (Rogelj et al. 2010; den Elzen et al. 2009a, b).

It is clear that without ambitious abatement targets for the Annex I and non-Annex I groups of countries, and specific measures to limit the trade and carry-over of surplus AAUs into future compliance periods, carbon credits will flood the market. This will result in low carbon prices and a widened emissions gap.

Literature Review

A number of studies were conducted to investigate the burden sharing options among Annex I and non-Annex I countries and to evaluate the extent and implications of the market power of Former Soviet Union (FSU) counties.

The IPCC (2007) estimated that in order to stabilise GHGs concentration at 450 ppm by 2100, Annex I countries would have to reduce GHG emissions by 25–40 % before 2020 and by 80–95 % before 2050 below the 1990 level, while developing countries need to abate GHGs by 15–30 % below their baseline projections until 2020 (den Elzen et al. 2008; den Elzen and Höhne 2010).

Over the past decades, computable general equilibrium (CGE) models have become a popular workhorse to quantify the impacts of policies whose effects are likely to propagate across multiple markets and affect industry competitiveness. These models operate large databases that consist of input–output tables of single or multiple regions. In contrast to input–output models, CGE models assign important roles to prices and values of elasticities that determine possibilities of substitution among different goods. These features are indispensable for the evaluation of carbon mitigation options that can be achieved by means of (a) a reduction in output of energy- and carbon-intensive goods, (b) substitution between energy goods that have different carbon intensities, and (c) substitution between energy and non-energy goods. Since CGE models capture all these mitigation options, they are widely employed for climate policy evaluation.

McKibbin et al. (2011) applied the dynamic CGE model G-Cubed to ten major regions of the global economy to evaluate the economic cost of the Copenhagen commitments. They found that although overall GDP losses reach 3.2 % of the baseline projections, the welfare impacts differ sharply across the countries. Welfare implications depend largely on the size and structure of each economy; a country's endowment of fossil fuels, structure of energy balance, technological possibilities to substitute among the fuels, and energy- and non-energy intensive goods. The important finding is that, due to spillover effects, domestic carbon price may not always define the welfare implications for climate policies. Therefore, irrespective of non-binding abatement targets and close to zero carbon prices, FSU countries face a 3 % decline in their GDP when other countries reach their targets.

Dellink et al. (2010) employed a dynamic CGE model to analyse the economic impact of attaining the Copenhagen pledges. The economic cost of action was estimated at 0.3 % of annual GDP for both Annex I and non-Annex I countries and 0.5–0.6 % of global real income. The study revealed that fossil fuel exporting countries, such as OPEC, Canada and Russia, would be affected by the falling fuel prices on the world market, as international climate policies lower demand and prices for fossil fuels. Unilateral implementation of carbon policies may actually have a smaller impact on these countries compared with the multilateral scenarios, as the former do not reduce international demand for hydrocarbons. However, the unrestricted trade of AAUs can reduce costs across the Annex I region, as Russia is able to sell emission permits at a low price.

Böhringer et al. (2007) applied a multi-country CGE model to investigate the implications of Russian market power for the international and European emissions markets. They found that Russia may have incentives to join the EU ETS as long as the latter remains separated from the international emissions market. In this case, Russia can use the monopolistic price discrimination between the two emissions markets, thereby maximising revenues from "hot air" sales. Compliance costs could be substantially reduced if emission trading is not restrained within the EU. However, part of the gains from extra-EU emissions trading may come at the expense of environmental effectiveness as more "hot air" will be traded.

Klepper and Peterson (2007) employed the CGE model DART to analyze the alternative institutional designs for permit allocation in the economies holding "hot-air" and to quantify the resulting effects with and without US participation. They found that the allocation mechanism for "hot air" in Eastern Europe influences the level of energy demand and energy prices. Propagation of these effects induces changes in the production structure of energy intensive goods and affects welfare, depending on whether a country is an energy net exporter or importer. Therefore, the repercussions of carbon policies are to a large degree determined by their effects on world energy demand and energy prices.

The other studies on multilateral climate policies that were based on CGE modelling include Alexeeva-Talebi and Anger (2007), Anger et al. (2009), van der Laan (2009), and Schleich et al. (2010).

This study builds on a previous perspective, aiming to find the balance between effective emission reduction and profits for the hot air-holding countries such as Russia and Ukraine.

The Methodology

The ICES (Intertemporal Computable Equilibrium System) model is a recursive dynamic CGE model developed at the Fondazione Eni Enrico Mattei to assess the economic impact of climate change on the world economy. The model description is available on the web-site http://www.feem-web.it/ices. The ICES structure is similar to the GTAP-E model (Burniaux and Truong 2002). The model is programmed in GEMPACK language (Centre of Policy Studies 2002). The current model version is calibrated to the 7th version of the GTAP database which consists of 57 commodities/sectors and 113 regions (Narayanan and Walmsley 2008).

For the purpose of this study, we used the aggregation which includes 9 sectors and 27 regions. The sectors' aggregation covers major dimensions in the analysis of carbon abatement options, such as differences in carbon intensities and possibilities for substitution between energy and non-energy goods. The choice of regional aggregation captures key players in international negotiations on climate change among the Annex I and non-Annex I parties to the UNFCCC.

The dynamics in the model are driven by the evolution of capital, foreign debt, natural resources, technological progress, demographic changes and fuel prices. The model baseline accounts for the recent economic crisis and its medium-term effects. The fuel prices were projected according to EURELECTRIC (2010). The possibility of carrying over emission permits from the Kyoto (2008–2012) to the Copenhagen commitment period (2013–2020), i.e. "banking", is not considered in this study as it would require zero carbon abatement from most Annex I countries.

Table 18.1 The model scenarios

	Base year or intensity target	AllDomestic		IntTrade		EUFSUTrade		EUDomestic	
		CO_2 reduction (%)	CO_2 trade	CO_2 reduction (%)	CO_2 trade	CO_2 reduction (%)	CO_2 trade	CO_2 reduction (%)	CO_2 trade
EU-27	1990	−20	Yes	−20	Yes	−20	Yes	−20	Yes
US	2005	−17	No	−17	Yes	0	No	0	No
Russia	1990	−20	No	−20	Yes	−20	Yes	0	No
Ukraine	1990	−20	No	−20	Yes	−20	Yes	0	No
RoA1[a]	2005	−31	No	−31	Yes	0	No	0	No
China	CO_2/GDP	−40	No	−40	Yes	0	No	0	No
India	CO_2/GDP	−20	No	−20	Yes	0	No	0	No
Brazil	BAU	−5.32	No	−5.32	Yes	0	No	0	No
NonA1[b]	2005	−1.74	No	−1.74	Yes	0	No	0	No
ROW[c]	0	0	No	0	No	0	No	0	No

[a]RoA1—the rest of Annex I developed countries (Australia, New Zealand, Japan, Canada, and EFTA)
[b]NonA1—the largest emitters among the non-Annex I countries with GHG abatement pledges (South Korea, Indonesia, Mexico, and South Africa)
[c]ROW—rest of the world aggregate

This study does not account for linking between the EU and Australian emissions trading systems. Although accounting rules for LULUCF emissions have not yet been decided, these offsets were not considered in this study.

The Model Scenarios

This study proposes four scenarios that reflect the low end of official COP 16 emission abatement targets. Since the size of post-2012 carbon markets and the rules that govern admission and transactions have not yet been confirmed, countries' commitments were evaluated for the range of alternative policy settings, which include international carbon trading, domestic actions, and linking of the EU ETS with Russian and Ukrainian carbon markets (see Table 18.1).

As shown in Table 18.1, the scenario AllDomestic reflects the case of carbon trade among EU countries when the other parties of the UNFCCC implement domestic actions to achieve their commitments. This policy is contrasted with the scenario IntTrade that verifies the economic cost of meeting the COP 16 obligations when countries that committed to abate emissions can trade them internationally. Under the scenario EUFSUTrade, the EU ETS is linked to Russian and Ukrainian carbon markets, and the emissions of other regions follow their baselines. The benefits of linking were evaluated against the EUDomestic scenario, where EU countries trade carbon permits to achieve their COP 16 targets and the emissions of other regions follow baseline projections. It was assumed in the model that multilateral carbon trading starts in 2013.

Table 18.2 CO_2 emissions and CO_2 price in 2020

	Baseline		AllDomestic		IntTrade		EUFSUTrade		EUDomestic	
	CO_2 (%) versus 2004	CO_2 price, €$_{2008}$/t CO_2	CO_2 (%) versus 2004	CO_2 price, €$_{2008}$/t CO_2	CO_2 (%) versus 2004	CO_2 price, €$_{2008}$/t CO_2	CO_2 (%) versus 2004	CO_2 price, €$_{2008}$/t CO_2	CO_2 (%) versus 2004	CO_2 price, €$_{2008}$/t CO_2
EU-27	–4.06	0	–13.16	30.88	–12.61	27.73	–8.51	7.64	–13.16	15.17
US	–6.80	0	–17.99	28.40	–19.01	27.73	–6.19	0	–6.19	0
Russia	36.51	0	22.39	16.47	10.69	27.73	27.22	7.64	37.88	0
Ukraine	45.98	0	69.00	0	–8.82	27.73	27.88	7.64	46.03	0
RoA1	–6.58	0	–29.47	68.62	–13.83	27.73	–5.88	0	–5.87	0
China	147.51	0	92.05	27.37	88.87	27.73	148.61	0	148.68	0
India	120.24	0	119.54	8.35	81.16	27.73	120.98	0	120.97	0
Brazil	24.08	0	19.77	26.33	17.55	27.73	24.70	0	24.66	0
NonA1	56.80	0	–0.16	100.30	38.57	27.73	57.73	0	57.78	0
ROW	55.04	0	73.88	0	71.50	0	56.10	0	56.05	0
Global CO_2, Gt	**41.22**		**35.96**		**35.87**		**41.00**		**41.01**	
Global CO_2 (%) versus 1990	**96.62**		**71.56**		**71.11**		**95.55**		**95.63**	

Discussion of the Results

Computer simulation demonstrated that, depending on model scenario, by 2020 global carbon emissions would grow by 71–96 % above the 1990 level. To reach a 450 ppm stabilisation target, in 2020 global GHGs should not increase by more than 15 % above their 1990 level (den Elzen et al. 2008; den Elzen and Höhne 2010), so severe cuts will be required after 2020. As shown in Table 18.2, the multilateral carbon policies AllDomestic and IntTrade, when all regions attain their COP 16 commitments, result in the lowest emission levels among the model scenarios (correspondingly, 35.96 and 35.87 Gt of CO_2 in 2020). However, the global emissions consequent from these policies overshoot the WEO's 450 ppm scenario (IEA 2010, 2011) by 5 Gt of CO_2, as it requires the peaking of global CO_2 at 30.9 Gt by 2020. Obviously the possibility to "bank" the AAUs will move the world further away from the GHGs' stabilisation target. Although both scenarios of multilateral policies reach similar emission levels, their allocative effects differ.

The AllDomestic policy which restricts emissions trading within the EU borders, results in higher CO_2 prices for the developed countries compared with the scenario of international carbon trading (IntTrade).

As shown in Table 18.2, only the possibility to export carbon permits (scenarios EUFSUTrade and IntTrade) can incentivise strong emission cuts in "hot-air"-holding countries such as Russia and Ukraine. As we can see, the extent of carbon abatement is directly proportional to the size of emissions market. Due to high energy intensity and low energy efficiency, Russian and Ukrainian marginal GHG abatement costs are much lower than in the EU. The possibility to trade

Table 18.3 GDP increase in 2020 (%) versus 2004

	Baseline	AllDomestic	IntTrade	EUFSUTrade	EUDomestic
EU-27	19.68	19.91	19.81	19.48	19.24
US	18.83	18.59	18.54	18.85	18.85
Russia	78.80	75.90	73.90	76.70	78.91
Ukraine	71.87	77.19	61.72	66.53	72.03
RoA1	20.89	19.83	20.85	20.95	20.95
China	273.93	266.52	263.89	274.29	274.3
India	232.69	238.19	227.47	233.04	233.1
Brazil	70.46	70.47	70.35	70.52	70.52
NonA1	117.97	104.76	116.32	118.16	118.2
ROW	113.48	116.99	116.34	113.78	113.7
Global GDP (%) versus 2004	**49.78**	**48.95**	**49.19**	**49.75**	**49.71**

carbon stimulates "hot air" holders to reduce their emissions in order to maximize proceeds from carbon exports. If profits from carbon trade are invested in energy efficiency improvements, that would lead to higher carbon abatement and more profits in the future.

The important finding is that for the European Union it is more profitable to trade emissions only with Russia and Ukraine (as in the EUFSUTrade scenario) than internationally. The EUFSUTrade policy enables the EU to absorb the Copenhagen surplus of Russian and Ukrainian carbon permits rather than sharing it with other countries as happens under the scenario IntTrade. Consequently, for a small market with a large surplus of permits, the carbon price is low (7.64 $€_{2008}$/t CO_2), and the EU reduces its emissions the least, by 8.51 % below the level of 2004, compared with the other scenarios (see Table 18.2).

Computer simulation reveals that some climate free-riders can actually benefit from strict carbon policies in other countries. Indeed, as the multilateral implementation of abatement pledges lowers demand and prices for fossil fuels, they are absorbed by the ROW countries (AllDomestic and IntTrade), Ukraine and India (AllDomestic), so both the GDP and emissions of these countries grow. At the same time, multilateral policies harm the exporters of fossil fuels, such as Russia and OPEC, but reward energy-intensive production in countries with non-binding abatement targets.

Even though the possibility to trade emissions internationally provides climate free-riders with incentives to cut GHGs, at the global scale such policy only lowers the cost of carbon abatement, without meeting the GHG stabilisation goal (Tables 18.2 and 18.3).

The modest impact of climate policies on global GDP is explained by the relatively small amount of abated carbon. The interesting finding is that, due to spillover effects, the domestic carbon price does not necessarily correlate with the welfare implications of carbon policies. Under the unilateral carbon policy, when only the EU achieves its COP 16 commitments (EUDomestic scenario), the GDP in the non-EU countries does not deviate much from the baseline, as they can still import energy-intensive goods and fossil fuels at relatively low prices.

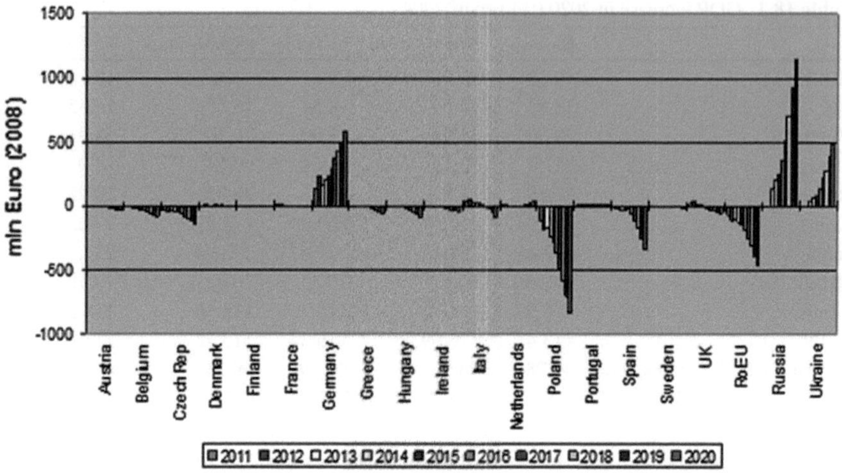

Fig. 18.1 Volumes of internationally traded CO_2 (scenario EUFSUTrade), million €$_{2008}$, (*plus*) stands for sale and (*minus*) for the purchase of permits

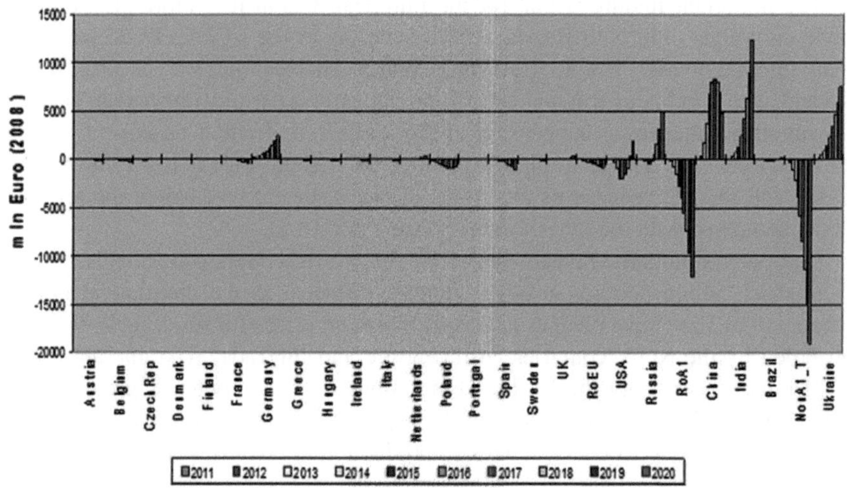

Fig. 18.2 Volumes of internationally traded CO_2 (Scenario IntTrade), million €$_{2008}$, (*plus*) stands for sale and (*minus*) for purchase of permits

On the joint EU, Russian and Ukrainian carbon market (EUFSUTrade) the largest sellers are Russia, Ukraine and Germany (Fig. 18.1).

Although Germany generates surplus permits due to the high rate of energy efficiency improvements assumed in the baseline, and not because of the lenient abatement target, for Russia and Ukraine the situation is the opposite.

On the global carbon market China, India, Russia, and Ukraine are the largest sellers (Fig. 18.2).

Table 18.4 Fossil fuel consumption per GDP in 2020 (%) versus 2004

	Baseline	AllDomestic	IntTrade	EUFSUTrade	EUDomestic
EU-27	−22.27	−27.14	−26.78	−24.89	−28.02
US	−18.32	−27.71	−28.35	−17.87	−17.84
Russia	−26.20	−31.34	−36.87	−30.21	−25.65
Ukraine	−30.30	−11.51	−46.00	−37.96	−29.42
RoA1	−22.87	−41.60	−28.90	−22.32	−22.33
China	−32.32	−46.25	−47.02	−31.98	−31.96
India	−32.21	−30.53	−45.24	−31.80	−31.80
Brazil	−26.62	−29.51	−30.71	−26.32	−26.33
NonA1	−27.91	−54.17	−35.92	−27.49	−27.46
ROW	−27.46	−20.59	−21.37	−27.03	−27.08

As expected, the revenues of Russia and Ukraine are much higher under the unrestricted global trade (IntTrade scenario) than under the linking with the EU ETS (EUFSUTrade).

China and India have considerable potential for carbon trade because of low GHG marginal abatement costs and lenient emission reduction targets.

Countries that form the RoA1 and NonA1 aggregates are the largest buyers on the international carbon market since the purchase of carbon permits reduces the need for domestic abatement to achieve their pledges (Table 18.1; CAT 2012).

As shown in Table 18.4, fossil fuel consumption negatively correlates with the stringency of abatement targets. Indeed, multilateral policies (IntTrade and AllDomestic) that lower the demand for fossil fuels result in the most pronounced improvements in fuel intensity of GDP.

Computer simulation demonstrates the significant structural changes towards the less energy-intensive sectors under the carbon constraints (Table 18.5).

As shown in Table 18.5, industry producers in the European Union benefit less from the unrestricted global carbon trade than from the inclusion of Russia and Ukraine in the EU emissions market. Indeed, under the EUFSUTrade scenario production levels of European sectors closely follow the baseline, because the bulk of abatement occurs in Russia and Ukraine (see Tables 18.1, 18.2, and 18.5). At the same time carbon trade with the EU has a milder effect on energy-intensive production in Russia and Ukraine, compared to unrestricted international trade (scenario IntTrade), see Table 18.5.

Since Russia is a major exporter of hydrocarbons, it may favour a small carbon market rather than the multilateral implementation of carbon policies, since the latter lowers demand and prices of fossil fuels.

Although the Ukrainian economy is heavily reliant on imported oil and gas, and on exports of energy-intensive goods, absenteeism from participation in multilateral carbon policies seems to be favourable for its production and GDP. Indeed, stringent carbon policies in the rest of the world would discourage energy-intensive production, and lower fuel prices, therefore rising demand for Ukrainian products.

Table 18.5 Domestic production of selected goods in 2020 (%) versus baseline scenario

	EU-27	US	Russia	Ukraine	RoA1	China	India	Brazil	NonA1	ROW
AllDomestic										
Agriculture	−0.8	−1.2	0.3	5.2	−2.5	−0.6	0.6	−1.8	−1.2	0.6
Electricity	−2.7	−5.7	−7.2	10.4	−5.4	−14.0	1.1	0.2	−29.3	6.0
Paper	−0.2	−0.7	5.2	6.9	−1.1	−2.8	1.4	−0.4	−6.2	4.4
Minerals	0.1	−0.1	−0.5	17.1	−1.6	−4.2	2.3	0.6	−15.0	6.9
Chemicals	1.6	−0.1	−2.8	44.1	−5.0	−6.8	7.0	−0.4	−21.3	19.0
Iron and steel	0.2	0.1	3.9	35.4	−2.6	−3.9	3.0	1.0	−17.9	18.3
Transport	0.0	−1.5	−1.7	10.9	−3.1	−2.9	3.1	−0.7	−14.1	5.6
ROI[a]	−1.0	−1.4	0.5	8.3	−1.2	−1.4	−1.0	−0.8	−2.0	3.1
Services	0.3	0.2	−3.2	2.5	−0.4	−2.3	2.7	0.5	−4.8	0.4
IntTrade										
Agriculture	−0.7	−1.1	−0.6	−6.1	−1.4	−0.8	−0.4	−1.6	−0.1	0.5
Electricity	−2.7	−5.9	−14.2	−11.4	−1.3	−15.4	−8.5	0.4	−7.6	5.3
Paper	−0.1	−0.6	6.4	−13.8	0.0	−3.7	−6.0	−0.4	−0.4	3.9
Minerals	0.0	−0.6	−3.4	−20.4	0.4	−5.2	−9.1	0.2	−2.3	5.9
Chemicals	1.0	−1.7	−19.0	−43.8	0.7	−9.6	−5.8	−1.0	−1.8	15.6
Iron and steel	1.2	−0.2	−6.8	−28.8	2.2	−5.4	−8.7	0.6	−0.5	17.8
Transport	−0.4	−2.0	−5.1	−13.9	−0.7	−4.1	−1.9	−1.5	−2.4	4.3
ROI	−0.6	−0.9	−3.8	−15.3	−0.5	−1.9	−3.3	−0.4	0.4	3.1
Services	0.2	0.1	−3.9	4.6	0.1	−2.9	−0.6	0.3	−0.9	0.3
EUFSUTrade										
Agriculture	−0.1	0.0	−0.5	−2.5	0.0	0.0	0.0	0.0	0.0	0.1
Electricity	−1.2	0.2	−5.4	−5.0	0.2	0.4	0.2	0.1	0.4	0.6
Paper	−0.2	0.0	0.2	−5.0	0.0	0.1	0.1	0.0	0.1	0.2
Minerals	−0.3	0.1	−2.0	−7.9	0.0	0.2	0.2	0.1	0.2	0.4
Chemicals	−0.6	0.2	−9.8	−18.3	0.3	0.3	0.4	0.2	0.3	1.0
Iron and steel	−0.2	0.2	−5.9	−11.8	0.4	0.4	0.5	0.5	0.5	1.6
Transport	−0.5	0.2	−2.1	−4.8	0.1	0.2	0.2	0.2	0.3	0.4
ROI	−0.1	0.0	−2.8	−4.7	0.0	0.1	0.0	0.0	0.1	0.2
Services	−0.1	0.0	−1.1	−0.4	0.0	0.1	0.1	0.0	0.1	0.0
EUDomestic										
Agriculture	−0.25	0.00	0.24	0.42	0.00	0.04	0.05	−0.06	0.05	0.05
Electricity	−2.52	0.30	0.71	0.70	0.22	0.40	0.24	0.12	0.39	0.55
Paper	−0.34	0.00	0.45	0.34	0.00	0.13	0.19	0.00	0.09	0.18
Minerals	−0.73	0.08	0.45	0.92	0.08	0.18	0.19	0.07	0.20	0.33
Chemicals	−1.38	0.38	2.53	2.63	0.44	0.45	0.55	0.20	0.40	0.86
Iron and steel	−1.39	0.08	1.80	1.94	0.08	0.19	0.29	0.15	0.25	0.78
Transport	−1.13	0.25	0.46	0.89	0.16	0.31	0.22	0.28	0.33	0.47
ROI	−0.26	0.00	0.76	0.36	−0.09	0.03	0.00	0.00	0.05	0.09
Services	−0.17	0.00	−0.15	0.06	0.00	0.12	0.15	0.00	0.09	0.00

[a]ROI—rest of the world aggregate

However, access to the international carbon markets incentivises Russia and Ukraine to abate emissions far below their baseline, even when they take lenient GHG abatement pledges. This conclusion is in line with the results obtained by Böhringer et al. (2007), Alexeeva-Talebi and Anger (2007), and Anger et al. (2009).

Conclusions

This paper aims to find the balance between the effective reduction of emissions and profits for hot air-holding countries such as Russia and Ukraine. If the possibility of carrying forward emission permits from the Kyoto to the Copenhagen commitment period is eliminated, by 2020 global carbon emissions would grow by 71–96 % above the 1990 level, depending on the model scenario. This means that in 2020 the global CO_2 level will exceed the CO_2 level recommended in the IEA's World Energy Outlook 450 ppm scenario by 5–10 Gt. Multilateral implementation of carbon policies when all regions abate CO_2 according to their COP 16 goals leads to the smallest increase in global emissions among the model scenarios. The important finding is that Russia and Ukraine could exercise market power to achieve increased environmental effectiveness as they will have incentives to reduce emissions in order to benefit from carbon price discrimination between the domestic and global markets. Carbon trade with Russia and Ukraine is also profitable for the EU, because it halves the European carbon price compared to when the EU implements its emission policy domestically. For the European Union the compliance cost is lower when Russia and Ukraine join the EU carbon market than in the case of a global carbon trade. Inclusion of Russia and Ukraine into the EU carbon market allows the EU to absorb their surplus of carbon permits rather than sharing it with other countries. However, Russian and Ukrainian revenues from carbon exports are much higher in the case of global trade. Global emission trading equalises carbon prices among countries, ensuring the steepest decline in carbon emissions of "hot air" holders such as Russia and Ukraine. However, at the global scale such policy would only lower the carbon price, and not meet the temperature stabilisation goal. Ambitious abatement targets, along with a large emissions market, would make climate policy effective.

Acknowledgments The research leading to these results has received funding from the European Community's 7th Framework Programme FP7/2007–2013 under the Grant Agreement No. PIIF-GA-2008-220852. The author would like to thank Dr. Francesco Bosello (Fondazione Eni Enrico Mattei) for his valuable comments and suggestions.

References

Alexeeva-Talebi V, Anger N (2007) Developing Supra-European emissions trading schemes: an efficiency and international trade analysis. ZEW discussion papers, No. 07–038

Anger N, Brouns B, Onigkeit J (2009) Linking the EU emissions trading scheme: economic implications of allowance allocation and global carbon constraints. Mitig Adapt Strat Glob Change 14(5):379–398

Böhringer C, Moslener U, Sturm B (2007) Hot air for sale: a quantitative assessment of Russia's near-term climate policy options. Eur Assoc of Environ and Resource Econ 38(4):545–572

Burniaux J-M, Truong T (2002) GTAP-E: an energy-environmental version of the GTAP model. GTAP technical papers 923, Center for Global Trade Analysis, Purdue University

CAT (2012) Climate action tracker. http://www.climateactiontracker.org/

Centre of Policy Studies (2002) An introduction to GEMPACK. Faculty of Business and Economics at Monash University. http://www.monash.edu.au/policy/gpdoc.htm
Dellink R, Briner G, Clapp Ch (2010) Costs, revenues, and effectiveness of the Copenhagen Accord. Emission pledges for 2020. Environment working paper no. 22, OECD
den Elzen MGJ, Höhne N (2010) Sharing the reduction effort to limit global warming to 2 °C. Clim Policy 10(3):247–260
den Elzen MGJ, Lucas P, van Vuuren D (2008) Regional abatement action and costs under allocation schemes for emission allowances for achieving low CO_2-equivalent concentrations. Clim Change 90(3):243–268
den Elzen MGJ, Roelfsema M, Slingerland S (2009a) Too hot to handle? The emission surplus in the Copenhagen negotiations. http://www.pbl.nl/en/publications/2009/Too-hot-to-handle-The-emission-surplus-in-the-Copenhagen-negotiations
den Elzen MGJ, Mendoza Beltran MA,van Vliet J, Bakker SJA, Bole T (2009b) Pledges and actions. A scenario analysis of mitigation costs and carbon market impacts for developed and developing countries. http://www.ecn.nl/docs/library/report/2009/o09034.pdf
EEA (2013) Atmospheric greenhouse gas concentrations. European Environment Agency—assessment published in Jan 2013. http://www.eea.europa.eu/data-and-maps/indicators/atmospheric-greenhouse-gas-concentrations-2/assessment
EURELECTRIC (2010) Power choices pathways to carbon-neutral electricity in Europe by 2050. www.eurelectric.org/PowerChoices2050/
European Commission (2009) Towards a comprehensive climate change agreement in Copenhagen. Communication from the Commission to the European Parliament, the Council, the European Economic and Social Committee and the Committee of the Regions. http://eur-lex.europa.eu/LexUriServ/LexUriServ.do?uri=COM:2009:0039:FIN:EN:PDF
IEA (2010) International Energy Agency—World Energy Outlook 2010. http://www.worldenergy outlook.org/publications/weo-2010/
IEA (2011) International Energy Agency—World Energy Outlook 2011. http://www.worldenergy outlook.org/publications/weo-2011/
IPCC (2007) Fourth assessment report. http://www1.ipcc.ch/ipccreports/assessments-reports.htm
Kartha S, Erickson P (2011) Comparison of Annex 1 and non-Annex 1 pledges under the Cancun agreements. Stockholm Environment Institute (SEI). http://seiinternational.org/mediamana ger/documents/Publications/Climate/seiorkingpaperus-1107.pdf
Klepper G, Peterson S (2007) Distribution matters: Taxes vs. emissions trading in post Kyoto. Kiel working paper 1380, Kiel Institute for the World Economy
McKibbin WJ, Morris A, Wilcoxen PJ (2011) Comparing climate commitments: a model-based analysis of the Copenhagen Accord. Clim Change Econ 02(02):79–103
Narayanan GB, Walmsley T (2008) The GTAP 7 data base. West Lafayette/Indiana: Center for Global Trade Analysis Purdue University. https://www.gtap.agecon.purdue.edu/databases/v7/
Rogelj J, Nabel J, Chen C, Hare W, Markmann K, Meinshausen M, Schaeffer M, Macey K, Höhn N (2010) Copenhagen Accord pledges are paltry. Nature 464(7292):1126–1128
Schleich J, Duscha V, Peterson EB (2010) Environmental and economic effects of the Copenhagen pledges and more ambitious emission reduction targets. http://www.uba.de/uba-info-medien/3998.html
UNEP (2010) The emissions gap report: are the Copenhagen Accord pledges sufficient to limit global warming to 2 °C or 1.5 °C. United Nations Environment Programme (UNEP). www.unep.org/pub lications/ebooks/emissionsgapreport/pdfs/The_EMISSIONS_GAP_REPORT.pdf
UNEP (2011) Bridging the emissions gap. United Nations Environment Programme (UNEP). www.unep.org/pdf/UNEP_bridging_gap.pdf
UNFCCC (2010) Copenhagen Accord, 21 April 2010. http://unfccc.int/meetings/copenhagen_dec_2009/items/5262.php

UNFCCC (2011) Submission to the AWG-LCA and AWG-KP. Information relating to possible quantified emissions limitation and reduction objectives as submitted by Parties. http://unfccc.int/files/kyotoprotoc_ol/application/pdf/awgkpjointqelrosubmission091009.pdf
van der Laan (2009) Impact of the Post-2012 expansion of the EU emission trading scheme. A computable general equilibrium assessment. ECORYS research programme. http://www.ecorys.com/%20sites/default/files/files/Pillar%201%20CD17500%20-%20EU%20Emission%20Trading%20Scheme.pdf

Chapter 19
Perceptions About Climate Change in Sidama, Ethiopia

Seyoum Hameso

Abstract This research reports on how farmers perceive and adapt to climate change in three agro-ecological zones of Sidama, South Ethiopia. The main aim is to increase understanding about smallholder farmers' perspectives by documenting and analysing local people's perception of climate change. The research was held as part of a large project investigating vulnerability and adaptation to climate change. The researcher spent six months (from January to May 2012) in the field undertaking, among other things, focus group discussion (FGD) meetings from March to May 2012. Findings revealed that farmers clearly perceive climate change based on their lived experience and knowledge of their local environment. They identify shifting seasons, increased aridity, drought, erratic rainfall, floods, extreme heat and the emergence/spread of diseases such as malaria as indicators of change. Yet their perception of the causes of climate change varied: deforestation, God's wrath, abandonment of past traditions/practices, and overpopulation. They also assigned important role to religious beliefs and government authority to address the problems engendered by what they refer to as "changed times." Since the problem is rightly recognised, government policy needs to focus on aligning local knowledge and values with scientific information for adaptation to climate change.

Keywords Adaptation • Agriculture • Climate change • Coffee • Drought • Enset or Wesse • Ethiopia • Focus group • Perceptions • Sidama • Smallholder farmers • Vulnerability

---

S. Hameso (✉)
School of Law and Social Sciences, University of East London, Docklands Campus, University Way, London E16 2RD, UK
e-mail: hameso@gmail.com; seyoum@uel.ac.uk

W. Leal Filho et al. (eds.), *International Perspectives on Climate Change*, Climate Change Management, DOI: 10.1007/978-3-319-04489-7_19,

Introduction

"We are living in changed times," reported a smallholder farmer in a Sidama village, in an apparent reference to climate change.[1] A matching narrative among scholars elsewhere makes similar point. Nordhaus (2007) argued that "we live in the age of global warming and global warming studies." Other scholars underscore that climate change is not only a reality but also one of the most urgent challenges of our time (Stern 2007; UNDP 2007; Giddens 2009). Far more important are the implications of climate change on economic growth, food security, and people's survival—especially those living in the poorest communities in developing countries. The issues facing smallholder farmers who largely rely on rain-fed agriculture for their livelihood are deeply complex. Here, as in most of Africa, smallholder farmers are the majority in agricultural sector and form the backbone of agricultural production in Africa (Dixon et al. 2004).

Addressing challenges emerging from climate change requires in-depth knowledge about how the problem of climate change is perceived by different stakeholders. Thus it is essential to explore how scientific position on climate change is shared and rationalised by common people. The motive for doing research of this nature is based on its timeliness and importance across communities, space, and disciplines. It is recognised that climate change is the major environmental problem facing the world.

For developing countries such as Ethiopia, research on perceptions about climate change is vital on both practical and theoretical levels. Practically, fast evolving and emerging knowledge in the field is required to inform policy. Theoretically, such research offers deep insight into the relationship between vulnerability, resilience and adaptive capacity of people and ecosystems under environmental risk, thus contributing to on-going debate on a topical and important problem of our times.

A growing body of literature exists on perceptions of farmers on climate change (Byg and Salick 2009; Crate and Nuttall 2009; Moghariya and Smardon 2012; Mortreux and Barnett 2009; Petheram et al. 2010; Roncoli et al. 2002); yet Ethiopia is least studied (Deressa et al. 2009) and the case of Sidama is even more wanting. This study fills the gap in the know and paucity of research.

Perceptions are grounded on complex set of social, political and environmental settings and they are context-specific. The IPCC (2007) recognised that perception is about human behaviour, which is one of the least understood components of the climate system. It is known that people's perception about climate change, and specially its causes are filtered through local knowledge, values and moral norms; hence they could diverge which may not necessarily conform to expert or scientific views.

[1] According to the Intergovernmental Panel on Climate Change (IPCC 2007), climate change is a variation in the mean state of the climate persisting for an extended period (typically decades or longer) and resulting from anthropogenic greenhouse gas emissions.

Studies on rural people's beliefs and understandings of climate change in Western India suggested that most rural respondents have not heard about the scientific concept of climate change, but they have detected changes in the climate (Moghariya and Smardon 2012). A study with East Tibet villages indicated that people were not aware of the global phenomenon of climate change and they assumed the changes were local (Byg and Salick 2009). Another study with coffee producing communities of Mexico found notable lack of recognition by farmers about the impacts of climate change for their coffee trees in comparison to changes of coffee prices, which they indicated was more significant (Eakin et al. 2005; see also Bacon 2005, for the case of Nicaragua). On the other hand, studies on Africa noted that farmers were aware of warmer temperatures and decline in precipitation (Maddison 2006; Juana et al. 2013). In Ethiopia, research on the Nile Basin reported differential perceptions among farmers living in the highlands and in the lowlands or the midlands (Deressa et al. 2010: 28). These findings are explored in the context of a coffee and Wesse (Enset) growing society of Sidama.

Study Area

Sidama is located in Southern Ethiopia. Ethiopia is currently structured into 9 regions and 2 city administrations (Addis Ababa and Dire Dawa). Sidama is found in one of the 9 regions known as Southern Nations, Nationalities and Peoples Region (SNNPR). According to Central Statistical Agency of Ethiopia, the total population of Sidama stood at 3.4 million. With the population density of 506 people per square kilometres, Sidama is one of the highest in the country, the Ethiopian average being 83 (CSA 2011) (Fig. 19.1).[2]

Rural residents in Sidama account for over 94 % of the population. As in much of the rural Ethiopia, farming is predominantly based on smallholdings as the dominant livelihood activity for the majority of people (Devereux and Guenther 2007: 2). In terms of administration, the Sidama Zone is divided into 19 rural districts and 527 rural villages or 'Kebeles' in Amharic. In terms of geography (topography), the elevation of the area ranges between 501 and 3,000 m above sea level. Annual mean temperature ranges between 10 and 27 °C and the annual mean rainfall ranges between 801 and 1,600 mm (Fig. 19.2).

Sidama has the total land size of 6,972 km^2, of which, 26.8 % belong to lowlands, 45.5 % to midlands and 27.7 % to highlands.

The midlands are known for coffee which has been the main cash crop and crucial to local livelihoods but under threat from increasingly irregular rainfall patterns and rising temperatures. Sidama is also home for Wesse plant that is instrumental for food security. Wesse (an indigenous tuber also known as Enset and false banana) is dubbed as drought resistant tree (Brandt et al. 1997).

[2] Ethiopia's population in 2011 was 85 million.

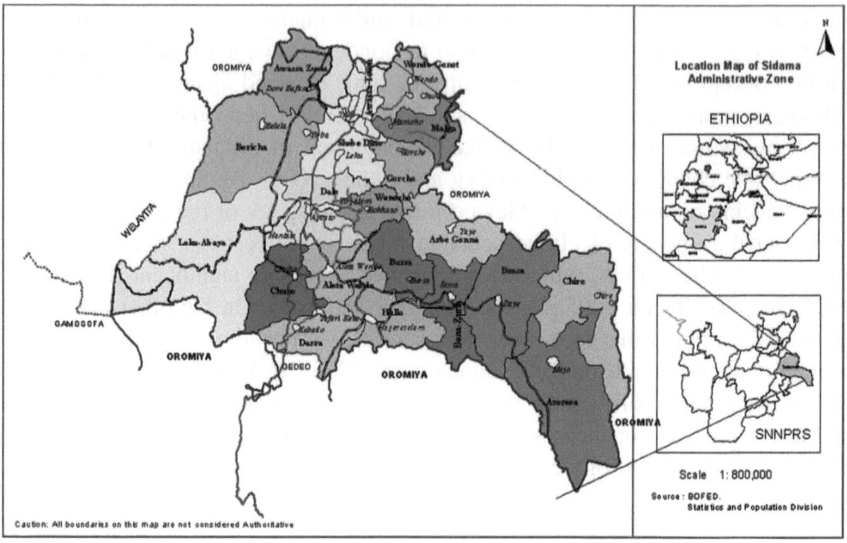

Fig. 19.1 Map of Sidama, in SNNPRS, Ethiopia (*Source* SNNPR)

Fig. 19.2 Map of Sidama, SNNPR, Ethiopia (*Source* Anonymous)

The research sites representing three agro-ecological zones were selected because the impacts and adaptation to climate change vary from place to place. The focus on specific coffee-Wesse livelihood system was based on the recognition that future ability to cope with climate change will be supported by specific knowledge about where, when, and how much climate change will affect human communities (Wheeler 2011, p. 4).

Ethiopia is country long accustomed to the problems of famine and food insecurity. Recurrent droughts and resultant famine and dependence on food aid have become most pronounced features of the country for a long time. In the past four decades alone, the country has suffered from serious droughts, some resulting in highly publicised famines and death of millions of people. The brunt of the catastrophe was born by farming and pastoralist communities. In order to address these issues, the impacts of climate change in the form of temperature rise and precipitation need proper exploration.

Methods

The study utilised comparative cross-sectional research design of livelihood systems in three ecological zones to gain better understanding of smallholder farmers' understanding and perception of climate change. Fieldwork took place between January and June 2012. Like other programmes of field research, the study involved site selection, sampling, and negotiating entry to samples for data collection (Bailey 2007). Research sites were selected on the basis of agro-ecological zones and availability of meterological stations[3] that have rainfall and temperature data for 20 years. These stations include Hula, Yirgalem and Hawassa. The environs in close proximity to these stations are Xexicha, Awaada, and Jara Galalcha, respectively. The geographic basis of sampling was neighbourhood (also known as Kebeles[4]) consisting of at least five hundred families, or the equivalent of 3,500–4,000 persons).

Participants to the FGD were key informants who lived in the area for two or more decades and from different age and sex groups. Each FGD involved 10 farmers, including 2 women, the main researcher and research assistant. The identification and selection of participants was facilitated by local extension workers and officials. Three meetings took place at village communal sites and offices. Farmers were asked to discuss openly about their perception of climate change. Some of the questions included:

- How do describe the climate of your area in the last 10 years?
- What do you think are the causes of the climate change?
- Can you explain any impacts on your livelihood and problems you face as the result of the climate change?
- What do you do to cope or adapt to climate change?

The meetings were facilitated and conducted in Sidamuafo or Sidama language. The researcher's knowledge of the local language helped with communication and direct

[3] Depending on the instruments they use and weather elements they record, weather stations are categorized in Ethiopia into four classes. First class stations or Synoptic stations (only 2 exist in the SNNP region, namely, in Hawassa and Arba Minch), record all weather elements and receiving satellite data. Second class or Principal or Indicative stations record temperature, rainfall, humidity, sunshine and some other weather elements. Third class or Ordinary stations record temperature and rainfall. Fourth class record only rainfall (SNNPR; NMA; http://www.ethiomet.gov.et/stations/regional_information/).

[4] Kebele is similar to a ward, a neighbourhood or a localized group of people. It is part of a woreda, or district, itself usually part of a Zone, which in turn are grouped into one of the Regions (or Kilil) that comprise the Federal Democratic Republic of Ethiopia.

observation of cultural sensitivities and subtleties. The role of research assistants and their knowledge of the local conditions were also important. The researcher travelled to rural farming sites to explore land use and landscape to account for land degradation, soil erosion and population settlement or density. Responses from FGDs were taped with digital recorders which were latter transcribed for analysis. The data collected in this way resulted in voluminous audio recordings and digital pictures.

At the time of the field research, data was also collected from secondary sources such as published documents on Ethiopian legislature, policies, laws and practical measures relating to climate change in the country. Limited amount of data in the form of participant and field observation was also gathered which was recorded during or soon after the field visits.

The unit of analysis for this study were individual heads of households. Nearly all of the participants were adults, i.e., household heads which included women. Households were selected, as they were basic economic units making decisions on production and consumption in rural settings. The following are findings about perception, causes and actions on climate change from the three sites of interest.

Results and Discussion of Findings

FGD responses were analysed using Nvivo 10 Qualitative Data Analysis Software. Analysis of qualitative data explored key emerging themes, meanings and discourses of climate change, through the deployment of Grounded Theory (Glaser and Strauss 1967; Fleming and Vanclay 2011). Analysis started with depth reading of transcripts from FGD meetings. In line with research ethics and to maintain participant privacy, the names of participants were rendered anonymous. The transcripts were then loaded into NVivo software to develop codes to assist with the analysis of qualitative data.

One of the initial insights into the research was realisation of the lack of direct linguistic equivalent of the term *climate change* in Sidamuafo. The words "*dilallote soorro*" refer to change in wind or weather conditions or what farmers referred to as changed times. The word *dilallo* literally means wind and the word *soorro* means change. Respondents tended to conflate the concepts of weather and climate. This is not surprising as Rudiak-Gould (2012) noted that cultures might not have two distinct words that unambiguously distinguish between western concepts of climate and weather. Such a process of translation necessarily entails transformation of the climate change concept and we ventured into the business of translation knowing full well that no translation is perfect.

Thus the researcher prioritised the clarification of the concept of climate change and the purpose of the study in all meetings. Responses to the questions varied in content and context with striking similarities of themes and issues across agro-ecological zones.[5] The following table (Table 19.1) provides an overview of emerging themes on perception, vulnerability and adaptation to climate change.

---

[5] Here the names of the locations Xexicha, Awaada and Jara Galalcha represent highlands, midlands and lowlands, respectively.

Table 19.1 Summary of perception about climate change, causes and adaptation actions

Climate change	Xexicha (highland)	Awaada (midland)	Jara Galalcha (lowland)
Perceptions	Climate change exists	Climate change exists	Climate change exists
Indicators	Higher temperature, change in wind direction, erratic rainfall pattern, higher disease incidence	Higher temperature, heat wave, erratic rainfall pattern, higher disease incidence	Higher temperature, drought, heat wave, erratic rainfall pattern, higher disease incidence
Causes	Deforestation, planting non-native trees, God's wrath, weakened indigenous practices and values	Deforestation, God's wrath, weakened indigenous practices and values	Deforestation, God's wrath, weakened indigenous practices and values
Vulnerability and impacts	Food insecurity, poverty, uncertainty, decline in production, overpopulation	Food insecurity, poverty, uncertainty, decline in production, overpopulation	Food insecurity, poverty, uncertainty, decline in production, overpopulation, shortage of grazing land, shortage of water, flooding
Adaptive actions	Resort to faith, fatalism, diverse livelihoods	Resort to faith, fatalism, diverse livelihoods	Fatalism, diverse livelihoods, use of hybrid seeds and fertiliser

Lay Perceptions of Climate Change

As the Table 19.1 above shows, farmers noticed climate change in the form of rising temperature, shifting seasonal distribution of rains, erratic rain, heat waves and emergence of health hazards such as malaria. But these indicators of change had varying weight in different altitudes and socio-economic settings.

Farmers in the drought prone, low-lying areas expressed higher understanding about causes and adaptive measures of climate change than in the highlands. This is in agreement with findings elsewhere, in the case of West India (Moghariya and Smardon 2012). However, throughout the research sites, Sidama farmers reported changes in broader sense. One farmer claimed: "We entered a new era. We are living in a new era. Old things have changed. … While the habitual calendar of months remains the same, the weather is changing and we don't know why it is so" (Farmer from highlands).

The same participant added:

> *…there is climate change at [different levels]. What we know to happen in the past and what is happening now are different. In the past, people knew what to expect. They knew what happens during the autumn…. People knew what happens in those months. Then they prepare for farming activities. With full knowledge of the events, people organise their life and prepare for work. They produce foodstuff for their own consumption and for the market. Then [they] improve their living standards. … In recent times, however, an unexpected event emerged. While changes in the past were imposed on people, such as revolutionary change in politics, now there is a change in climate. It has unexpectedly engulfed people without their knowledge …* (Farmer from highlands).

The challenge for farmers lies not only in the changed climate but also in the irregularity and unpredictable nature of the changes in the form of rising temperature, heat waves, late onset of rain, and unpredictable seasonal distribution of rain.

Increased Temperature

In the highlands of Xexicha, participants noted that their land is gradually converted to midland conditions. For instance, a participant noted that "the sun has turned our highland to dry midlands. Things that never grow in our lands are currently growing due to climate change. We are now growing things that used to grow the in the lower altitudes" (Farmer from highlands).

In the midlands of Awaada, farmers felt that drought was getting stronger and on-going with significant impact on their livelihood including loss of crops. Comparing appearances to the reality of lived experience of heat waves, an elderly farmer said: "… with the changed times, all people are in trouble (…) When people are seen walking on the streets, it looks as if they are well and comfortable, but underneath we think that survival is at stake. Life with heat waves is very difficult" (Farmer from midlands).

Alongside drought, farmers mention decline in river levels and underground water. In the lowlands, there are palpable evidence of aridity. For example, farmers of Jara Galalcha were forthcoming to linking temperature rise in their locality to climate change than the other localities surveyed. According to one farmer "Temperature has increased. Why all this happens? It is climate change. The temperature has increased. It has increased specially in the last three years. While rain is decreasing, temperature is rising" (Farmer from lowlands).

Erratic Rainfall Patterns

The perception of climate change as manifested in erratic rains is common to all ecological zones under study. Farmers reported that rain used to start in January or February in the past, but it moved to April–May in recent times. According to a farmer from Awaada, fellow farmers used to plant around mid-February in the past. "These were years of prosperity, where we get plenty, use and take good care of our families" (Farmer from midlands).

Shifting cropping season was also emphasised by participants. Indicating late onset of rain and changes in cropping calendar, a farmer from Awaada told: "While we used to plant Wesse and maize in January and February in the past, today planting the same moved to April or June" (Farmer from midlands).

Change of direction and extent of wind is also reported in one instance. A farmer told:

> In highly unknown ways, change happened in manners that harm our area. The climate has changed significantly. It does not wind the way it used to (…). It does not wind the way we knew. In the past, the wind was heavy. It can carry a person. But it was not harmful (…) It would not destroy anything. It sweeps away peacefully. Today's wind causes damage to people. Its behaviour is not known. When rain appears on the horizon without signals, the wind spreads it across the land and we don't enjoy abundant rainfall (Farmer from highlands).

An elder man in Jara Galalcha, reported that indigenous knowledge established regular seasons. "Rain comes on its time," he said. "The spring rain comes on time. In the interim, there is a flowering time which happens without changing its time" (Farmer from lowlands).

The above description of climate change as perceived by farmers is compared with climate data obtained from the southern branch of the National Meteorology Agency.

Climate Data Analysis

In terms of agro-ecological characteristics of the study sites, data is aggregated to show average annual maximum (and minimum) temperature and rainfall (see Table 19.2). Accordingly, Hagereselam with the highest elevation of 2,759 metres above sea level (masl), recorded milder temperature (19 °C) and higher rainfall.

Table 19.2 Agro-ecological zones of study sites (1992–2011)

Survey site	Main crops/livelihood system	Elevation, metres above sea level	Max annual mean temperature (°C)	Minimum annual mean temperature (°C)	Average annual rainfall (mm)
Hawassa (Jara)[a]	Maize and livestock	1,665	27	13	81
Yirgalem (Awaada)[a]	Coffee, enset and livestock	1,750	27	12	101
Hagere Selam (Xexicha)[a]	Enset and livestock	2,759	19	–	111

Source National Meterological Agency
[a]Sites in close proximity to weather stations or about 10–20 km. For example, Awaada site is close to Yirgalem town where weather data is collected by the Metrology Agency

Both Yirgalem and Hawassa with relatively lower altitude, recorded higher temperature and lower precipitation. Furthermore, the low lying areas of including Hawassa cover part of the Great Rift Valley system and hence they are warmer and drier in comparison with the mid- and highlands.

Differences in the livelihood systems surveryed is predicated on spatial and temporal differences. For instance, the midlands specialise on the production of crops such as coffee for export and domestic consumption.The highlands rely predominantly on enset and livestock while the lowlands grow maize alongside livestock production.

Among climate change indicators, drought poses significant impact on the livelihood systems of Ethiopia. Even though nationwide research on impacts has remained fragmentary (Conway et al. 2007), available climate data shows that temperature has increased by 0.37 °C every ten years for the last fifty years, 1951–2006 (Tadege 2007; MoWRNMA 2007). Within this national framework, the Southern Nations and Nationalities and People's Region (where Sidama Zone is subsumed) has experienced an average temperature increase of 0.4 °C (NREPA 2012) which is above the national 0.37 °C. This shows the seriousness of the problem in the region, with high risk of drought.

A recent study by Viste et al. (2013) listed nation-wide drought episodes in Ethiopian occurred in 1972–1975, 1984, 1987, 1990–1992, 1999–2000, 2002–2003, and most recently in 2008–2011. Accordingly, 2009 was the second driest year, surpassed only by 1984. Spring droughts have occurred more frequently in all parts of Ethiopia during the last 10–15 years. Since rising temperature and erratic rainfall account for the drought conditions, it is relevant to explore the climate data for Sidama in detail.

Temperature and Rainfall Data

The main factor influencing temperature in Sidama is elevation. As elevation increases air temperature falls. The lowlands are thus the warmest parts of the country, with annual mean temperatures ranging from 20 to 25 °C (Coppock 1994). As

Table 19.3 Station temperature data (Cumulative average for 1992–2011)

Station/ Month	Jan	Feb	Mar	Apr	May	Jun	Jul	Aug	Sep	Oct	Nov	Dec
Hawassa												
Av. MaxT[a]	28.8	30.1	30.0	28.6	27.3	25.8	24.6	24.8	25.7	26.9	28.1	28.3
Av. MinT	11.7	12.0	13.0	14.2	14.3	14.3	14.5	14.6	13.8	12.8	10.5	10.4
Av. RF[a]	33.4	33.3	77.0	104.9	115.0	102.4	122.0	123.8	116.0	75.9	39.6	24.7
Yirgalem												
Av. MaxT	28.3	29.0	29.0	27.5	26.0	25.1	23.9	24.3	24.7	25.6	26.8	27.4
Av. MinT	10.4	10.7	11.8	12.6	12.7	12.6	12.9	12.6	12.1	11.6	10.6	9.7
Av. RF	35.5	28.1	87.4	166.9	174.5	102.6	110.3	125.6	137.9	169.5	46.3	25.2
Hagereselam												
Av. MaxT	19.9	20.7	20.5	19.1	18.6	17.9	16.6	16.8	17.4	18.0	18.7	19.4
Av. MinT	7.2	7.8	8.3	8.6	8.5	7.8	7.3	7.7	7.6	7.5	7.3	6.8
Av. RF	53.1	57.6	88.3	158.5	191.5	118.2	100.8	159.6	128.1	184.2	65.1	32.4

[a]Temperature is measured in °C and rainfall is measured in millimetres

the Table 19.3 shows, Hawassa recorded the highest temperature, an average of 28.3 °C in the year 2009 followed by 1994/1995 and 2001/2002 (27.8 °C). This trend corresponds with the national trend by Viste et al. (2013).

Average minimum temperature is at its lowest (6.8 °C) in December in Hagereselam followed by 10.4 °C in December in Yirgalem and 10.5 °C in November in Hawassa. This coincides with farmers' report to the researcher about dew or cold conditions for the months of November in the past (2–3 decades ago) but this conditions is moving to February and early March in recent times, due to climate change. Average maximum temperature is at its highest (30.1 °C) in February in Hawassa. The same is true of Yirgalem for February and Hagereselam.

In terms of rainfall, the area has a bimodal rainfall pattern with three seasons in a year, namely, two rainy periods and one dry period. The Belg season is characterised by small rains while the Kiremt is the main rainy season running from June to September. Bega is the dry season running from October to February. However, Sidama farmers refer to two main seasons: *hawado* and *arro*, or wet and dry seasons, respectively.

Rainfall data from the study sites demonstrates a mixed picture. For example, the Table 19.3 above shows that average rainfall is the highest (191.5 mm) in the highlands for the month of May which does not reflect the national seasonal classification. Yet for Yirgalem, the classification meets the national classification. In August (which is part of the main rain season/Kiremt), Yirgalem recorded 169 mm in October. This is not in line with the national classification that posits October as the dry season/Bega. Quite interestingly, the month of December in all the three agro-ecological zones showed the lowest rainfall, namely 24.7, 25.2 and 32.4 mm for Hawassa, Yirgalem and Hagere Selam, respectively. It means that December is the month of lowest precipitation throughout Sidama. This is confirmed in FGD meetings and the climate data corresponds with farmers' responses about their experience with climate change.

Perception of Causes of Climate Change

Existing literature established that the warming of the climate system is unequivocal (IPCC 2007). Climate change is caused by human activities that have grown since pre-industrial times. Almost all of the increase in greenhouse gases in the atmosphere is accounted by 70 % increase between 1970 and 2004 (IPCC 2007). These greenhouse gases (GHGs) are generated from the burning of fossil fuels such as coal, oil and gas as well as deforestation. Carbon dioxide (CO_2) contained in these fuels is one of the main gases which contribute to global warming—one of the key aspects of climate change.

One would expect components of the above scientific explanations to filter into discussions related to causes of climate change with rural communities. Yet the connection between CO_2 concentration and global warming or climate change was not directly established during the FGD meetings. For instance, the burning of fossil fuels or even the use of chemical fertiliser or dung as contributing to GHGs is hardly mentioned as a cause of climate change, given a rural context that hardly uses the fossil fuel products. In all the sites, wood is used for cooking and light.

Discussions around the causes of climate change were rather focused around religious explanation. Reference to God appeared in all FGDs. In one meeting, a farmer summed it thus: "This climate change, when we think about it, is an act of God" (Farmer from highlands). Yet a participant who argued that climate change is the work of man and not God held quite an opposite position. With reference to cutting of trees or deforestation, he noted: "… the mistake is ours only. What God is supposed to do?" (Farmer from midlands). This argument links well with findings from the study of Eastern Tibet where people had a strong belief that human actions are the root cause of climate change, directly or indirectly by angering or weakening gods or deities (Byg and Salick 2009).

This is further demonstrated by a study of rural Indian communities which showed that people "correctly detect climate change, have good understanding about causes and solutions, but also have views about climate change causation that are different from scientific perspectives" (Moghariya and Smardon 2012). The same authors noted that this is in part due to the limited extent of formal schooling among many rural people who often lack the background information necessary to process detailed scientific views on climate change" (Moghariya and Smardon 2012).

Debate also emphasised shifting inter-generational attitude towards forest: "People in the past pray to God and pay sacrifice under a tree. Today's people hurriedly cut trees and devastate native forest" (Farmer from highlands). Economic imperatives, increased population and the introduction of markets are behind the shift. The shift also led to declining importance once attached to community forest resource management and indigenous knowledge. Sacred places that were once filled with sycamore and other native trees have declined given enormous pressure emanating from economic, social and political changes. Eucalyptus trees are pervasively present through Sidama with vital importance on livelihoods.

Noting the conversion of fertile highlands to semi-arid conditions and that human action contributed towards climate change, a farmer argued: "it is wrong to blame God for lack of rain after planting trees in a wrong place. The land was fertile even when there were no rains. Planting wrong trees ruined it … Similarly, we plant eucalyptus tree where it should not be planted and we bring poverty upon ourselves. What I mean is, we dried up the rain by inappropriately cutting down indigenous trees, which were balancing our climate" (Farmer from highlands). This perception corresponds to recognition that deforestation and land use changes contributed to climate change. Yet the links are not directly established by farmers whose knowledge is limited to the local world.

> *In the beginning, it is not development workers who planted trees. It is not government. In the days when neither development workers nor government existed, God had balanced our climate on this land with forest, with water (…) and everything. We messed up and we complain that God has withheld rain (…) As the saying goes: a cloth owner at fault blames her cloth. After causing the problem, after failing to conserve trees, after planting trees in inappropriate places, we complain of land aridity (…) Without taking care of the land, we denounce government and God. This kind of thinking caused problems* (Farmer from highlands).

The same farmer went further and noted:

> *There is something which harms our way of life, our pleasant way of life. It is coming from wealthy people. It is the heat produced by their activities. The heat is causing us problems. For example, if a man installs a coffee processing machinery, that could benefit him and his children, but it can cause problems to other people. The residue of coffee berries pollutes the river. Other people, who live downstream, wash and drink from this polluted water. Both people and animals drink it. Similarly, what rich people are doing, what the government is doing (…) has impact on our land; it is changing our climate* (Farmer from highlands).

Perceptions of Vulnerability and Impact of Climate Change

Farmers describe their vulnerability to climate change in terms of food insecurity, poverty, overpopulation and general uncertainty. The sense of decline in livelihood is captured when they compare and contrast the past with the present: "We lived in affluence. Today carrying through a year is a struggle. Everything is on decline" (Farmer from midlands). Dependence on food alms (albeit in the form of food-for-work package) is mentioned as evidence of this:

> *… since time immemorial, Sidama people have never seen a problem of such a magnitude … and since 1999 we are overwhelmed with problems. In the past, people are used to work, use rain water for farming. They never look for government hand-outs. They would want to help the government instead. That was all in the past* (Farmer from midlands).

People in the highlands shared similar anxiety. In the words of one farmer, "people are troubled by this situation. People are troubled by uncertain situation. Climate change has imposed another change on people and people are seeking answers" (Farmer from highlands).

In a mixed farming-livestock economy, the shortage of land impacted not only on farming but also on grazing. The story is stark in the lowlands as the following complaint from a farmer demonstrates: "We wonder why the government which cares for wild animals in parks is not looking after the needs of our livestock which provides us with milk and butter" (Farmer from lowlands). The same farmer applauded Ethiopian government's efforts at improving health access and coverage in the country, but blamed inflation as a problem.

> *We were pleased with the departure of the Derg [the military regime that ruled the country from 1974–1991]. Our children have milk. Mothers are getting attention. Babies are weighed ... they are cared for while they are in the womb; they are delivered safely without problems. They grow without problems. Now what is a problem is inflation, the cost of maize seeds and fertiliser has skyrocketed* (Farmer from lowlands).

The combined impact of rain shortage and rising cost of hybrid seeds and fertiliser is raised quite often. A farmer recounts:

> *... what people produce gets damaged. It means climate change. People do things. Some would borrow from government [banks] to buy fertiliser and hybrid maize seeds. When crops fail, a farmer is left with empty hands. Wealthy farmers, who have their own cash, buy their own seeds. This group will also lose due to extreme weather. Farmers are now saying that it is not possible to develop by farming land; therefore they use their land to grow feed for livestock and continue on that path. People are pursuing such alternative paths—paths of laziness* (Farmer from highlands).

In semi-arid lowlands, a link is made of speedy aging of people and untimely growth of grey hair with increased temperature and heat waves. They compare this situation with slow aging process in higher altitudes where temperature is cooler.

Vulnerability of farmers in semi-arid lowlands is at its highest during the dry season when the shortage of water is at its peak. In the months of February and March, the researcher observed water being carried in yellow jerry cans (water containers) by carts, or even trucks, for sale to households (Fig. 19.3).

People of Jara Galalcha are close to Lake Hawassa, they use water from the lake which contains fluoride and nitrate acids with negative damage to teeth and bone. The damage was widely visible on young children who have brown teeth, often disfigured (Fig. 19.4).

In more than one instance, participants displayed a sense of despair as a revelation of vulnerability. The following quotes are testimony to this: "We are only looking up to God and government. God wouldn't let this to continue" (Farmer from lowlands). "If God and the times do not permit," added another farmer, "nothing can be done. Unless there is rain, what are we expected to do? Except perhaps using Woyma river or digging underground water and wells!" (Farmer from midlands).

There is the recognition that increasing population size with declining land holding and primeval farming technology, contributed to decline in livelihood. A farmer recounted this: "Human birth in the past was small. As human birth has increased without commensurate increase in land size, there are now many problems for people. Too many problems" (Farmer from highlands). Such explanation underscores the Malthusian worldview which is shared by participants.

Fig. 19.3 Dry weather and water scarcity near Hawassa, Feb 2012 (*Source* Author)

Fig. 19.4 Land degradation and flooding, Boricha, near Hawassa, Feb 2012 (*Source* Author)

Adaptive Actions

People admit that they cope with changed times in different ways ranging from fatalism to proactive measures. A meeting in the lowland noted: "As the saying goes, God that gave you an itching skin disease (*bijajo*), does not deny you nails (*culunqa*). When a problem appears, God offers the means to resolve it" (Farmer from lowlands).

The sense of fatalism and helplessness is also observed. An elderly participant lamented that "What is a solution if wesse is not productive anymore? What will happen to me? I would pray to God. I would ask God: what say you? Earth: what say you? I appeal: God, don't abandon me. Earth, don't abandon me. What else can I do?" (Farmer from midlands).

On the other hand, farmers reported deploying a variety of mechanisms to cope with changed times. Among the foremost are diversifying and growing more crops, including vegetables and edible trees (such as avocado and papaya trees), instead of sole reliance on traditional wesse and coffee. A farmer said: "Sidama's wealth in the past was Wesse … Our wealth is Wesse and coffee followed by harcourt beans, potatoes, and green paper. The latter are not things I am used to by tradition. They are things I adopted in recent times" (Farmer from lowlands).

The measures farmers take in the face of climate change are haphazard and less systematic. Farmers mention the support and awareness raising efforts from development workers (extension agents). Every Kebele or site surveyed is staffed with three development workers, each focusing on (a) Crops, (b) Livestock and (c) Natural resources management and environmental protection, respectively. They are often government employees whose remit sometimes expands beyond their titles and roles. For example, over two years ago they were expected to collect land tax and arrears of fertiliser sales from farmers. This put them in conflict with the advisory and consultancy role to farmers with impact on relationships of trust. The practice was suspended later.

Mainstreaming Climate Change to Development Policy

There is important role for public policy in "changed times". In this regard, Ethiopia has made important inroads in terms of policy orientation to address the challenges of climate change including adaptation and mitigation of climatic risks. In 2011, it announced a policy framework that integrates climate change into development policy. Known as the Climate Resilient Green Economy Initiative, the ambition is to build a climate resilient green economy by 2025. However, the efforts are still ill-equipped to deal with the magnitude of the problem.

Since climate change is an important challenge, policy makers need to develop deeper understanding of how climate change is perceived by farmers. Such understanding forms the basis for successful intervention to deal with the impacts of climate change. It also means that coordinated action is needed in policy terms to help farmers adapt to climate change. The silo or functional nature of current sectoral governmental service design will mean that one only deals with one's 'piece' of the problem. Such an approach cannot address the complex interactions between climatic, economic, political, institutional, social and technological processes.

To cope with the challenges in ever-increasing population and inevitable climate change, there is need to develop a policy environment that incentivises the production of early maturing and drought tolerant crop varieties and animal breeds that can withstand the effect of climate change. There is need to make good use of waste by soil improvement through recycling of organic materials from farms.

It should also be understood that public understanding of climate change is based not only on scientific information of its physical causes and solutions, but also on local knowledge, values, moral responsibilities (Bulkeley 2000), and cultural interactions with nature. This suggests that there is a need to increase the emphasis on analysis of public understandings in terms of underlying thoughts and ideas, instead of merely a quantitative assessment of climate change knowledge (Moghariya and Smardon 2012).

Policy interventions will have to explore ways of enhancing provision of water, irrigation agriculture, diversification of activities and research into affordable inputs (HYVs and fertiliser). Policy should address provision of credit and capital for energy efficiency and land use patterns, through processes which recognize public understanding of climate change (Bulkeley 2000). Moreover, public policy need to correct possibilities of maladaptation where the human response actively undermines the capacity of society to cope with climate change or further contributes to the problem; for example, the breakdown of institutionalized relationships (Bulkeley 2000).

Conclusions and Recommendations

This research set out to explore farmers' perceptions about climate change and their lived experience about a challenge much broader than their scope, both is space and time—global environmental challenge. It explored how farmers perceive and act upon key indicators of climate change such as change in seasonal distribution of rainfall (flooding or drought), changing wind conditions and temperature variations (drought), and changes in the ecosystem (deforestation, soil degradation).

By using qualitative method involving FGD with smallholder farmers, it noted that times have indeed changed for smallholder farmers in Sidama. These people are never new to changes. They had experienced a host of social, political and economic changes through generations, some more tumultuous than others. Yet none of the above mentioned changes have come any close in equivalence to climate change which they described to be beyond their control. However, they have clearly noticed that climate change existed and they tried to make sense of the world around them. Their explanation of the causes of climate change is based on their local beliefs, values and norms. Like most farmers, elsewhere in the world, they formulate expectations from observation of natural phenomena; and they have shared and selected repertoire (Roncoli et al. 2002).

It was striking throughout the FGDs that the world of farmers was quite local or parochial, or what I may call hyper-local. Their perception of the nature of the problem under scrutiny (climate change) and its causes and adaptive actions are largely limited to their local world. There were hardly times when discussions linking climate change with global warming, green house such as CO_2 emissions, except mention of pollution of rivers caused by coffee processing plants.

On all occasions, farmers made reference to God as the cause and as solution to the problem of climate change. I would link such preponderance of faith-based explanations and ever growing faith-based networks to what I refer to spiritual capital, which can complement the social capital in sustainable livelihood framework. One can link religion and development to develop the notion of spiritual capital.

In sum, there is broader agreement that climate change is a reality. The causes of climate change remain matters for debate. Some believed it has all to do with nature and supernatural. Other cite deforestation and changing land use as factors. Some participants associated cultural aspects with climate change such as declining morality, the collapse of customs and traditions or "God" as other reasons for climate change. Yet the reality of climate change is rightly perceived; so is its impact. The latter is more evident with the livelihood of communities solely dependent on rain-fed farming and livestock. The systems of food production of smallholder farmers are adversely affected by the variability in timing and amount of rainfall, heat stress and outbreak of diseases.

The implications for policy intervention are broad ranging from ensuring food security and promotion of sustainable development. There is a need for concerted public policy effort to mainstream climate change into development policy. Without this, some areas, particularly the lowlands, would gradually be converted to semi-desert conditions.

There is also a need to synchronise scientific information and knowledge on climate change to farmers' understanding of climate change impacts and efforts at adaptation. Enhancing public awareness plays an important part. Adaptation to climate change requires that farmers need to notice that the climate has changed, and then identify useful adaptations measures and implement them (Maddison 2006). Moreover, measures for adaptation on the local need to be complemented by support from various levels of the regional and national governance systems. Most certainly, in uncertain futures, changed times require changed strategies, policies and approaches.

Acknowledgments I am grateful for financial support for the field-work from the NORAD, Hawassa University and the University of East London. Thanks also to two anonymous referees for useful comments on the first draft of the paper. The views expressed and any errors or omissions are my own.

References

Bacon C (2005) Confronting the coffee crisis: can fair trade, organic, and specialty coffees reduce small-scale farmer vulnerability in Nicaragua? World Dev 33(3):497–511

Bailey C (2007) A guide to qualitative field research, 2nd edn. Pine Forge Press, London

Brandt SA, Spring A, Hiebsch C, McCabe T, Tabogie E, Diro M, Wolde-Michael G, Yntiso G, Shigeta M, Tesfaye S (1997) The "*Tree Against Hunger*": Enset-based agricultural systems in Ethiopia. American Association for the Advancement of Science. Available at http://www.aaas.org/international/africa/enset/enset.pdf. Accessed on 26 Jan 2013

Bulkeley H (2000) Common knowledge? Public understanding of climate change in Newcastle, Australia. Public Underst Sci 9:313–333

Byg A, Salick J (2009) Local perspectives on a global phenomenon: climate change in eastern Tibetan villages. Glob Environ Change 19(2):156–166

Conway D, Schipper L, Yesuf M, Kassie M, Persechino A, Kebede B (2007) Reducing vulnerability in Ethiopia: addressing the implications of climate change: integration of results from phase I. University of East Anglia, Norwich

Coppock L (ed) (1994) The Borana Plateau of Southern Ethiopia: synthesis of pastoral research, development and change, 1980–91. International Livestock Centre for Africa (ILCA), Addis Ababa, Ethiopia

Crate SA, Nuttall M (2009) Anthropology and climate change: from encounters to actions. Left Coast Press, Walnut Creek

CSA (Central Statistical Authority) (2011) Tables: section-b population. Available at http://www.csa.gov.et/index.php?option=com_rubberdoc&view=doc&id=226&format=raw&Itemid=561. Accessed on 21 Mar 2013

Deressa TT, Hassan RM, Ringler C, Alemu T, Yesuf M (2009) Determinants of farmers' choice of adaptation methods to climate change in the Nile Basin of Ethiopia. Glob Environ Change 19(2):248–255

Deressa TT, Ringler C, Hassan RM (2010) Factors affecting the choices of coping strategies for climate extremes: the case of farmers in the Nile Basin of Ethiopia. IFPRI Discussion Paper 01032, Nov

Devereux S, Guenther B (2007) Social protection and agriculture in Ethiopia. Country case study paper prepared for a review commissioned by the FAO on 'social protection and support to small farmer development'. Institute of Development Studies, University of Sussex. 30 Nov

Dixon J, Tanyeri-Abur Y, Wattenbach H (2004) "Smallholders, globalization and policy analysis". Agricultural management, marketing and finance service occasional paper 5. Agricultural Support Systems Division, FAO, Rome. Available at http://www.fao.org/docrep/007/y5784e/y5784e02.htm#bm02. Accessed on 9 Sept 2013

Eakin H, Tucker C, Castellanos E (2005) Market shocks and climate variability: the coffee crisis in Mexico, Guatemala, and Honduras. Mt Res Dev 25(4):304–309

Fleming A, Vanclay F (2011) Farmer responses to climate change and sustainable agriculture. In: Lichtfouse E et al (eds) Sustainable agriculture, vol 2. Springer, Dordrecht, pp 283–293

Giddens A (2009) The politics of climate change. Polity, Malden

Glaser BG, Strauss A (1967) The discovery of grounded theory: strategies for qualitative research. Aldine, Chicago

IPCC (2007) Climate change 2007: synthesis report. Cambridge University Press, Cambridge

Juana SJ, Kahaka Z, Okurut FN (2013) Farmers' perceptions and adaptations to climate change in Sub-Sahara Africa: a synthesis of empirical studies and implications for public policy in African agriculture. J Agric Sci 5(4):121–135

Maddison D (2006) The perception of and adaptation to climate change in Africa. CEEPA discussion paper no. 10, Centre for Environmental Economics and Policy in Africa (CEEPA), Pretoria

Moghariya D, Smardon R (2012) Rural perspectives of climate change: a study from Saurastra and Kutch of Western India. Public Underst Sci. doi:10.1177/0963662512465698

MoWRNMA (2007) Climate change national adaptation programme of action (NAPA) of Ethiopia. Ministry of Water Resources and National Meteorological Agency, Addis Ababa

Mortreux C, Barnett J (2009) Climate change, migration and adaptation in Funafuti, Tuvalu. Global Environ Change 19(1):105–112

Nordhaus W (2007) The challenge of global warming: economic models and environmental policy. New Haven. Available at http://nordhaus.econ.yale.edu/dice_mss_072407_all.pdf. Accessed on 30 Aug 2011

NREPA (2012) Climate change adaptation guideline—Draft document, Hawassa: SNNPR. Natural Resources and Environmental Protection Agency

Petheram L, Zander KK, Campbell BM, High C, Stacey N (2010) "Strange changes": indigenous perspectives of climate change and adaptation in NE Arnhem Land (Australia). Glob Environ Change 20(4):681–692

Roncoli C, Ingram K, Kirshen P (2002) Reading the rains: local knowledge and rainfall forecasting in Burkina Faso. Soc Nat Resour 15(5):409–427

Rudiak-Gould P (2012) Promiscuous corroboration and climate change translation: a case study from the Marshall Islands. Glob Environ Change 22(1):46–54

Stern N (2007) The economics of climate change: the Stern review. Cambridge University Press, Cambridge

Tadege A (ed) (2007) Climate change national aptation programme of action (NAPA) of Ethiopia. Ministry of Water Resources and National Meteorological Agency. Federal Democratic Republic of Ethiopia, Addis Ababa

UNDP (2007) Human development report 2007/08. Fighting climate change: solidarity in a divided world. United Nations Development Programme, Cambridge

Viste E, Korecha D, Sorteberg A (2013) Recent drought and precipitation tendencies in Ethiopia. Theor Appl Climatol 112(3–4):535–551

Wheeler D (2011) Quantifying vulnerability to climate change: Implications for adaptation assistance. CGD working paper 240. Centre for Global Development, Washington, D.C. Available at http://www.cgdev.org/content/publications/detail/1424759. Accessed on 18 Aug 2011

Chapter 20
Perceptions of and Attitudes Toward Climate Change in the Southeastern United States

David Himmelfarb, John Schelhas, Sarah Hitchner, Cassandra Johnson Gaither, KathErine Dunbar and J. Peter Brosius

Abstract Despite a global scientific consensus on the anthropogenic nature of climate change, the issue remains highly contentious in the United States, stifling public debate and action on the issue. Local perceptions of and attitudes toward climate change—how different groups of people outside of the professional climate science community make sense of changes in climate in light of their personal experiences and social, political, economic, and environmental contexts—are critical foci for understanding ongoing conflicts over climate change. Contributing to a growing body of literature on the social science of climate change, we use an ethnographic approach to examine these perceptions and attitudes in three sites in Georgia across the urban–rural continuum. Our research demonstrates that the way people view the concept of climate change, its potential effects, and mitigation strategies are mediated by a range of factors, including political and religious affiliation, race and ethnicity, personal experience, economic status, environmental context, media exposure, and sense of community and place. We argue that an ethnographic approach that explores the perceptions

D. Himmelfarb (✉) · S. Hitchner · K. Dunbar · J. P. Brosius
Center for Integrative Conservation Research, University of Georgia, Athens, GA, USA
e-mail: daveh@uga.edu

S. Hitchner
e-mail: slhitchn@uga.edu

K. Dunbar
e-mail: kdunbar@uga.edu

J. P. Brosius
e-mail: pbrosius@uga.edu

J. Schelhas · C. J. Gaither
USDA-Forest Service, Athens, GA, USA
e-mail: jschelhas@fs.fed.us

C. J. Gaither
e-mail: cjohnson09@fs.fed.us

W. Leal Filho et al. (eds.), *International Perspectives on Climate Change*,
Climate Change Management, DOI: 10.1007/978-3-319-04489-7_20,

and attitudes of specific communities in detail can add nuance to the broad-scale surveys that have dominated the field to date.

Keywords Climate change • Perceptions • Social dimensions • Southeastern US • Georgia

Introduction

Climate scientists overwhelmingly agree that the climate is changing in ways unprecedented in climate cycles over the past 600,000 years and that such change is in part anthropogenic (IPCC 2007). However, climate change remains a contentious issue, especially in the United States (Antonio and Brulle 2011). Mainstream news outlets present conflicting views about the reality of climate change, and surveys indicate that in the US, belief in climate change has decreased slightly over the past decade (Boykoff 2011; Hulme 2009; Malone 2009). Hoffman (2012: 33) observes that the increased polarization of attitudes toward climate change has stifled the robust public debate necessary for mitigating, coping with, and adapting to the predicted changes in climate.

The fields of social science are uniquely positioned to contribute to an understanding of contemporary climate debates (Barnes et al. 2013; Crate and Nuttall 2009; Kempton et al. 1996). Local perceptions of and attitudes toward climate change—how different groups of people outside of the professional climate science community make sense of changes in climate in light of their personal experiences and social, political, economic, and environmental contexts—are critical foci for climate change research (Crate 2011; Roncoli et al. 2009). Vedwan and Rhoades (2001: 109) argue that "understanding local perceptions of the relative amounts, direction, and impact of climate change are key to arriving at an understanding of patterns in human responses."

Recent investigations of perceptions of climate change have used surveys to examine the views of large populations (e.g. Leiserowitz and Smith 2012). While highly useful, by nature, these surveys cannot capture the nuance of social context and sense of place, which researchers have found influence people's understandings of climate change (Crate and Nuttall 2009; Strauss and Orlove 2003). Anthropologists have suggested that an ethnographic approach can contribute the nuance absent from larger-scale methods (Crate 2011; Roncoli 2006). To date, ethnographic studies of climate change have focused primarily on rural contexts in the Global South (e.g. Carey 2010; Strauss and Orlove 2003; Vedwan and Rhoades 2001). However, researchers have recently expanded their view to include urban and peri-urban environments in the Global North (Crate 2011; e.g. Norgaard 2011; Schelhas et al. 2012; Stone 2012).

Anthropologists can contribute to an understanding of how people perceive and respond to climate change by elucidating the ways that personal observations are shaped by cultural values, political relations, power dynamics, and social status and by examining the ways that scientific understandings are transformed when they enter other social contexts (Barnes et al. 2013). For example, earlier research by Kempton et al. (1996) found that publicity about global warming provided people with a framework through which they interpreted their personal observations of weather such that they often reported personal observations of global warming even when actual changes could not be observed in weather records. More recent research, however, focuses on climate change denial and how it may be socially organized to protect individual identity, sense of empowerment, and the maintenance of culturally produced conceptions of reality (Antonio and Brulle 2011; Norgaard 2011). Even natural resource-dependent communities, who are often seen as being keenly aware of their environment, see weather changes through cultural lenses and draw on diverse cultural models from domains beyond the strictly environmental, for example media discourses or popular culture, to interpret these observations (Barnes et al. 2013; Roncoli et al. 2009). Norgaard (2011) emphasizes the social nature of these processes, highlighting the importance of studying perceptions in a community context rather than focusing on individuals through surveys. An ethnographic approach to perceptions of climate change can reveal the ways perceptions are shaped by these diverse social, cultural, and political interactions and thereby guide constructive intervention in societal debates (Barnes et al. 2013).

This study builds on recent ethnographic analyses of climate change in the complex social landscapes of the US and the Southeast in particular, a region largely absent from the climate change literature (cf. Crane et al. 2010). We present the preliminary phase of a broader project to integrate qualitative and quantitative approaches to social vulnerability and climate change in Georgia. We focus on three sites that represent much of the social, economic, and environmental diversity found in the contemporary Southeast. In the urban Cascade Road Corridor of Atlanta, suburban Norcross (adjacent to Atlanta), and rural Jasper County communities, we examine: (1) climate change belief and skepticism; (2) observed changes in weather and climate; (3) effects experienced and anticipated; (4) perceived causes; and (5) attitudes toward mitigation.

As survey-based studies on public perceptions of climate change have found (Leiserowitz and Smith 2012), we also find a continuum of acceptance of the scientific explanations for climate change, from "believers" to "skeptics" to "deniers." In keeping with the large-scale surveys, we find that these categories of belief in climate change are likely to fall along political and social lines, with more acceptance among Democrats and racial and ethnic minorities and less among Republicans and the racial majority. However, our research demonstrates that attitudes toward the concept of climate change and its potential effects are also mediated by personal experience and religious affiliation, environmental context, media exposure, and sense of community and place. These influences not captured to date in public surveys or explored in depth in climate research in the Global North, and they complicate the generalizations of the categories of belief. Further, we find that skepticism about or rejection of the concept of anthropogenic climate change is not a result of lack of knowledge about the scientific explanation of climate change, so simply providing more climate education is not likely to increase belief in climate change or spur individual or communal actions to mitigate it. Many intertwined factors influence public perceptions of climate change, and our preliminary results show that an ethnographic approach can reveal connections between experiences, beliefs, and ideas that can complement and add nuance to survey results.

Methods

This qualitative research project used semi-structured interviews, focus groups, and, to a lesser extent, participant observation. Sites were selected to represent the range of urban to rural environments that characterize Georgia. Within each site, we endeavored to reach a heterogeneous group of participants based on recommendations from key informants and referrals from other interviewees. Initial contacts were made with local non-profits, government representatives, and churches. We followed up with focus groups and interviews aimed at representing the diversity in each community. The same interview guide was used at all sites, but the ways that we met with and talked to people were adapted to local conditions and opportunities. Interviews and focus groups began by eliciting impressions and histories of participants' communities, followed by questions about resiliency, awareness, and understanding of weather and climate change and its effects, and sources of information about climate change. To gain perspective into community relations and to recruit interview and focus group subjects, participant observation was carried out at church ceremonies, at informal gatherings in mobile home parks, and community organization meetings. We recorded focus groups and interviews where possible and took notes. Transcripts and field notes were coded, organized, and analyzed thematically using Nvivo qualitative analysis software.

Site Descriptions

The Cascade Road Corridor follows an 11-mile stretch of road in southwest Atlanta. It begins close to the city center in the West End neighborhood and extends on a diagonal through business and residential areas, from lower-income rental properties and public housing, to small residences, and multi-million dollar mansions in gated complexes. The 14 census tracts encompass approximately 32 neighborhoods and subdivisions with a population of approximately 58,000 (US Census Bureau 2010a). The Cascade Corridor population is 97 % African American and economically diverse, with the poorest residents closest to the city center and the wealthiest out toward the intersection of Cascade Road and Interstate 285, or "The Perimeter" (US Census Bureau 2010a). We conducted six focus groups and interviewed 30 individuals. All but three participants identified as African American.

The city of Norcross lies in western Gwinnett County in metropolitan Atlanta. The city transitions from suburban developments and shopping centers in the west to large-lot subdivisions in the east. Norcross has a diverse population of just under 10,000 inhabitants, about 40 % of whom are Latino, 25 % white, 18 % black, and 15 % Asian (US Census Bureau 2010b). Our research focused on the Latino population, with sites including a mobile home park, an apartment complex, two elementary schools, and a community-based Latino organization. We conducted six focus groups and 15 interviews, which were conducted in Spanish where necessary.

Jasper County is located about 60 miles southeast of Atlanta. While there are some bedroom housing developments in the northern part of the county, from which some residents commute to jobs in Atlanta, the remainder of the county is rural with two small towns, Monticello and Shady Dale. The agrarian economy focuses on cattle and timber/pulp production. The population of 14,000 is approximately 75 % white, 22 % black, and 2 % Latino (US Census Bureau 2010c). We conducted one focus group and 18 interviews in Jasper County.

Climate Change Belief and Skepticism

In the United States, where climate change has become a politically polarized and contentious issue, any discussion of how people perceive local changes in weather and environment must begin with an examination of people's attitudes toward the concept of climate change advanced by mainstream science. We found that avowed belief or skepticism in the concept of climate change influenced where participants sought information about climate and provided a lens through which they interpreted their personal observations, perceptions of risk, and attitudes toward climate change mitigation.

All but one of our participants in Cascade and Norcross expressed at least some belief in the scientific concept of climate change.[1] While some voiced a lack of specific scientific knowledge regarding climate change, most people—especially, but not limited to, those with college and post-graduate levels of education—seemed comfortable using the basic terms of climate science, making reference to the greenhouse effect, ozone layer, and greenhouse gases. Those who subscribed to scientific explanations connected their personal observations with broader scale phenomena such as melting glaciers and sea level rise, which they understood as the products of global fossil fuel consumption. They asserted the authoritativeness of scientific information, which they gleaned from television, newspapers, the Internet, and personal research.

[1] It is important to note that the widespread belief in and concern for climate change in Cascade and Norcross, and to a lesser degree Jasper County may not be representative of the populations of these areas as a whole. Participation in the research was based on personal interest, which may have skewed the sample. Participants often remarked that their level of concern was not shared by many in their communities.

For most Cascade residents and some in Jasper County, climate change belief had strong political associations with the political left through the work of former Vice President Al Gore. The population of Cascade is overwhelmingly aligned with the Democratic Party and participants told us that such connections encouraged their acceptance of the climate change concept. Political affiliation did not seem to influence residents of Norcross, who tended to view themselves (perhaps because of their often ambiguous immigration status) outside of the electoral political sphere.

We encountered the most striking division on the concept of climate change in Jasper County, where five out of 18 interviewees and five out of seven focus group participants opposed the standard scientific explanation of climate change. For most, though not all participants, their beliefs had strong political associations (with believers, like those in Cascade, tending to identify with the political left and skeptics with the right). Two skeptics in Jasper fell outside the political divide, preferring to attribute any potential change to divine intervention or the manifestation of Biblical prophesy. Some skeptics denied any change while others felt that if change was occurring (and they were not definitively saying that it was), then it was likely due to natural, not anthropogenic, causes. "They're trying to blame that [climate change] on fossil fuels, which I don't agree with," one farmer told us. He and others referenced politically conservative talk shows and news programs that supported the skeptical perspective. Among several of the skeptics, the concept of climate change was inextricable from what they perceived as unpalatable social, ethical, and economic ideologies and policies prevalent in more urban (and implicitly more politically liberal) communities, including environmentalism, animal rights, social welfare, and economic regulation.

Observed Changes in Weather and Climate

While pointing out the difficulties of identifying broad trends and the challenges of disentangling personal experience from media exposure, residents across sites described similar changes in weather patterns over their lifetimes. From the concrete city-center neighborhoods and verdant subdivisions of Cascade, to the trailer parks and strip malls of Norcross, and the cattle pastures and pine plantations of Jasper County, participants described rising temperatures, weather and seasonal unpredictability, increased drought, and more frequent severe storms.

As examples of change, Cascade residents pointed to the loss of clear divisions between seasons, the increasing absence of iconic flora and fauna, and increased incidence of storms. In Norcross, most participants had migrated to the area within the last 10–20 years from a variety of locations in Latin America and noted that conditions in Norcross paralleled the increasing unpredictability and warmer temperatures occurring in their countries of origin. Many emphasized the diminishing frequency and intensity of cold weather to illustrate their descriptions of the weather unpredictability and warming trend. Norcross residents also described increased incidence and severity of droughts, as manifested by shrinking lakes and tree and vegetation die-off around their homes as well as an increase in extreme weather events such as tornadoes that threatened their trailer homes.

Climate believers in Jasper County also emphasized changes in seasonality. Drawing on dates when certain plants bloomed or leafed out, they observed that summer was becoming longer and that there was less distinction between seasons. Respondents also identified the increased incidence and severity of both drought and tornadoes in recent years. Local officials said that between the lack of rain and increased commercial and household water use, they were facing serious challenges in managing a dwindling water supply. As evidence of the drought, participants like those in Norcross pointed to the drying of creeks and reservoirs, the lack of sufficient forage for cattle, and the necessity of drilling deeper to find water for wells. Several participants felt that tornadoes posed an increasing threat to the county, especially as public resources have dwindled.

Skeptics in Jasper County did not necessarily deny the presence of unusual weather conditions, but were largely unwilling to connect those conditions with broader trends, pointing to times in the past that were hotter or drier than the present to illustrate natural climate variability. As one government official observed, "People notice differences in the weather, but they don't marry it with the concept of climate change." Some felt that if change was occurring, then it was likely due to natural, not anthropogenic, causes.

Effects Experienced and Anticipated

Respondents expressed varying levels of concern about the effects of these changes on their lives and livelihoods in the present and in the future. While few participants felt that their lives had been deeply affected by climate change to date, participants in all three sites recognized the potential of climate change to transform their daily lives and economic opportunities over time, exacerbating existing economic disparities. Though participants shared some concerns, site-specific economic, environmental, and housing conditions tended to influence the range of effects and concerns identified.

For residents in Cascade, Norcross, and Jasper County, an increase in energy bills was the most commonly observed present effect of higher temperatures. During times of extreme heat, all respondents noted that they depended on their air conditioning units to stay cool. While higher energy expenses were not of major concern for upper-middle class or wealthy individuals, those with more limited means had less capacity to absorb the extra expense. Not only do lower income households have fewer resources to pay the bills, but their consumption costs are higher because they tend to live in poor quality, energy inefficient housing.

Most Cascade participants reported changing their daily schedules and activities to avoid heat exposure. For some, heat was merely a discomfort. Elderly respondents and those with health problems were more seriously affected. Several older people noted that the heat prevented them from leaving the house. Heat also exacerbated the experience of allergies and asthma, common afflictions in urban areas such as Atlanta, posing severest risk for the very young, elderly, and poorest segments of the population.

Sustained drought, elevated temperatures, and heavy storms were viewed as threats to the Cascade Corridor's iconic tree canopy, widely valued for its contribution to residents' sense of place and community. Vegetable gardens, important dietary supplements for several participants, have also fallen victim to increasingly dry and hot conditions.

Increased frequency of tornadoes and strong storms that felled trees and caused flooding held particular menace for those living in mobile homes in Norcross. Several respondents described harrowing encounters with severe storms, where strong winds tore away pieces from their trailers and floods threatened to carry their homes away. Government officials in Jasper County similarly expressed their fears regarding severe storms. The local government's dwindling budget in recent years, they explained, increased Jasper County's vulnerability to tornadoes. An official said, "A tornado would kill us…This is a small, poor town. We're on a thin edge."

Residents in Norcross and Jasper County expressed concern that erratic temperatures, droughts, and extreme weather events would impact employment opportunities and production in the construction and agricultural sectors. For now, participants in Norcross felt that such effects were minimal, but parents were worried for their children's prospects. Climate believers in Jasper County, an area reliant on agricultural production, felt that changes in climate were already seriously affecting agricultural production. "We're not producing the crops like we [do] normally," one woman observed. Between the changes in temperature and season, ongoing drought, and the arrival of new invasive weed and pest species, the mainstays of the economy—row crops, cattle, and timber and pulp production—were becoming increasingly precarious.

The climate skeptics took a different perspective on the phenomena that concerned climate believers. Without belief that the climate would continue to get hotter and drier, skeptics tended not to be as worried about the future. The farmers among them did not deny the effects that intense heat or drought might have. Instead, they preferred to take a deeper historical view: "There's nothing new that's [not] been going on for tens of thousands of years." Several challenged the notion of climate change by referring to times of similar hardship in the past, indicating that they had seen weather variability long before the popularization of climate change.

Perceived Causes

Consistent with the standard scientific explanation of climate change, participants in Cascade, Norcross, and climate believers in Jasper County viewed the changes they had observed as anthropogenic, or as one person in Cascade put it, "things we're doing as humans." Participants in all three sites agreed that such factors were occurring at multiple scales.

On the individual or household scale, people noted that heavy reliance on cars and air conditioning and lack of household recycling contributed to climate change-inducing pollution. Some respondents felt that the culture of consumption and waste in the US influenced their peers' ecologically damaging behaviors. One respondent in Norcross originally from Mexico observed that the "US consumes all the natural resources they can;" another concurred, "In my country water is very valued, but here it is wasted."

On a local scale, participants in Cascade pointed to recent residential development as a contributor to rising temperatures. While earlier subdivisions left some trees standing to preserve the urban canopy, developers now "clear out all the trees." A subdivision resident described how 10 years ago she used to be able to detect a drop in temperature on her drive home from work in the city-center to her well-wooded neighborhood on the outskirts of Cascade. Now that many of the trees have been replaced with houses, she can no longer feel the change.

To most participants across sites, however, the most significant contributor to climate change came from large-scale industrial emissions. Participants in Cascade and Norcross mentioned the factories and power plants that are scattered across the urban landscapes surrounding their homes. Some in Cascade and Norcross connected industrial production with the degradation of the ozone layer. Emphasizing the disproportionate contribution of emissions by large corporations, respondents in Cascade, an area noted for its historical role in the Civil Rights Movement, framed climate change as a social justice issue. Referencing the history of corporations concentrating environmental hazards in marginalized communities and lax regulatory policies associated with the political right, an urban food activist described climate change as "not something that's just happening to us. It's being done to us."

In two focus groups and two interviews in Cascade, older participants proposed the notion that various kinds of scientific intervention had contributed to climate change, a view we did not encounter in Norcross or Jasper County. An older woman explained her view that "The climate now is not natural…They've [scientists] been experimenting for years on how climate can be manipulated in warfare." Space exploration and cell phone technology were also seen as possible influences on climate. Those who voiced such views qualified them as speculation, were vague as to the specific nature of the experiments or technological mechanisms, and were not confident in their scientific literacy.

Attitudes Toward Mitigation

Participants indicated that accepting the scientific concept of climate change entailed embracing the anthropogenic nature of such change and, by extension, human capacity to mitigate the causes of such change. Aside from recycling, reducing waste, and reducing their driving, participants identified few options for climate change mitigation on the individual scale, and most expressed little inclination to dramatically change their lifestyles to address climate change.

Despite the relatively high degree of awareness and concern about climate change, participants expressed a feeling of disconnection between their understanding of climate change and their personal lifestyle decisions. Wealthier individuals observed that they were able to draw on their personal financial resources to insulate themselves from the effects of climate change. Though they were uncertain of the future, "until it gets extreme," as one Cascade woman said, she and her peers were unlikely to substantially reduce their consumption. Another participant in Jasper agreed, noting that "It's a hard sell to ask people to change their creature comforts."

For lower-income individuals in all three sites, respondents said climate change and other environmental concerns were beyond the purview of their daily economic struggles. One woman who lived in a low-income housing complex in Cascade said, "people are so caught up in 'I have to eat today,' so global warming is. That's off. I gotta eat today, I gotta put gas in my car, I gotta get my kids to school. I gotta make sure no one hits me over the head." Environmental concerns like climate change, in the words of one Cascade social activist, were "for rich white folks who don't have anything else to do."

While respondents in Cascade, Norcross, and Jasper County reported taking steps to reduce their carbon footprints, they questioned the impact potential of their actions considering that the majority of emissions came from industrial activity. They argued that meaningful change must happen at a larger political scale. As a Norcross woman put it, "It doesn't matter what we do individually if the big companies do not change their ways. The change needs to be more dramatically on their part since the changes are rapid and dramatic." Left-leaning individuals doubted that the US government would be able to initiate large-scale industrial regulations, citing the close relationship between corporations and government actors.

For the climate change skeptics in Jasper County, the path ahead was straightforward: nothing needed to be done to mitigate climate change. Issues of climate, in their view, were beyond human control, for better or for worse. One farmer said: "I don't think that it's a thing we should be concerned about at this point… We don't call the shots. I don't think we can change it." Other religiously inclined skeptics took comfort in this lack of human influence, noting that, as one interviewee put it, "The good Lord is in control."

Conclusions

As Braman et al. (2011) observe, the enduring controversy over climate change is commonly attributed to the public's limited knowledge of science (e.g. Ungar 2000). This preliminary investigation in three locations in Georgia suggests that, while many were uncertain of the specifics, most people we spoke to across social, economic, and urban–rural continua possessed a basic understanding of climate change as elaborated by climate scientists. Even those who rejected the concept of climate change expressed familiarity with the scientific argument. Further, our findings support recent social science research that has shown how attitudes toward the concept of climate change and people's willingness to adopt mitigation strategies are not a simple function of scientific literacy, but are also dependent upon both the social and natural context of community (see Barnes et al. 2013; Crate 2011; Braman et al. 2011; Roncoli et al. 2009). In terms of adopting mitigation strategies, our data suggest that Cascade Corridor and Jasper County residents, in particular, are likely to face constraints in adopting mainstream prescriptions. Well-intentioned interest groups, for instance, prescribe alternatives to the private automobile—walking or public transportation—as mitigation tactics. However, these options may either not be available in rural settings (Jasper County) or deemed unacceptable by middle-class, urban residents because of their association with crime. While a large literature has addressed climate change perceptions and awareness, relatively little research has focused on place-based factors that may encourage or limit people's ability to respond to climate change at the local level, particularly when these actions take place outside of the house and involve transportation alternatives.

In all three sites, individuals described how their understandings were formed through the interplay between their personal experiences and observations and information derived from a variety of sources, including community leaders, social relations, and broadcast media. Nevertheless, belief/disbelief in the scientific concept of climate change did seem to act as a powerful frame for how people evaluated their experiences and other information sources. Respondents throughout Cascade, Norcross, and Jasper County, regardless of their position on climate change, identified the same recent weather phenomena: high temperatures, unpredictable seasonality, drought, and severe storms. Whether they viewed such phenomena as part of global-scale trends, how they attributed causation, and what they felt should be done seemed to depend on their belief in the scientific concept of climate change. In Cascade and Jasper County, climate belief was incorporated into broader social and political issues, where taking one stance or the other was fraught with cultural significance.

Due to the time and resource constraints of this exploratory phase, we were not able to represent the full diversity of views in the large and diverse research sites or put participant observation to its fullest use. The opportunistic way participants were solicited may have resulted in the under- or over-representation of certain views or certain categories of people. Nevertheless, this research forms an important basis for future research in the Southeast. Our findings suggest the diversity of viewpoints found within the climate believer/skeptic categories and the ways that climate belief and skepticism become intertwined with pre-existing cultural narratives and conflicts, in particular, deserve further scrutiny. As policy makers and scientists devise responses to climate change, we argue that failure to recognize and grapple with the complex social processes that produce public understandings of climate change risks exacerbating the cultural polarization that has to date prevented meaningful public debate and action.

References

Antonio RJ, Brulle RJ (2011) The unbearable lightness of politics: climate change denial and political polarization. Sociol Q 52:195–202

Barnes J, Dove M, Lahsen M, Mathews A, McElwee P, McIntosh R, Moore F, O'Reilly J, Orlove B, Puri R, Weiss H, Yager K (2013) Contribution of anthropology to the study of climate change. Nat Clim Change 3:541–544

Boykoff MT (2011) Who speaks for the climate? Making sense of media reporting on climate change. Cambridge University Press, Cambridge

Braman D, Kahan DH, Wittlin M, Slovic P, Ouellette LL, Mandel GN (2011) The tragedy of the risk-perception commons: culture conflict, rationality conflict, and climate change. GW Law Faculty Publications & Other Works, Paper 213

Carey M (2010) In the shadow of melting glaciers: climate change and Andean society. Oxford University Press, Oxford

Crane TA, Roncoli C, Paz J, Breuer N, Broad K, Ingram KT, Hoogenboom G (2010) Forecast skill and farmers' skills: seasonal climate forecasts and agricultural risk management in the Southeastern United States. Weather Clim Soc 2:44–59

Crate SA (2011) Climate and culture: anthropology in the era of contemporary climate change. Annu Rev Anthropol 40:175–194

Crate SA, Nuttall M (eds) (2009) Anthropology and climate change: from encounters to actions. Left Coast Press, Walnut Creek

Hoffman AJ (2012) Climate science as culture war. Stanford Soc Innovation Rev Fall 2012:30–37

Hulme M (2009) Why we disagree about climate change: understanding controversy, inaction, and opportunity. Cambridge University Press, Cambridge

IPCC (2007) Climate change 2007: impacts, adaptation and vulnerability. Working group II summary for policymakers. IPCC Secretariat, Geneva

Kempton W, Boster JS, Hartley JA (1996) Environmental values in American culture. MIT Press, Cambridge

Leiserowitz A, Smith N (2012) Knowledge of climate change across global warming's six Americas. Yale University, New Haven

Malone EL (2009) Debating climate change: pathways through argument to agreement. Earthscan, London

Norgaard KM (2011) Living in denial: climate change, emotions, and everyday life. MIT Press, Cambridge

Roncoli C (2006) Ethnographic and participatory approaches to reserach on farmers' responses to climate predictions. Climate Res 33(1):81–99

Roncoli C, Crane T, Orlove BS (2009) Fielding climate change in cultural anthropology. In: Crate SA, Nuttall M (eds) Anthropology and climate change: from encounters to actions. Left Coast Press, Walnut Creek, pp 87–115

Schelhas J, Hitchner S, Johnson C (2012) Social vulnerability and environmental change along urban-rural interfaces. In: Laband DN, Graeme Lockaby B, Zipperer W (eds) Urban-rural interfaces: linking people and nature. American Society of Agronomy, Soil Science Society of America, Crop Science Society of America, Madison, pp 185–200

Stone G (2012) The city and the coming climate: climate change in the places we live. Cambridge University Press, Cambridge

Strauss S, Orlove BS (eds) (2003) Weather, climate, culture. Berg, Oxford

Ungar S (2000) Knowledge, ignorance and the popular culture: climate change versus the ozone hole. Public Underst Sci 9(3):297–312

US Census Bureau (2010a) 2010 census. T54. Race, Prepared by social explorer. http://www.socialexplorer.com.proxyremote.galib.uga.edu/pub/reportdata/htmlresults.aspx?ReportId=R10475610. Accessed 15 April 2013

US Census Bureau (2010b) 2010 census. T54. Race, Prepared by social explorer. http://www.socialexplorer.com.proxy-remote.galib.uga.edu/pub/reportdata/htmlresults.aspx?ReportId=R10475624. Accessed 15 April 2013

US Census Bureau (2010c) Norcross quick facts. Accessed 15 April 2013

Vedwan N, Rhoades RE (2001) Climate change in the Western Himalayas of India: a study of local perception and response. Clim Res 198:100–117

Chapter 21
The Role of Higher Education in Institutionalising Climate Change in Bangladesh

Gazi Mahabubul Alam and Abul Quasem Al-Amin

Abstract Since the inception of higher education (HE) over one hundred years ago, its purpose has been to cater for the religious and social leaders who will be competent enough to run the country. Paradigm shifting in HE has made society realise that every profession demands competent professionals to provide balanced development across all educational sectors in order to see national development progress. As a result, universities now provide a wide range of training and education in the arts, sciences, social sciences and commerce, which may have caused the recent IT and e-commerce revolution. However, the primary contribution of HE is still seen as supporting economic development because it contributes less than other developmental phenomena. In developing nations, the issue of climate change is yet to be perceived either as an economic or as a social developmental agenda item. Consequently, HE provides less attention to climate change. However, climate change has a serious impact not only on economic development but also on social development. Therefore, HE needs to discover a concrete and substantial way in which to handle climate change. This novel piece of research, which is the first of its kind in Bangladesh, used qualitative methods in order to outline the policy direction that HE may adopt to make its role more distinct in handling climate change.

Keywords Climate change • Higher education • Sustainability in education Policy • Bangladesh

G. M. Alam (✉)
Academic Performance Enhancement Unit, University of Malaya,
59990 Kuala Lumpur, Malaysia
e-mail: gazimalamb@yahoo.com

A. Q. Al-Amin
Faculty of Economics and Administration, University of Malaya,
50603 Kuala Lumpur, Malaysia
e-mail: amin_cant@yahoo.com

W. Leal Filho et al. (eds.), *International Perspectives on Climate Change*,
Climate Change Management, DOI: 10.1007/978-3-319-04489-7_21,

Introduction

The criteria of education and their schemata have evolved from historical social practices. The interpretation of social events is guided and constrained by the prevailing rationality, which itself reflects the dominant constellation of power (Alam 2009). Currently, every provision, type, level and pattern of education is experiencing a serious experimental process (Alam 2008). Adherents of this experimental philosophy argue that in order to cope with modernisation and to ensure timely education, the philosophy of education outlined in earlier times cannot be followed in this contemporary world (Alam et al. 2010). They further argue that the philosophy of education needs to be revised in a way in which it can respond to the needs of the 21st century (Alam et al. 2011). On the other hand, opponents advocate that in this era of 'commodification', the education system is overly responsive towards programmes that can offer early economic benefits to its 'customers' (Brennan 2012; CHE 2000). According to Alam (2013) and Konrad (2013), this attitude forces the education system to ignore many fundamental and pertinent components connected to social development, especially those programmes related to sustainability in education. However, sustainability programmes in education with a spotlight on climate change programmes provide a wider value in terms of both economic and social returns[1] (Laessoe et al. 2013; Lemons 1995, 2011; Feinstein 2009; IALEI 2009; Rowe 2007; Bowers 2008; Kerr 2001; Weis et al. 1992; Aldrich and Kormondy 1972).

Primary education is seen as an important component for the civic and ethical development of children. Secondary and tertiary education is considered to be the main weapon to develop 'human capital' with an assurance of providing the necessary citizenship skills to students (Dalton and Crosby 2010). The core business of higher education (HE) is considered to be a production house of knowledge through research, and this knowledge is fundamentally disseminated to the community for wider benefit (Boulton and Lucas 2011). HE should partially be responsible for the dissemination of knowledge; however, the prime responsibility is to discover a substantial mechanism both for primary and for secondary provisions that can ensure the decent and pragmatic dissemination of knowledge invented by HE (Kerr 2001).

Climate change is a contemporary topic, especially in the area of research (Laessoe et al. 2013; Lemons 2011). This topic falls under the scope of sustainability programmes in education (Laessoe et al. 2013; IALEI 2009). Even though the topic of climate change has been gaining increased popularity, universities and research institutions have mainly considered this to be an issue exclusively for research. Therefore, a complete academic-taught programme deliverable from primary education to HE is missing (Laessoe et al. 2013; O'Brien 1998). In the

[1] If the world vanishes because of climate change, what would be the point in having economic gains.

current situation, primary education in some countries may have some content on climate change as a part of its social sciences programme. Meanwhile, secondary provision via science departments may deliver little content on climate change, particularly related to awareness programmes, compared with the multidisciplinary climate change programmes at university (Vincent 2010; Rowe 2007). Although theoretically claimed as multidisciplinary, every single faculty conducts research or offers courses at a small scale in the area of climate change without being rigorously involved in inter-faculty collaboration and linkage (Fahey 2012; Filho 2010). This is primarily seen as the root cause of the sluggish progression of the climate change topic to become a full-fledged subject of study (Reed and Mera 2011). It is fair to assume that because of lower achievement in HE, the climate change topic is yet to receive high attention from primary and secondary provisions (Fahey 2012). Therefore, although the research impact on climate change may theoretically be high, the overall impact on society to date is insignificant because of the low level of participation by both primary and secondary provisions in the due course of dissemination (Fahey 2012; Reed and Mera 2011).

A good number of studies are ongoing around the world in research institutions and higher academia, but the current evidence on climate change issues and impacts and vulnerability issues indicates that the world is still behind in terms of effective actions and appropriate awareness programmes for dealing with climate change matters (Al-Amin and Filho 2011). Keeping this view as a research issue, this novel piece of research aims to discover how HE can play an important and constructive role in institutionalising climate change as a core subject of study using Bangladesh as a case study. This may also help researchers find a way for other subjects in the area of sustainability in education to institutionalise this as a core subject of study.

Conceptual Development

Climate change and its nature vary because of a number of parameters. However, the causes of climate change can broadly be differentiated into two directions: (1) human activities and the factors for transformation towards modernisation and globalisation and (2) nature-driven causes (Ritter 2009).

Human Factors

Climate change is a repercussion of globalisation and modernisation; consequently, within current trends, its impact is inevitable (IPCC 2007; Stern 2007). Lately, a reasonable number of scientific studies have evidenced the repercussion effects of globalisation (Ritter 2009; Oreskes 2004). It is thus straightforward to see the linkage between climate change and modernisation by human activities.

In exchange for the complete cessation of modernisation, some degree of climate change could be prevented; however, the existence of human beings on the planet will continue to cause climate change (IPCC 2007; Wigley 2001). Ceasing modernisation would simply be impossible. Climate change caused through modernisation is a human-made concept that is not feasible to prevent completely (Stern 2007). However, a greater degree of awareness of the people involved in different stages would help in reducing the volume of climate change and damage without interrupting modernisation too much (Singh 2013). Well-evidenced arguments confirm that industrialised countries are the principal contributors to climate change, while developing countries are victims because of their relatively low bargaining power (Sheila 2003). Extensive knowledge and awareness both at macro and at micro levels may help. Consequently, education and effective knowledge on awareness can play a major role in change. It is a fundamental responsibility for legislators to discover a substantial mechanism that can answer: what can universities do to address the issue of climate change?

Natural Factors

In addition to the human-led causes of climate change, natural causes are also influential. The climate of the world can be affected by natural factors that are external to the climate situation, such as changes in solar output, volcanic activity and the Earth's orbit around the sun (GOV 2012). These factors primarily influence the incoming energy, which is important to the Earth's energy balance. Volcanic eruptions are episodic and have relatively short-term effects on climate change. Over the past century, although changes in solar irradiance have contributed to climatic trends, increasing greenhouse gases caused by the Industrial Revolution has affected the atmosphere by about 10 times as much. However, both human and natural causes are interrelated (GOV 2012). A better way of handling the issue and building awareness is to offer a key solution to the problem, which can mainly be acquired through institutionalising such knowledge.

Scenario of Climate Change in Bangladesh

The issues of climate change and its vulnerabilities and impacts are not new concerns in Bangladesh (Rahman and Alam 2003). Climate change causes vulnerability to the demography and physiography of the national economy for which development is affected on a regular basis (INC 2002; Saleemul et al. 2003; Ahmed 2008; Mazumder 2010). Hence, climate change is no longer only an ecological problem but also a fundamental issue for future development (INC 2002). The increases in temperature, inadequate rainfall, irregular weather pattern and short winter and long summer are causing catastrophes (Rahman and Alam 2003;

Saleemul et al. 2003; Mazumder 2010). In particular, according to INC (2002), IPCC (2007) and UNFCCC (2007), 8 % of the low-lying lands in Bangladesh may become inundate, 8 million people could be affected by drought and 70 million people could suffer from floods by 2,050. The probable impacts of climate change are already well known and changing climate patterns are adversely affecting plant, water, soil, animal, agriculture and ecosystems (Al-Amin et al. 2010; Al-Amin and Alam 2011; Al-Amin and Filho 2011).

Since the economy of Bangladesh depends on climate-sensitive sectors (such as fisheries, livestock and agriculture), any alteration in the climate causes more vulnerability compared with many other counterparts. In particular, the poor, who essentially depend on a sector controlled by sustainable conditions in nature, are the main victims of climate change (INC 2002; Saleemul et al. 2003; UNFCCC 2007). Bangladesh lacks long-run climatic national plans. The sustainable development of natural resource management to handle climate change is yet to be developed in Bangladesh. It rather translates and orients short-versioned action plans that only deliver short-term solutions (Reza 2004). Environmentalists and related agencies, however, have recently started looking for mechanisms, especially awareness programmes, to handle these issues. However, owing to budgetary constraints and a lack of institutional support, short- or long-term action-oriented tools to deal with future impacts are lacking. This undoubtedly restricts the ability to cope with potential vulnerabilities. Political commitments, international pressure and an awareness of the impact on development have all helped make some legislative changes in the past two years. However, sincere attention is now required for periodic reviews of legislative provisions on HE, environmental policy aims and a redefinition of sustainable development (Saleemul et al. 2003). Bangladesh is yet to realise that building awareness is a key solution to the problem, which can mainly be acquired through institutionalising such effective knowledge.

HE: A Paradigm Shift

The criteria of HE and their schemata have evolved from historical social practices. Evidence confirms that the first university in the world, Al-Azhar University in the Middle East established in the 970s, was the exclusive world leader in knowledge for a long period. The activities carried out and role played by Al-Azhar University primarily shaped the philosophy and role of HE in general (Altbach 1999, 2004). Later, universities in Europe especially in the United Kingdom, took over the leading role of HE with some amendments made both to philosophy and to the role of HE (Ball 1993). The US system gradually began to compete through large investment in both public and private provisions (Symes 1996, 1998; Ball 1993). After the rigorous involvement and drastic shift in paradigm (which was primarily created by the US system), global HE has enjoyed significant successes and tensions (Altbach 2004; Symes 1996).

After examining the regime of HE, let us now examine the role of HE in different eras. Until the 1500s, HE followed the concepts prescribed by Plato and Aristotle, namely it set out to form the intellect and character of students, with character defined to include moral and civic aims (Alam et al. 2011). From the late 1500s to the 1700s, HE consisted of seminaries to prepare religious leaders and liberal arts colleges to prepare the wealthy to become the leaders of the society. It was intended for the few and the elite, so that they could lead the uneducated masses. In the late 1700s, Oxford, Cambridge and Harvard advocated that the purpose of a state university was to promote social improvement and individual happiness (Nigel 2005). Newman in the 19th century argues that the purpose of a university in the early 1800s to the 1900s was to train elite scholars in order to render them competent enough to work for the public service that ran the country (Alam 2009). From the 1900s to the 1960s, multidimensional views asserted that universities do far more than what Newman initially envisaged. Universities recognised that every profession demands competent professionals (human capital) that countries need to bring about balanced development across all educational sectors in order for national progress (Alam 2009).

As a result, universities now provide a wide range of training and education in the arts, sciences, social sciences and commerce, which may have caused the recent IT and e-commerce revolution. Well-developed communication systems in the ether and across landscapes remind us that universities are centres for sharing and exchanging knowledge, and excellence in scholarships (Alam 2009). From the late 1960s, HE has traditionally been considered to be for both the ‘public’ and the ‘private’ good. It delivers a unique product—that of knowledge—and the credentials to apply that knowledge in a modern society. It provides the necessary skills for individuals to raise their income levels and to follow prestigious careers (Alam 2009). At the same time, HE improves the condition of the human resources needed for societal growth and the operation of a modern economy (Alam 2009). Since the late 1980s, the concept of commodification in HE has taken place both in public and in private institutions, especially in private provision (Alam 2009). These days, buyers, consumers, suppliers and providers see HE as a commodity that should provide financial gain to each party or stakeholder involved (Alam 2013).

Lately, discussion on HE often generates more heat than constructive criticism. The adherents of different thoughts have been putting a significant amount of effort into proving their excellence (Alam 2009). Consensus is yet to be placed because most states and regions of the globe need a distinct nature of HE suited to their economic, social, cultural and religious patterns (Alam 2013). While the area of science in HE may have a common international phenomenon with little adjustment in light of local needs, the areas of social sciences, arts and humanities simply cannot progress if there is no distinction based on the local context (Nigel 2005). However, distinction concocted for the local context always faces recognition and validation challenges internationally (Alam 2013). Developing and underdeveloped countries always struggle to set their education systems to reflect their own needs and demands because of budgetary and cultural constraints and an ostensible colonial mental setup in both ruling and ruled countries (Alam 2009). This always forces them to breathe in a colonial pattern imposed by the prevailing dominants. Bangladesh is no exception to this because of its economic backdrop and long heritage of being colonised.

Some debated questions have lately developed from the dramatic paradigm shift in the arena of HE. Should HE institutions follow the philosophy of HE determined earlier? Should they be included in the philosophy of HE whatever they have lately been doing? Should HE institutions have a predetermined role? Should all the activities carried out by institutions of higher learning be expected of HE? Common answers are yet to be framed. However, the drastic paradigm shift in HE is lately driving us to realise that each state or region should adopt a distinct nature of HE aligned with a global prototype, thus ensuring a full-fledged course for the overall education system connected to sustainability programmes in education towards a better state. This also shows us that sustainability programmes in education should receive higher attention.

HE in Bangladesh

Dhaka University (DU) is the premier university established in the early 1900s for training the elite intelligentsia to run a flawless bureaucracy. Later in the 1960s, challenging DU's concept of a 'Man as Man' knowledge delivery method, the concept of 'Land-Grants delivery' mode came to be known as the Bangladesh Agriculture University. Simultaneously, the Bangladesh University of Engineering and Technology started operations with its 'Factory/Industry Grants' delivery mode.

The total number of public and private universities in Bangladesh is 42 and 73, respectively. Four out of the 42 public universities are traditional such as DU, while the rest are specialised. Lately, some traditional universities have started offering additional programmes taught in specialised universities and vice versa. On the contrary, most private universities are confined to either a business or an IT school. Universities provided access only to 4 % of qualified secondary graduates until the 1980s. With the affiliated institutions of HE of National University of Bangladesh (NUB) and the private HE sector, this number has increased dramatically. Currently, HE provides access to 42 % of qualified secondary graduates. However, Alam (2013) argues that most secondary graduates are only theoretically qualified to be enrolled into HE and that the quality of secondary graduates is sacrificed for the sake of a quantity-friendly market. This has benefitted both rich customers and the private sector.

HE received 20–25 % of the total budget in education until the late 1990s. This was curtailed to 8–12 % after the massive growth of the private sector. The education sector is still one of the highest priorities when allocating the national budget, and the reduction of allocation from HE increases the allocation for primary and secondary provisions. Until the late 1970s, the employment of university graduates was limited to only public service commission. Therefore, many graduates from the area of agriculture and engineering have been working as administrators under the Bangladesh public service commission. From the late 1980s, the development of NGOs and the private sector has increased dramatically, which has provided remarkable employment for HE graduates. However, there is a mismatch between the degree and available jobs, resulting in a huge number of HE graduates remaining either underemployed or unemployed.

Academics in the public HE sector receive the same salary as their counterparts in other public sectors. However, the counterparts working in the public service enjoy some additional benefits such as transport and a travelling allowance, while academics enjoy only flexitime. The private HE sector is yet to follow a prescribed salary structure; hence, some academics in the private sector may enjoy huge salary packages, while the rest are badly underpaid. Academics from business and IT in public universities earn significantly higher because of their part-time involvement in the private sector. Despite high qualifications, academics are always paid less in comparison to their colleagues in the private sector.

Research Design

This article aims to generate discourse through the analysis of secondary data and intellectual debates. In order to obtain information, a number of related official websites, scholarly sites hosted by different organisations and blog sites were browsed. Moreover, data received through fieldwork were also supplemented. Many of the arguments were also made through the analysis of the data received through fieldwork. Qualitative methods were used that allowed interviewees to express their views in a free and personal way, giving as much prominence as possible to their thematic associations.

Semi-structured interviews by using a qualitative approach were held with the personnel at the Ministry of Education, Ministry of the Environment and Forestry, Department of the Environment, University Grants Commission (UGC) and management, staff and students at both public and private universities. Focus group discussions were conducted with teachers, administrators and students in secondary and primary provisions. The paper also concentrated on the use of data collected from a document review and observations.

Data Collection, Analysis, Confidentiality and Limitations

The document review provided us with an excellent opportunity to preview the overall scenario and prepare the semi-structured interviews to generate in-depth understanding. The objectivity of the study was achieved through the triangulation of samples. As suggested by Bell (1999), we conducted trial interviews with colleagues, which guided us to realise the need to ask further questions based on their answers in the final interviews. Therefore, the final interviews with respondents were semi-structured. Interviews were segmented into four phases. While the first phase included students, the final phases covered key legislators. This sequence guided us to ask the relevant questions on a phase-to-phase basis. Since legislators are the key designers of strategies, they were interviewed last to communicate the relevant issues from the other phases. A second session of interviews was conducted with some respondents for cross-checking purposes.

A briefing on the purpose, focus and confidentiality of the research was made before the final interviews. Each interview lasted 35–45 min compared with legislators' interviews, which lasted around an hour. We asked several indirect questions and respondents' answers led us to ask further questions. We enquired if they could tell us more that was not covered but relevant to the research. This prompted respondents to disclose additional useful information. The exact sequence of the interview questions was altered to maintain a friendly discussion. Our discussion ended with a healthy rapport by thanking all participants for their constructive, positive and critical feedback with an assurance of security and confidentiality regarding information. We sought their permission to record the interviews to which most of them agreed, leading us to record, transcribe and listen to the interviews for our analysis.

As researchers from educational management, we took care with all our cultural baggage and aimed to be entirely objective in our data collection by being aware of possible positional power issues that might arise from power differences. The findings of this qualitative research project are limited to the affiliated HE institutions in Bangladesh and cannot be generalised. They can, however, be a pathway for further research in other developing nations.

Findings and Discussion

Before narrowing the topic to the role of HE on climate change education and research, let us state some basic statistics and facts on HE in Bangladesh.

Genesis of HE and Public Universities

Offering informal HE was started by the visiting religious and spiritual leaders from the Arab World. In July 1921, DU, the first official institution of HE, started its operations as a British colonial university (DU 2012). This university was the witness and main activist of many revolutions over the period. Consequently, DU is seen as the apex of the knowledge world in the region (Alam 2008). Until 1971, the growth of universities and HE institutions was unexpectedly slow. After independence in 1971, even though legislators realised the need for the expansion of

HE, owing to budgetary constraints, the expansion did not take place immediately. From the early 1980s, the government considerably expanded HE by establishing new universities and institutions or transforming institutions into universities (Alam et al. 2007).

Altogether, 34 public universities have been actively involved in providing HE. Out of these 34, 32 are conventional and two are unconventional universities. Of the latter, the Bangladesh Open University mainly covers the dropped out graduates from the secondary system in order to give them some basic training and secondary education through open and distance modes. On the other hand, National University, Bangladesh is not directly involved in teaching and research. Through its affiliated institutions, National University only provides teaching to undergraduates and master's students without any involvement in research. This university shares 90 % of the total enrolment in public sector. Of the 32 conventional universities, a significant number are engineering universities, while a few are for agriculture and medicine.[2] Universities rarely engage in research. Some academics of these universities may be involved in research through their individual consultancy services for different agencies and companies.

Private Universities

In the early 1990s, Bangladesh HE passed a new milestone as a GATS (General Agreement on Trade in Services) member country by opening its HE for the private sector. After the commencement of the private sector, its expansion was dramatic, rapid and politically motivated (Alam 2008). This attitude developed a quantitative sector, thereby sacrificing its qualitative nature (Alam 2013). These days, private counterparts are challenging the public sector from multiple positive and negative angles. Since challenges from negative angles are dominant, it is no surprise to see the quality of both public and private universities declining, while the growth in the private sector is mushrooming (Alam 2013). In total, 73 private universities are currently recognised by the UGC; however, there are few ostensible franchises universities in the market without a valid operating licence from concerned regulatory bodies (UGC 2010).

Owing to the regular entrance of newcomers and sudden collapse of private HE institutions, providing an accurate figure is impossible. According to the UGC, the private sector currently shares more than 65 % of total university enrolment but only offers very limited popular programmes. These programmes are seen as Western, especially American, programmes. Without identifying local needs, the private sector mainly offers business, IT and other low-cost programmes (Alam 2013).

---

[2] A number of public medical colleges are affiliated with different universities. Medical colleges provide education in medicine.

Role of HE in the Education and Research of Climate Change

The following subsections examine the specific role that HE plays in different areas of climate change.

Education Programmes

According to the information available on the websites of the universities, neither a single public nor private university offers a complete programme on climate change. It is surprisingly observed that none of these universities offers a complete programme (undergraduate or postgraduate) connected to a course in the area of sustainability in education. However, data gained from telephone conversations with the management of public and private universities testify that even though there are no such programmes on climate change, universities have lately started offering sustainability programmes in education (e.g. environmental science, urban and rural planning). In this regard, the statement by a Vice Chancellor of a reputable university is worthy of note:

"As we hardly design any programme or course curriculum in Bangladesh, we are overly dependent on the programmes, courses and curricula of Western universities. Given this nature, the current market of higher education and operational bureaucracy, it is somewhat impossible to offer sustainability programmes in education, especially programmes such as climate change and disaster management. However, we are slowly entering into these by providing some tiny courses alongside with our different programmes, although it is very insignificant."

The statement above implies many constraints that Bangladesh HE may face. However, it also shows the hope and willingness for sustainability programmes in education, especially on climate change. By using this scope, a well-implementable decent policy and action plan on climate change education may ensure a better education, which the country needs.

Research on Climate Change

HE plays a significant role in research on climate change. Interviews with the academics and students of public universities confirm that a few public universities have recently produced a remarkable number of PhD theses and master's dissertations in the area of climate change. All this research is devoted to discovering insights into climate change that are applicable regionally. These attempts are encouraging. Some development partners (i.e. DFID, British Council, Action AID and USAID) have been working in the areas of both research and field implementation to address climate change. Academics from both public and private counterparts are involved in research on climate change. In this regard, one of the officials of a development partner shares her experience:

"Initially, academics in Bangladesh used to blame industrialised countries for the adverse situation that Bangladesh was experiencing due to climate change. Fortunately, they now realise that this is a grave concern and, without finding the faults of others and making further delays, everyone should contribute their best to tackle it. I am very happy that academics in Bangladesh are now contributing remarkably to research outputs on climate change."

An academic of a public university observes:

"Since the fund for climate change research comes from development partners, the findings may be prejudiced to protect the funder's objectives, which also fail to provide proper recommendations to protect the interests of our country."

Currently, the British Council in Bangladesh is providing support, mainly logistics, to academics to conduct research on climate change and to discover pragmatically implementable projects for society that could help reduce both human and natural causes.

Climate Change Education for Primary and Secondary Levels

Before stating the role of HE in climate change education for primary and secondary levels, let us quote a few lines of a public speech given by the Minster of the Environment and Forestry: "Higher education can only play a think-tank role for climate change. It is the primary and secondary graduates who can really bring the best outcomes for climate change. They are the largest group who are involved with both human and natural causes of climate change."

This is inevitably a real concern. However, we are well aware that primary and secondary provisions do not live in isolation. It is HE that conducts research and designs the programmes, courses and syllabuses applicable both for primary and secondary provisions. In this regard, an observation of an academic can well interpret the situation:

"A programme may initially be started either in higher education or in primary and secondary; however, if we want to give sustainability for the said programme and course, it should be readily available for every provision. If it is only available for higher education, the effectiveness of the course will ultimately be insignificant since higher education graduates use the degrees as paper qualifications to obtain a job. Thus, for real changes and the grounded impact on society, both primary and secondary graduates should be included."

Hence, the role of HE is manifold. In addition to conducting research and discovering a programme on climate change suited to students pursuing HE, HE needs to discover a wide range of programmes on climate change suitable for various target groups studying at different levels of primary and secondary education.

Social Awareness of Climate Change

Lately, HE has conducted seminars and symposiums on climate change, which have created a degree of social awareness. The messages delivered through these seminars and symposiums are only communicated to the education community. Bangladesh has a significant population who are still out of reach of the messages usually provided by these seminars and symposiums. Development partners are striving to knock door-to-door in order to ensure that the messages are conveyed. However, isolation from the education system and industries is keeping the programmes carried out by development partners far behind reaching the targeted group.

Strategic Planning and Policy

In order to describe the strategic plan and future policies, the following key findings are hereby discussed. The topic of climate change is living within the research boundary of HE. The programme is still an unpopular subject to be taught. This is partly because the psychological and social attitudes of students and parents always favour a degree that provides quick employment with a reasonable private benefit. Although development partners advocate topics on climate change, the graduates employed with them have science, engineering or social science backgrounds. Consequently, they do not raise their voices in order to institutionalise climate change as a complete academic programme. It is an easy assumption that because of the non-supply of climate change graduates, development partners are forced to employ others to act for the climate change issue. However, if development partners make climate change knowledge a compulsory requirement of obtaining the job, this would gradually motivate students, parents and suppliers to ensure a proper supply of graduates specialising in climate change. This will be easier now since private HE institutions have become more responsive to job market patterns for survival.

Currently, academics are enjoying a greater private benefit through their private consultancy on topics of climate change. This is why they may be somewhat reluctant to materialise this topic into an academic programme. However, the long-term benefit will be erased if climate change is not institutionalised. Moreover, an academic programme is not a competitor of private consultancy. They would rather supplement each other. Currently, both the Ministry of Education and the Ministry of the Environment live in isolation, which hinders the institutionalisation of climate change as an academic programme. Owing to the urgency of the issue, the UGC and Department of the Environment should start collaborative work to open the academic gate for sustainable programmes on climate change as an academic programme with collaboration among different stakeholders (i.e. industry, development partners, the ministries and agencies concerned for sustainability programmes such as climate change, disaster management, water and natural resources).

Conclusion

The outcome of this article is fundamental to dealing with the role of HE in institutionalising climate change. Academics recognise the lack of HE policy in Bangladesh. As climate change is a real concern there, HE should be seen as supporting the climate change concern with other economic development; however, unfortunately, it is yet to be perceived effectively as a developmental phenomenon. We understand that HE needs to be discovered in a substantial and concrete way to handle climate change. We note that legislators should re-examine the roles and responsibilities of every sector by reflecting on the possible ways in which to dissolve the differences that can complement each other. This may contribute to having a complete academic programme on climate change to support national development and international competitiveness on this vital issue and thus to change education in Bangladesh.

References

Ahmed S (2008) The status and challenges of water infrastructure development in Bangladesh. The first regional workshop on co-efficient water infrastructure in Asia, Nov 11, Seoul, Republic of Korea

Alam GM et al (2010) Who gains from EFA –State business of education or private higher education business in developing nation: a study to understand the policy impact in Bangladesh? Afr J Bus Manage 4(5):770–789

Alam GM et al (2011) National development and student politics in Bangladesh. Afr J Bus Manage 5(15):6044–6605

Alam GM (2013) Excellence in private sector higher education: a challenge to public sectors in developing nations. British Council, Colombo, Sri Lanka

Alam GM (2008) Impact of the private universities on Bangladeshi Education System. VDM, Verlag, Germany

Alam GM (2009) Can governance and regulatory control ensure private higher education as business or public goods in Bangladesh? Afr J Bus Manage 3(12):890–906

Alam M et al (2007) Private higher education in Bangladesh. International Institute for Educational Planning, Bangladesh

Al-Amin AQ, Alam GM (2011) The impacts of climate change on animal health and economy: a way forward for policy option. Asian J Anim Vet Adv 6(11):1061–1068

Al-Amin AQ, Filho WL (2011) An overview of prospects and challenges in the field of climate change in Malaysia. Int J Glob Warming 3(3):390–402

Al-Amin AQ et al (2010) Climate change mitigation and policy concern for prioritization/privatization. Int J Clim Change Strat Manage 2(4):418–425

Aldrich JL, Kormondy EJ (1972) Environmental education: academia's response. Bioscience 22:353–354

Altbach PG (1999) Private higher education: themes and variation in comparative perspective. Prospects 29(3):311–323

Altbach PG (2004) Globalisation and the university: myths and realities in an unequal world. Tert Educ Manag 10(1):3–25

Ball SJ (1993) Education markets, choice and social-class: the market as a class strategy in the UK and the USA. Br J Sociol Educ 14(1):3–19

Boulton G, Lucas C (2011) What are universities for? Chin Sci Bull 56(23):2506–2517

Bowers CA (2008) University reform in an era of global warming. Ecojustice Press, Ypsilanti

Brennan J (2012) Is there a future for higher education institutions in the knowledge society? Eur Rev 20(2):195–202

CHE (2000) Towards a new higher education landscape: meeting the equity, quality and social development imperatives of South Africa in the 21st century. Council on Higher Education Size and Shape Task Team, Pretoria, South Africa

Dalton JC, Crosby PC (2010) How we teach character in college: a retrospective on some recent higher education initiatives that promote moral and civic learning. J Coll Charact 11:1–10

DU (2012) News achieves, University of Dhaka, Dhaka, Bangladesh. Available online 17 July 2012. http://www.du.ac.bd/news_archive.php. Accessed 17 July 2012

Fahey SJ (2012) Curriculum change and climate change: inside outside pressures in higher education. J Curriculum Stud 44(5):703–722

Feinstein N (2009) Education for sustainable development in the United States of America. In: Laessøe J, Schnack K, Breiting S, Rolls S (eds) Climate change and sustainable development: the response from education. Danish School of Education University of Aarhus, Aarhus

Filho WL (2010) Climate change at universities: results of a world survey. In: Filho WL (ed) Universities and climate change. Springer, Berlin

GOV (2012) Canada's action on climate change. Government of Canada. Available online 16 July 2013. http://www.climatechange.gc.ca/default.asp?lang=En&n=65CD73F41#X201208011007431

IALEI (2009) Climate change and sustainable development: the response from education. The Danish School of Education. Aarhus University, Aarhus

INC (2002) Initial National Communication to UNFCC. Ministry of Environment and Forest. Government of the People's Republic of Bangladesh, Dhaka

IPCC (2007) In climate change: 2007, mitigation of climate change, contribution of working group III to the 4th assessment report of intergovernmental panel on climate change, annex 1. Cambridge University Press, Cambridge

Kerr C (2001) The uses of the university. Harvard University Press, Cambridge

Konrad O (2013) The role of higher education institutions in promoting sustainable development. Available online 1 July 2013. http://www.ias.unu.edu/sub_page.aspx?catID=705&ddlI=887

Laessoe J et al (2013) Environmental education policy research challenges and ways research might cope with them. Environ Educ Res 19(2):231–242

Lemons J (1995) Sustainable development and environmental protection: a perspective on current trends and future options for universities. Environ Manage 19:157–165

Lemons J (2011) The urgent need for universities to comprehensively address global climate change across disciplines and programs. Environ Manage 48(3):379–391

Mazumder KM (2010) Climate change, environment and Bangladesh's development. Bangladesh Development Forum (BDF), 15–16 Feb, Dhaka, Bangladesh

Nigel T (2005) Philosophy's higher education. Springer, Berlin

O'Brien GD (1998) All the essential half-truths about higher education. The University of Chicago Press, Chicago

Oreskes N (2004) Beyond the ivory tower: the scientific consensus on climate change. Science 306(5702):1686

Rahman A, Alam GM (2003) Mainstreaming adaptation to climate change in least developed countries (LDCS). Working paper 2: Bangladesh Country Case Study, The International Institute of Environment and Development, London, UK

Reed J, Mera E (2011) Developing integrated studies programs in "sustainability" through international exchange, experience based learning, and multi-disciplinary collaboration. In: Torres IC, Chova LG, Martinez AL (eds) Proceedings of 4th international conference of education', research and innovation (ICERI), Madrid, Spain, 14–16 Nov 2011, pp 6266–6271

Reza AH (2004) Natural resources management in Bangladesh: linking national priority to global perspective. Tigerpaper 31(2):10–16

Ritter SK (2009) Climate debate: what's warming us UP? Human activity or mother nature? global: warming and climate change. Chem Eng News 87(51):11–21

Rowe D (2007) Education for a sustainable future. Science 317:323–324

Saleemul H et al (2003) Mainstreaming adaptation climate change in least developed countries (LDCS). The International Institute for Environment and Development (IIRD), Nottingham, UK

Sheila P (2003) Developing countries: victims or participants-their changing role in international negotiations. Overseas Development Institute, London, UK

Singh P (2013) Environmental education: enhancing learning and awareness through assessment. Syst Pract Action Res 26(3):299–314

Stern N (2007) The economics of climate change: the stern review. Cambridge University Press, UK

Symes C (1996) Selling futures: a new image for Australian universities? Stud High Educ 21(2):133–147

Symes C (1998) Education for sale: a semiotic analysis of school prospectuses and other forms of educational marketing. Aust J Educ 42(2):133–152

UGC (2010) Annual report: University Grants Commission of Bangladesh. Dhaka, Bangladesh

UNFCCC (2007) Climate change: impacts, vulnerabilities and adaptation in developing countries. United Nations Framework Convention of Climate Change (UNFCCC)

Vincent S (2010) Interdisciplinary environmental education on the nation's campuses: elements of field identity and curriculum design. National Council for Science and the Environment, Washington

Weis JS et al (1992) Report of the workshop on undergraduate environmental science education. Environ Prof 14:363–370

Wigley T (2001) The science of climate change: 2001. Claussen E (ed) Climate change: science, strategies, and solutions. Pew Center on Global Climate, Arlington, VA

MIX
Papier aus verantwortungsvollen Quellen
Paper from responsible sources
FSC® C105338

If you have any concerns about our products,
you can contact us on
ProductSafety@springernature.com

In case Publisher is established outside the EU,
the EU authorized representative is:
Springer Nature Customer Service Center GmbH
Europaplatz 3, 69115 Heidelberg, Germany

Printed by Libri Plureos GmbH
in Hamburg, Germany